VOLUME ONE HUNDRED AND NINETY FIVE

ADVANCES IN IMAGING AND ELECTRON PHYSICS

Logarithmic Image Processing: Theory and Applications

VOLUME ONE HUNDRED AND NINETY FIVE

ADVANCES IN IMAGING AND ELECTRON PHYSICS

Logarithmic Image Processing: Theory and Applications

MICHEL JOURLIN

Laboratoire Hubert Curien
University of Saint-Etienne
France

ELSEVIER

AMSTERDAM • BOSTON • HEIDELBERG • LONDON
NEW YORK • OXFORD • PARIS • SAN DIEGO
SAN FRANCISCO • SINGAPORE • SYDNEY • TOKYO
Academic Press is an imprint of Elsevier

Cover photo credit:
Metrics Based on Logarithmic Laws
Advances in Imaging and Electron Physics (2016) 195, pp. 61–114
Dynamic Range Expansion, Night Vision. Stabilization, Centering. Industrial and Biomedical Applications
Advances in Imaging and Electron Physics (2016) 195, pp. 115–164

Academic Press is an imprint of Elsevier
50 Hampshire Street, 5th Floor, Cambridge, MA 02139, United States
525 B Street, Suite 1800, San Diego, CA 92101-4495, United States
The Boulevard, Langford Lane, Kidlington, Oxford OX5 1GB, United Kingdom
125 London Wall, London, EC2Y 5AS, United Kingdom

First edition 2016

ISBN: 978-0-12-804813-9
ISSN: 1076-5670

For information on all Academic Press publications visit our website at https://www.elsevier.com/

Publisher: Zoe Kruze
Acquisition Editor: Poppy Garraway
Editorial Project Manager: Shellie Bryant
Production Project Manager: Radhakrishnan Lakshmanan
Cover Designer: Mark Rogers

Typeset by SPi Global, India

CONTENTS

PREFACE

It is a pleasure to welcome Michel Jourlin back to these Advances. With his colleagues, he has already written extensively on the logarithmic image processing (LIP) model, which he originated. In this volume, he brings together both the theoretical background of the model and the motivation for using it, together with a host of applications in different subject areas. The book begins with an account of the origins of the model and the ways in which it has developed, which includes much useful advice for new generations of scientists and mathematicians embarking on such subjects. I say no more about this here, for I would merely be repeating his vividly written Foreword. He also presents the contents of each chapter there and draws attention to novel features.

I am confident that readers will appreciate this full and very lucid new account of the LIP model, and I am delighted to include it in these Advances.

Peter Hawkes

FUTURE CONTRIBUTIONS

S. Ando
Gradient operators and edge and corner detection

J. Angulo
Mathematical morphology for complex and quaternion-valued images

D. Batchelor
Soft x-ray microscopy

E. Bayro Corrochano
Quaternion wavelet transforms

C. Beeli
Structure and microscopy of quasicrystals

C. Bobisch, R. Möller
Ballistic electron microscopy

F. Bociort
Saddle-point methods in lens design

E. Bosch, I. Lazic
High-resolution STEM and related developments

K. Bredies
Diffusion tensor imaging

A. Broers
A retrospective

N. Chandra, R. Ghosh, (vol. 196)
Quantum entanglement in electron optics

A. Cornejo Rodriguez, F. Granados Agustin
Ronchigram quantification

C. Edgcombe
Electron phase plates

J. Elorza
Fuzzy operators

R.G. Forbes
Liquid metal ion sources

P.L. Gai, E.D. Boyes
Aberration-corrected environmental microscopy

M. Haschke
Micro-XRF excitation in the scanning electron microscope

R. Herring, B. McMorran
Electron vortex beams

M.S. Isaacson
Early STEM development

K. Ishizuka
Contrast transfer and crystal images

K. Jensen, D. Shiffler, J. Luginsland
Physics of field emission cold cathodes

U. Kaiser
The sub-Ångström low-voltage electron microcope project (SALVE)

T. Kirk, (vol. 196)
The near-field-emission scanning electron microscope

A.I. Kirkland, R. Clough, J. Mir, (vol. 196)
Electron detectors

C.T. Koch
In-line electron holography

O.L. Krivanek
Aberration-corrected STEM

M. Kroupa
The Timepix detector and its applications

I. Lazic, E. Bosch
High-resolution STEM

B. Lencová
Modern developments in electron optical calculations

H. Lichte
Developments in electron holography

A. Lubk
Fundamentals of Focal Series Inline Holography

M. Matsuya
Calculation of aberration coefficients using Lie algebra

J.A. Monsoriu
Fractal zone plates

L. Muray
Miniature electron optics and applications

S.A. Nepijko, V.G. Dyukov, G. Schönhense, (vol. 196)
Voltage contrast mode in a scanning electron microscope and its applications

M.A. O'Keefe
Electron image simulation

V. Ortalan
Ultrafast electron microscopy

D. Paganin, T. Gureyev, K. Pavlov
Intensity-linear methods in inverse imaging

N. Papamarkos, A. Kesidis
The inverse Hough transform

H. Qin
Swarm optimization and lens design

Q. Ramasse, R. Brydson
The SuperSTEM laboratory

B. Rieger, A.J. Koster
Image formation in cryo-electron microscopy

P. Rocca, M. Donelli
Imaging of dielectric objects

J. Rodenburg
Lensless imaging

J. Rouse, H.-n. Liu, E. Munro
The role of differential algebra in electron optics

J. Sánchez
Fisher vector encoding for the classification of natural images

P. Santi
Light sheet fluorescence microscopy

R. Shimizu, T. Ikuta, Y. Takai
Defocus image modulation processing in real time

T. Soma
Focus-deflection systems and their applications

I.J. Taneja
Inequalities and information measures

T. Tanigaki
Aberration-corrected high-voltage electron microscopy

J. Valdés
Recent developments concerning the Système International (SI)

ACKNOWLEDGMENTS

I want first to say that this book would never be written without the unfailing support and confidence of Dr. Peter Hawkes. Thus, I am deeply indebted to him, even if the last 3 months were entirely dedicated to the logarithmic imaging process (LIP), which became the obsession of my days and my nights. I hope that the achievement of this work will signify a return to normal life.

I also take this opportunity to express my gratitude to

- a number of my students for the pleasure it was to work with them: Pierre Guilbert, Victor Deshayes, Bassam Abdallah, Joris Corvo, Enguerrand Couka, Josselin Breugnot, Maxime Carré, Frédéric Itthirad, and earlier Freddy Darsonville, Pascal Michoud, Nathalie Montard, among others.
- my colleagues whose honesty and open minds made the collaboration both enjoyable and fruitful: Jean-Marie Becker, Marie-José Labouré, Isabelle Fillère—with whom we animated a working group focused on geometry, metrics, and shapes—Thierry Fournel, Guillaume Noyel, Guy Courbebaisse, Laurent Navarro, and earlier to Richard Grisel.
- the researchers who shared my passion of image processing: Dominique Jeulin, Françoise Préteux, Jacques Demongeot, Isabelle Magnin, among others who always accepted my invitations to PhD juries, Congresses, etc.
- and those who contributed to the LIP developments, including the teams of Prof. Cahill, Buzuloiu, and Agaian.

Writing all these names, I am very aware how truly lucky I was to meet these people.

Michel Jourlin

FOREWORD: SHORT HISTORY OF THE LIP MODEL

1. JUSTIFICATION OF THE MODEL

The first question we will answer is: *Why the logarithmic image processing (LIP) model?*

In fact, when we consider preexisting logarithmic look-up tables (LUT) we should ask what novel logarithmic tools can really bring to image processing. Moreover, when we analyze the content of the many papers dedicated to the LIP framework, we notice that their contribution directly relies on the model's nonlinearity. Particularly, most of them point out the ability of LIP tools to enhance or process dark images or dark regions inside an image. In fact, such tools are now well known to be practically insensitive to lighting variations or lighting drift.

In this book, a number of applications we will expose also rely on these capabilities of the LIP model. Nevertheless, using the strong mathematical and physical properties of the model, we will try to understand and justify the power of logarithmic operators. This way of reasoning permitted us to propose many novel tools and results. For example, we invite the readers to focus their attention on these sections:

Chapter II

- The physical meaning of the logarithmic additive contrast and its link with the classical Michelson contrast
- The logarithmic multiplicative contrast and its strong sensitivity in bright regins

Chapter III:

- The metrics associated with logarithmic contrasts
- Novel notions: the various Asplünd's metrics and particularly the additive Asplünd's metric, whose insensitivity to exposure time and, at a lesser level, to diaphragm aperture variations offers a new way of processing low-light images as well as images acquired under variable lighting.

Chapter IV:

- Based on the mathematical vector space structure valid on the space of images (Chapter "Gray-Level LIP Model. Notations, Recalls and First Applications"), the use of negative gray levels as light intensifiers permits the enhancement of very low-light images.

We have also taken care to:

- illustrate the main results with plenty of biomedical and industrial applications, and
- highlight the places where new research developments are required.

2. ORIGIN OF THE MODEL

In 1983, I attended a training week dedicated to image analysis at the "Ecole des Mines de Fontainebleau." I already worked in collaboration with this team, called the "Fontainebleau School," and understood the principles and tools of "mathematical morphology." Nevertheless, this training was an opportunity for me to enter the formulation of the main tools of mathematical morphology in depth, and I focused my attention on the gray-level mathematical morphology. Since my initial education was purely mathematics, I was particularly interested in the set of theoretical results developed by the Fontainebleau School on the basis of the Minkowski's addition $A \oplus B$ of two binary shapes A and B. I remarked that for dilation (resp. erosion) operator, noted $A \oplus B$ (resp. $A \ominus B$), the structuring element B and the studied shape A were of same nature (binary shapes). However, when transferred to functions, they produce dilation/erosion operators $f \oplus B$ and $f \ominus B$ where the structuring element B remains binary.

Thus, I thought that it would likely be interesting (at least intellectually) to imagine new operators $f\,''\oplus''\,g$ and $f\,''\ominus''\,g$ where g could play the role of a functional structuring element, ie, could be chosen as a simplified function dedicated to the study of an image f, as it is commonly done for binary shapes. Moreover, as a former mathematician converted to image processing, I had in mind that the laws $''\oplus''$ and $''\ominus''$ would not be limited to a mathematical formulation, but would possess a tangible usefulness. This is the reason why my first concern was to think about a realistic situation in image processing where the sum of two images could have a physical meaning. This reflection directed me toward images acquired in transmission (microscopy, tomography, scanner, etc.) where the observed object appears often as the superimposition of various components. For example, the simple observation with a microscope of the nucleolus of a cell consists of the stacking of the cell cytoplasm, of the nucleus and the nucleolus. The resulting image is then nothing but the addition, in transmission, of the images corresponding separately to each component.

In such conditions, I decided to define first the law $''\oplus''$ in the context of transmitted signal (images acquired in transmission), ie, when the observed object is located between the source and the sensor. I finally noted this addition law with the symbol ⨹ so as not to lead to confusion with the standard morphological operators.

The first research on this subject permitted me to effectively introduce an internal addition law ⨹ on the space $I(D, [0, M[)$ of images defined on the same spatial support D and with values in the same gray scale $[0, M[$. Such a law is the well-known transmittance law expressed in terms of gray levels (cf. following Chapter "Gray-Level LIP Model. Notations, Recalls and First Applications"). A classical reflex for a mathematician when he disposes of an addition law is to deduct a scalar multiplication ⨻ by stacking n times the object corresponding to the image *f*: $f ⨹ f ⨹ \cdots ⨹ f = n ⨻ f$. Once established for integers, this law is then extended to every real number λ. At this moment, we dispose of two laws, one permitting to add two images ($f ⨹ g$), the other to multiply an image by a real number ($\lambda ⨻ f$). Having done this "physical" reasoning based on the transmittance law, I returned to a mathematical point of view, which consisted of asking what the properties of these laws are in terms of associativity, neutral element, opposite, commutativity, distributiveness, etc.

All these properties are satisfied when the space $I(D, [0, M[)$ of images is extended to the space $F(D,]-\infty, M[)$ of functions defined on the same domain D with values in $]-\infty, M[$. In fact, the introduction of possible negative values allows us to define the opposite $⨺ f$ of an image *f*, representing the function, which results in the neutral element 0 when added to the image $f : f ⨹ (⨺ f) = 0$.

The set of these properties structures $F(D,]-\infty, M[), ⨹, ⨻)$ as a vector space, allowing the use of the numerous tools available in such spaces, such as interpolation, gauges, norms, scalar products, etc.

In conclusion, we dispose of a strong physical and mathematical framework to perform image processing in good conditions.

I do not know if the reader is unfamiliar with the LIP model considered at this step, or whether it is interesting to pursue these first developments, but I will say that when I proposed this subject to my PhD students, no one was enthusiastic. I most strongly insisted until one of them, J.C. Pinoli, demonstrated some interest. I must admit that nobody, not even myself, had the least idea of the subject's extent at this period.

After I published the first results with J.C. Pinoli and R. Zeboudj (1985-89), I had some contact with researchers interested in the LIP model. I remember particularly the visit in Lyon of Professor Cahill (La Trobe

University, Australia) who had the courtesy to ask my agreement before beginning a PhD on the LIP subject with his student, Dennis Deng.

To refer to the most important milestones punctuating the LIP progresses, I must cite:

- In the biomedical domain, the collaboration I developed with Professor Jörg Schüpbach (University of Zurich) concerning the three-dimensional (3D) visualization of T4-lymphocytes cultured in presence of HIV virus. T4-lymphocytes were removed from the culture and fixed in an epoxy block before being cut into very thin slices (thickness: 40 nm) with an ultramicrotome. Each slice was then imaged with an electron microscope in transmission and the role of the LIP was to interpret some missing cuts (logarithmic interpolation) and to normalize the lighting changing observed on the cuts due to thickness variations. This research took place between 1985 and 1989 and resulted in a scientific prize (Plano Prize, Salzburg, 1989).
- In the research domain, I had a lasting collaboration with the team created and led by Professor Vasile Buzuloiu (Polytecnica, Bucaresti), including the works of V. Patrascu, E. Zaharescu, C. and L. Florea, C. Vertan, among others.
- In research-development and industrial applications, I am working with a start-up called NT2I, which has integrated the set of logarithmic tools and tests the final tools as we proceed in their creation. I have headed two PhD theses prepared by research engineers from NT2I and we work together daily.

2.1 Spirit and Goals of the Book

It's a pleasure for me to recall the long and fruitful collaboration I developed with Peter Hawkes: thanks to him, it was possible to publish extended papers dedicated to the LIP model. I had already the occasion to emphasize the particular role of AIEP Serial in the "jungle" of scientific journals, where it is customary today to publish papers mostly devoted to abstract, introduction, notations, recalls, conclusion, perspectives, and references, with very little space for research itself. On the contrary, AIEP Serial permits me to present extended papers that allow in-depth study of a subject. With Peter Hawkes agreement, I decided that this book's contents would not be the simple aggregation of previous papers. It seemed more interesting to summarize the main existing results (which remain accessible to the interested reader) and to highlight the novel ones, as well as application examples.

However, the goal of the book is not to overview the abundant bibliography dedicated to the LIP model, because I take exception with explaining and appropriating results that are not mine. I recommend the same behavior to the young researchers working with me, contrary to what is commonly done: in my opinion, the first step to initialize a novel research subject is not to consult an extended bibliography concerning it. In fact, there is a serious risk that this could restrict the imagination because the researcher might believe that the main results have been already discovered. I consider it much more productive to work as if the subject was novel and afterward compare the obtained results with existing ones. In such a way, even in case of similar innovation, at least the subject will be enlightened with a novel point of view.

Being at the origin of the LIP model, I prefer to deliver the reader an authentic message: what is my own point of view on the model, what are the most significant theoretical results, illustrated by the best applications. I hope this book will aid novices in logarithmic processing and perhaps also the researchers who previously heard about it.

A last point must be underlined: I have chosen to focus the book on the gray-level LIP model. The main reason is that we previously published an extended paper dedicated to the logarithmic image processing for color images (*Advances in Imaging and Electron Physics*, Vol. 168, edited by Peter Hawkes), to which the interested reader can refer. Moreover, the logarithmic color tools are not as easy to understand as the gray-level ones. Nevertheless, we will sometimes propose simple extensions of our algorithms to color images by taking into account the three red, green, blue channels.

MICHEL JOURLIN

MAIN NOTATIONS

$\oplus$ mathematical morphology dilation
$\ominus$ mathematical morphology erosion
$\triangle\!\!\!\!+$ LIP addition
$\triangle\!\!\!\!-$ LIP subtraction
$\triangle\!\!\!\!\times$ LIP multiplication
$\boldsymbol{f, g, h}$ gray-level images
$\boldsymbol{\alpha, \beta, \lambda, \mu}$ scalar elements (real numbers, elements of $\mathbb{R}$)
$\boldsymbol{i, j, k}$ integers (elements of $\mathbb{N}$)
$\mathbb{Q}$ space of rational numbers
$\boldsymbol{D}$ spatial domain (definition domain of images)
$\boldsymbol{M}$ maximum value of an image ($M = 256$ for an 8-bits image)
$\boldsymbol{[0, M[}$ gray scale
$\boldsymbol{I(D, [0, M[)}$ space of gray-level images defined on D with values in $[0, M[$
$\boldsymbol{F(D,]-\infty, M[)}$ over-space of $I(D, [0, M[)$
$\triangle\!\!\!\!- f$ opposite function of image f
$\boldsymbol{T_f(x)}$ transmittance of an image f at $x \in D$
LUT look-up table
$\boldsymbol{C^{Mi}_{(x,y)}(f)}$ Michelson contrast between $f(x)$ and $f(y)$
$\boldsymbol{C^{\triangle\!\!\!\!+}_{(x,y)}(f) = LAC_{(x,y)}(f)}$ logarithmic additive contrast between $f(x)$ and $f(y)$
$\boldsymbol{C^{\triangle\!\!\!\!+}_{x}(f, g)}$ logarithmic additive contrast between $f(x)$ and $g(x)$
$\boldsymbol{N(x)}$ **(resp.** $\boldsymbol{N_8(x), N_4(x)}$**)** neighborhood of x (resp. 8-points and 4-points neighborhood)
Card $\boldsymbol{(E)}$ number of elements of a set E
$\mathbf{IC}^{\triangle\!\!\!\!+}_{\boldsymbol{R}}\boldsymbol{(f)}$ inner logarithmic contrast of region R
$\sum^{\triangle\!\!\!\!+}$ sum in the LIP sense
$\boldsymbol{B(t)}$ boundary generated by the threshold t
$\boldsymbol{C^K(B(t))}$ Köhler contrast associated to $B(t)$
$\boldsymbol{C^{K,\triangle\!\!\!\!+}(B(t))}$ LIP-Köhler additive contrast associated to $B(t)$
$\mathbf{AC}^{\triangle\!\!\!\!+}_{\boldsymbol{x}}\boldsymbol{(f)}$ average value of the logarithmic additive contrasts between x and its neighbors
$\mathbf{MC}^{\triangle\!\!\!\!+}_{\boldsymbol{(x)}}\boldsymbol{(f)}$ maximal logarithmic additive contrast at x
$\boldsymbol{C^{\triangle\!\!\!\!\times}_{(x,y)}(f)} = \mathrm{LMC}_{(x,y)}(f)$ logarithmic multiplicative contrast between $f(x)$and$f(y)$
$\boldsymbol{C^{\triangle\!\!\!\!\times}_{x}(f, g)}$ logarithmic multiplicative contrast between $f(x)$ and $g(x)$
$\boldsymbol{C^{\triangle\!\!\!\!\times N}_{(x,y)}(f)}$ normalized logarithmic multiplicative contrast between $f(x)$ and $f(y)$
$\boldsymbol{C^{\triangle\!\!\!\!\times N}_{x}(f, g)}$ normalized logarithmic multiplicative contrast between $f(x)$ and $g(x)$
$\boldsymbol{C^{K,\triangle\!\!\!\!\times}(B(t))}$ LIP-Köhler multiplicative contrast associated to $B(t)$
$\mathbf{AC}^{\triangle\!\!\!\!\times}_{\boldsymbol{x}}\boldsymbol{(f)}$ average of the logarithmic multiplicative contrasts between x and its neighbors
$\mathbf{MC}^{\triangle\!\!\!\!\times}_{\boldsymbol{x}}\boldsymbol{(f)}$ maximal logarithmic multiplicative contrast between x and its neighbors
$\mathbf{IC}^{\triangle\!\!\!\!\times}_{\boldsymbol{R}}\boldsymbol{(f)}$ inner multiplicative contrast associated to a region
$\boldsymbol{R, G, B}$ red, green, blue channels of a color image
$\boldsymbol{C^{\triangle\!\!\!\!+}_{R}(C_1, C_2)} = \mathrm{LAC}_{\boldsymbol{R}}(C_1, C_2)$ logarithmic additive contrast in red channel between C_1 and C_2

$\mathbf{ALAC}_R(f(x))$ red average logarithmic additive contrast between x and its neighbors
$\mathbf{MLAC}_R(f(x))$ red maximal logarithmic additive contrast at x
$d_1(f,g)$ "global" distance between two images
$d_\infty(f,g)$ "atomic" distance between two images
$d_\varepsilon(f,g)$ intermediate metric between "global" and "atomic" ones
$A\Delta B = A \cup B \setminus A \cap B$ symmetric difference between two sets A and B
$d_\Delta(A,B) = \text{area}(A\Delta B)$ symmetric difference distance between A and B
$d_H(A,B)$ Hausdorff distance between A and B
$d_1^{\triangle}(f,g)$ "global" distance between two images in the LIP sense
$d_\infty^{\triangle}(f,g)$ "atomic" distance between two images in the LIP sense
$d_{1,SupR}^{\triangle}(f,g)$ intermediate metric in the LIP sense
$d_{\mathrm{As}}(A,B)$ Asplünd's distance between two binary shapes A and B
$N_{\mathrm{As},\varepsilon}(A)$ set of shapes whose Asplünd's distance to A is less than ε
$d_{\mathrm{As}}^{\triangle}(f,g)$ multiplicative Asplünd's distance between two images
$f^{\triangle}$ equivalence class of f for the law $\triangle$
$I^{\triangle}(D,[0,M[)$ set of equivalence classes $f^{\triangle}$
$d_{\mathrm{As}}^{\triangle}(f^{\triangle},g^{\triangle})$ multiplicative Asplünd's distance between two equivalence classes
$f|_{D_t}$ restriction of image f to a subset D_t of D
$T_{\alpha,\beta,\varepsilon}(f)$ tolerance tube of f (region delimited by $\alpha \triangle f$ and $\beta \triangle f$ such that $Ln\left(\frac{\alpha}{\beta}\right) = \varepsilon$)
$d_{\mathrm{As},R}^{\triangle}(f,g)$ additive Asplünd's distance between f and g on a subset R of D
$f^{\triangle}$ equivalence class of f for the law $\triangle$
$I^{\triangle}(D,[0,M[)$ set of equivalence classes $f^{\triangle}$
$d_{\mathrm{As},R}^{\triangle}(f^{\triangle},g^{\triangle})$ additive Asplünd'sdistance between two equivalence classes
GAAM gradient additive Asplünd's map
$d_{\mathrm{As}}^{\triangle}(C_1,C_2)$ multiplicative Asplünd's distance between two colors
$d_{\mathrm{As},C,S}^{\triangle}(f,g)$ multiplicative Asplünd's distance between two colors images on a subset S of D
$d_{\mathrm{As},C}^{\triangle}(f,g)$ multiplicative Asplünd's distance between two colors images on D
$d_{\mathrm{As},C}^{\triangle}(C_1,C_2)$ additive Asplünd's distance between two colors
$(f|g)^{\Delta}$ scalar product between two images
j_A gauge associated to A
$\mathbf{DR}(f)$ dynamic range of image f
$h(f)$ histogram of image f
FDRE full dynamic range expansion
$\mathbf{GDR}(f)$ global dynamic range of a color image f
$\mathbf{GH}_f$ global histogram of a color image f
HAC hierarchical ascendant classification
d_e Euclidean distance
SNR signal-to-noise ratio

CHAPTER ONE

Gray-Level LIP Model. Notations, Recalls, and First Applications

Contents

1. BASIC NOTATIONS

Gray-level images will be generally noted $f, g, h\ldots$ A gray-level image f is defined on a domain D included in the plan $\mathbb{R}^2$ (or in the space $\mathbb{R}^3$ for 3D images) and takes its values (gray levels) in the gray scale $[0, M[$. For 8-bits images, $M = 256$ and the 256 gray levels are in the scale of integers $[0, \ldots, 255]$.

In such conditions, a gray-level image is associated with its representative surface. To illustrate this point, Fig. 1A displays a very simple image (near binary) and its representative surfaces depend on the meaning of the 0-value of the scale: classically, 0 corresponds to the *black* extremity (Fig. 1B), but it is possible to make the other choice, 0 being associated with the *white* extremity (Fig. 1C).

Advances in Imaging and Electron Physics, Volume 195
ISSN 1076-5670
http://dx.doi.org/10.1016/bs.aiep.2016.04.001

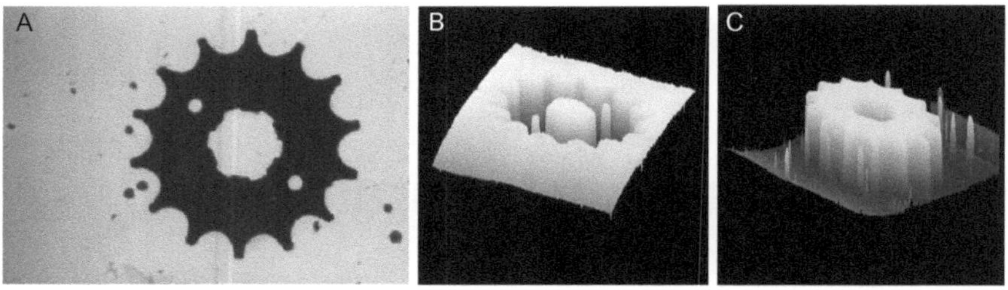

Fig. 1 Illustration of the correspondence between surface and gray-level image. (A) Gray-level image of a cogwheel. (B) Representative surface of image (A) (hypothesis 0 = "*black*"). (C) Representative surface of image (A) (hypothesis 0 = "*white*").

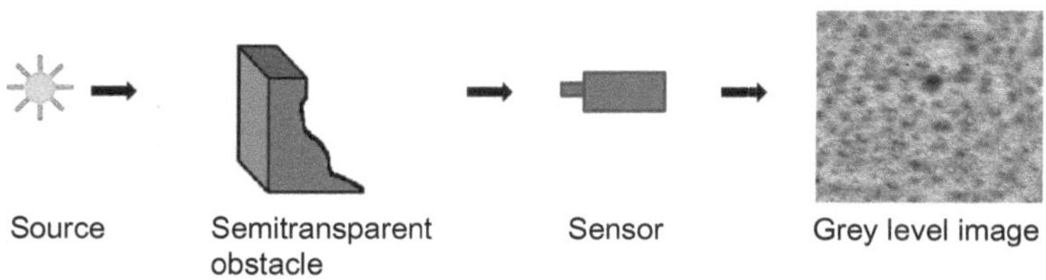

Fig. 2 Acquisition in transmission. (See the color plate.)

Based on Fig. 1, it is not surprising that in image processing and image analysis we often meet terms commonly used in topography, such as peaks, valleys, crest lines, watersheds, etc.

Coming back to basic notations, we will represent a gray-level image as a function defined on D with values in the gray scale $[0, M[$:

$$f : D \subset \mathbb{R}^2 \rightarrow [0, M[\subset \mathbb{R}$$

The space of gray-level images defined on the same domain D with values in $[0, M[$ is noted $I(D, [0, M[)$.

As cited in the Introduction, we consider in a first step *images acquired in transmission, ie, when the observed object is located between the source and the sensor* (cf. Fig. 2).

Comment on Fig. 2: In order to simplify notations, we will often confuse in a same letter "*f*" the gray-level image and the semitransparent object that generated *f*. In fact, under stable acquisition conditions (source intensity, exposure time, and diaphragm aperture of the sensor), the function associating to an object its gray-level image is obviously a bijection (one object corresponds to a unique image).

2. DEFINITION OF LOGARITHMIC LAWS ON THE SPACE OF IMAGES

The reader interested by initial papers concerning the LIP model can refer to Jourlin and Pinoli (1985, 1987, 1988).

Based on the transmittance law, the LIP model proposes two operations on images allowing the addition of two images and the multiplication of an image by a scalar, each of them resulting in a novel image.

The transmittance $T_f(x)$ of an image f at $x \in D$ is defined as the ratio of the outcoming flux at x by the incoming flux (intensity of the source). In a mathematical formulation, $T_f(x)$ may be understood as the probability for a particle of the source incident at x, to go through the obstacle (ie, to be seen by the sensor).

2.1 Definition of the Logarithmic Addition Law

The addition of two images f and g is interpretable as the superposition of the obstacles (objects) generating, respectively, f and g. The resulting image will be noted $f \triangle\!\!\!+ g$ and the equation corresponding to this addition will be deducted from the transmittance law:

$$T_{f \triangle\!\!\!+ g} = T_f \cdot T_g \tag{1}$$

Eq. (1) means that the probability for a particle emitted by the source, to go through the "sum" of the obstacles, f and g, equals the product of the probabilities to go through f and g, respectively. Jourlin and Pinoli (2001) established the link between the transmittance $T_f(x)$ and the gray level $f(x)$:

$$T_f(x) = 1 - \frac{f(x)}{M} \tag{2}$$

Replacing in Eq. (1) the transmittances by their values deducted from Eq. (2) yields:

$$f \triangle\!\!\!+ g = f + g - \frac{f \cdot g}{M} \tag{3}$$

Warning 1: within the LIP model, 0 corresponds to the "white" extremity of the gray scale, which means to the source intensity [ie, when no obstacle (object) is placed between the source and the sensor]. The reason for this gray-scale inversion is justified by the fact that 0 will appear as the neutral element of the addition law defined in Eq. (3). The other extremity M is a limit situation where no element of the source is transmitted (black value).

Warning 2: Classically, image sensors use a gray scale where 0 is "black." In order to not confuse a classical gray-level image $\tilde{f}$ and its corresponding function f as used in the LIP context, this last one is sometimes called a *gray-tone* function. Considering that we will work in the LIP framework throughout this book and $\tilde{f}$ is simply the negative picture of f (in the digital case: $f = M - 1 - \tilde{f}$ (cf. Fig. 1B and C), we made the choice to call f a gray-level image, knowing that we will notify the use of the classical scale $[0, M[$ when it will (rarely) happen.

Let us come back to the consequences of Eq. (3). When they dispose of an addition defined on a set, mathematicians generally look for properties of such a law, for example, in our case:

- Associativity: $\forall f, g, h \in I(D, [0, M[), (f \triangle\!\!\!\!+ g) \triangle\!\!\!\!+ h = f \triangle\!\!\!\!+ (g \triangle\!\!\!\!+ h)$
- Neutral element: $\exists n \in I(D, [0, M[), \forall f \in I(D, [0, M[), f \triangle\!\!\!\!+ n = f$
- Commutativity: $\forall f, g, \in I(D, [0, M[), f \triangle\!\!\!\!+ g = g \triangle\!\!\!\!+ f$

These properties obviously hold. The neutral element n is nothing but the image 0 (white in every point of D).

To structure the space of images $I(D, [0, M[)$ as a "*Commutative Group*," we must add a fourth property: the existence, for each image f of $I(D, [0, M[)$, of an opposite function, noted "$\triangle\!\!\!\!- f$" satisfying $f \triangle\!\!\!\!+ (\triangle\!\!\!\!- f) = 0$ at each point of the definition domain D.

2.1.1 Fundamental Remarks

Remark 1. The equality $f \triangle\!\!\!\!+ (\triangle\!\!\!\!- f) = 0$ is impossible inside the space $I(D, [0, M[)$, ie, if we expect that the opposite $\triangle\!\!\!\!- f$ remains an image. In fact, it would mean that:

The addition to f of an image $\triangle\!\!\!\!- f$ results in a "white" (transparent) image.

Such a property seems as marvelous as it is impossible.

Nevertheless, it is not so unrealistic if we accept working inside an over-space of $I(D, [0, M[)$, namely, the space $F(D,]-\infty, M[)$ of functions defined on D with values in $]-\infty, M[$. We have clearly the inclusion $I(D, [0, M[) \subset F(D,]-\infty, M[)$ and an obvious formal calculus based on Eq. (3) permits us to make explicit $\triangle\!\!\!\!- f$:

$$f \triangle\!\!\!+ (\triangle\!\!\!- f) = 0 \Leftrightarrow f + (\triangle\!\!\!- f) - \frac{f \cdot \triangle\!\!\!- f}{M} = 0$$
$$\Leftrightarrow \triangle\!\!\!- f = \frac{-f}{1 - \frac{f}{M}} \tag{4}$$

In the same way, the difference $f \triangle\!\!\!- g$ between two images exists and satisfies formally:

$$f \triangle\!\!\!- g = \frac{f - g}{1 - \frac{g}{M}} \tag{5}$$

Note that $f \triangle\!\!\!- g$ is an element of the space $F(D,]-\infty, M[)$, but is not an image, except when the inequality $g(x) \leq f(x)$ holds for every x in D.

Note also that $|f \triangle\!\!\!- g| \neq |g \triangle\!\!\!- f|$

Remark 2. From Eq. (4), we have a mathematical solution to make an image f transparent by adding its opposite $\triangle\!\!\!- f$. Such a property will take all its *physical* meaning when we will interpret the negative part of the interval $]-\infty, M[$ as *gray levels brighter than the gray-level 0 of the initial source* (ie, as light intensifiers). From this moment, it will become possible to *perfectly control image enhancement* (Chapter "Dynamic Range Expansion, Night Vision. Stabilization, Centering. Industrial and Biomedical Applications") and to *simulate images obtained by increasing the source intensity* (Chapter "Ability of the LIP Model to Simulate Variable Acquisition Conditions").

Why are these remarks fundamental? The reason is that they strongly exploit the two rigorous bases on which the LIP model is founded:

- on one hand the mathematical structure of Commutative Group of the over-space $F(D,]-\infty, M[)$ allows the use of negative gray levels
- on another hand the physical justification of the additive law $\triangle\!\!\!+$ is based on the Transmittance law

To be perfectly honest, the contents of these remarks and their theoretical and applicative consequences were not obvious to me from the beginning, but they appear essential now: they have determined the structure of the present book so that the reader will focus on our strongest results.

2.2 Definition of the Logarithmic Scalar Multiplication

When they dispose of an addition law (here $f \triangle\!\!\!+ g$), we already said that mathematicians have a natural reflex: they immediately try to deduct from it a scalar multiplication. In fact, in our case, if we add an image f to itself, we can write, according to Eq. (3):

$$f \triangle\!\!\!+ f = 2 \cdot f - \frac{f^2}{M} \tag{6}$$

In the case of transmitted light, it means that the obstacle generating f is stacked twice between the source and the sensor, which we will note $2 \triangle\!\!\!\times f$. This constitutes the first step to define a scalar multiplication. The next steps are:

- generalize Eq. (6) to every integer n: such an extension is performed by the transmittance law (Eq. 1). In fact, stacking n times the same obstacle f between the source and the sensor may be written in terms of transmittances:

$$T_{n \triangle\!\!\!\times f} = (T_f)^n$$

Replacing the transmittances in this equation by their expression given in Eq. (2) yields:

$$n \triangle\!\!\!\times f = M - M\left(1 - \frac{f}{M}\right)^n \tag{7}$$

- apply an inversion to Eq. (7):

$$\frac{1}{n} \triangle\!\!\!\times f = M - M\left(1 - \frac{f}{M}\right)^{\frac{1}{n}} \tag{8}$$

- establish the equation for every positive rational number $\frac{p}{q}$ (ratio of two integers); this is done based on Eqs. (7) and (8):

$$\frac{p}{q} \triangle\!\!\!\times f = M - M\left(1 - \frac{f}{M}\right)^{\frac{p}{q}} \tag{9}$$

- finally, knowing that the set of rational numbers is topologically dense in the real set $\mathbb{R}$, each real number λ may be written as the limit of a sequence of rational numbers: $\lambda = \lim_{n \to \infty}\left(\frac{p_n}{q_n}\right)$. The well-known continuity of the exponential function permits then to pass to the limit, which gives:

$$\lambda \triangle\!\!\!\times f = M - M\left(1 - \frac{f}{M}\right)^{\lambda} \tag{10}$$

Remark 3. Note that the previous reasoning establishes Eq. (10) for any positive real number λ. The extension to negative real numbers is achieved based on the obvious equality:

$$(-1)\triangle\!\!\!\!\times f = \triangle\!\!\!\!- f.$$

For this novel law, we proceed as for the additive one; listing the main mathematical properties it satisfies:

- Associativity: $\forall f \in I(D, [0, M[), \forall(\lambda, \mu) \in \mathbb{R}^2, \lambda \triangle\!\!\!\!\times (\mu \triangle\!\!\!\!\times f) = (\lambda \cdot \mu) \triangle\!\!\!\!\times f$
- Neutral element: The integer "1" answers obviously the question:

$$\forall f \in I(D, [0, M[), 1 \triangle\!\!\!\!\times f = f$$

- Double distributive law:

$$\forall f \in I(D, [0, M[), \forall(\lambda, \mu) \in \mathbb{R}^2, (\lambda + \mu) \triangle\!\!\!\!\times f = (\lambda \triangle\!\!\!\!\times f) \triangle\!\!\!\!+ (\mu \triangle\!\!\!\!\times f)$$
$$\forall (f, g) \in I^2(D, [0, M[), \forall \lambda \in \mathbb{R}, \lambda \triangle\!\!\!\!\times (f \triangle\!\!\!\!+ g) = (\lambda \triangle\!\!\!\!\times f) \triangle\!\!\!\!+ (\lambda \triangle\!\!\!\!\times g)$$

All these equalities are very easy to prove.

Fundamental result: The set of properties previously established for the two laws corresponds to the definition of a vector space structure (see, for example, Lang, 2002 or Schaefer, 1999). More precisely, the following results hold:

- The set $[F(D,]-\infty, M[), \triangle\!\!\!\!+, \triangle\!\!\!\!\times]$ is a *Real Vector Space*
- The space of images $[I(D, [0, M[), \triangle\!\!\!\!+, \triangle\!\!\!\!\times]$ is its *Positive Cone*

2.3 Physical Interpretation of These Logarithmic Laws

These laws automatically possess physical properties due to the Transmittance law because of the situation initially chosen to define the logarithmic laws (images acquired in transmission).

2.3.1 Addition Law

Remember that the logarithmic addition of two images f and g represents the superposition of the semitransparent objects generating f and g.

We can focus on the particular case where one function is a constant C, which means that it corresponds to a uniform obstacle. Such a constant may be added to f or subtracted from f (cf. Fig. 3).

The images displayed in Fig. 3E show the ability of the logarithmic addition law to *brighten/darken an image*. Such a property is not characteristic of the LIP framework: plenty of image processing operators are able to *visually* brighten or darken an image, generally without any physical justification.

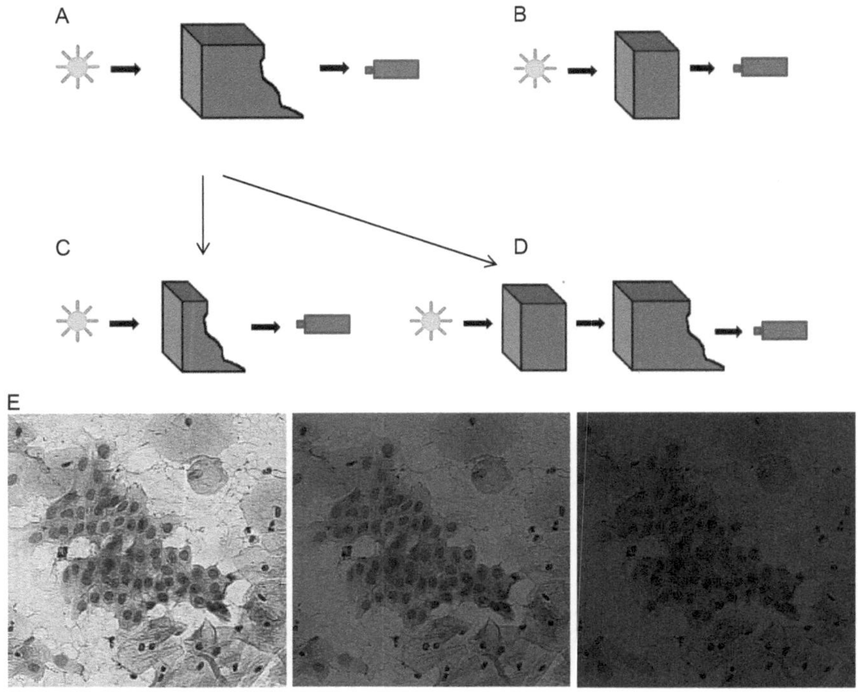

Fig. 3 Subtracting (resp. adding) a constant from (resp. to) an image. (A) Acquisition of initial image *f*. (B) Uniform image *C*. (C) Acquisition of $f \triangle\!\!\!\!-\, C$. (D) Acquisition of $f \triangle\!\!\!\!+\, C$. (E) Visualization of $f \triangle\!\!\!\!-\, C$, *f*, and $f \triangle\!\!\!\!+\, C$ (from *left* to *right*), case $C=128$. (See the color plate.)

In the case of the LIP model, such a property is strongly founded: not only based on the transmittance law, the logarithmic addition (resp. subtraction) of a constant to (resp. from) an image f will be demonstrated later (see Chapters "Dynamic Range Expansion, Night Vision. Stabilization, Centering. Industrial and Biomedical Applications" and "Ability of the LIP Model to Simulate Variable Acquisition Conditions") as a *perfect tool to simulate together exposure time changing, diaphragm aperture modification, and source intensity decrease (resp. increase).*

2.3.2 Multiplication Law

In order to justify Eq. (7), we have already explained that the image $n \triangle\!\!\!\!\times f$ was obtained by stacking n times the object generating f. When extending the equation to a real number λ, this property remains true—ie, $\lambda \triangle\!\!\!\!\times f$ is (virtually or physically) acquired by multiplying the thickness of f by λ. Thus the image $\lambda \triangle\!\!\!\!\times f$ will be darker than f if $\lambda \geq 1$ and brighter if $\lambda \in [0,1]$ (cf. Fig. 4).

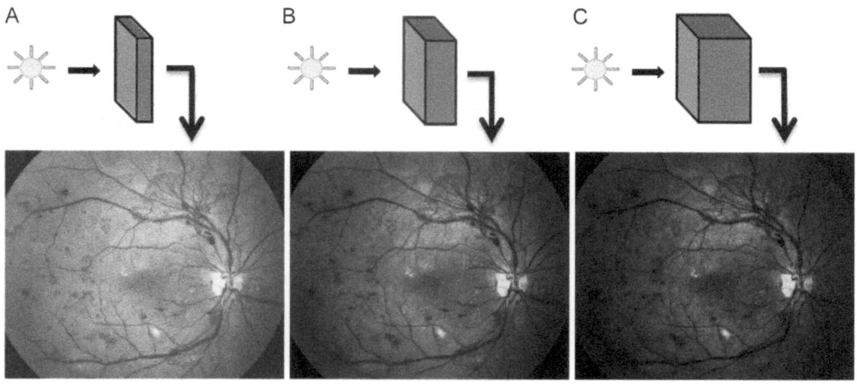

Fig. 4 Multiplying an image by a real number. (A–C) Visualization of $0.5 \triangle f$, f, and $2 \triangle f$ (from *left* to *right*). (See the color plate.)

Comment on Fig. 4: As for the logarithmic addition of a constant, the logarithmic scalar multiplication appears able to *brighten/darken an image*. Nevertheless, we can note that the two laws do not have the same effects. For example, the scalar multiplication better preserves the bright parts.

We will see (Chapter "Dynamic Range Expansion, Night Vision. Stabilization, Centering. Industrial and Biomedical Applications") the applicative interest of the vector space structure. In fact, to process very low-light images (near night vision), an efficient solution consists of using the negative part of the gray scale by multiplying the studied image by a negative real number λ. In my opinion, it is intellectually satisfactory to observe that the enhancement of an image by decreasing its thickness remains true when the thickness becomes negative.

2.4 Consistency of the LIP Model with Human Vision

This section will be very short, but its consequences are extremely important. In fact, we had noticed in our first papers introducing the LIP model (see Jourlin & Pinoli, 1988, for example), that the LIP model could be valid for nonlinear imaging systems, *human vision* for instance. In fact, the eye's response to light intensity variations is known to be logarithmic. It follows the Weber–Fechner law defined by:

$$W = \frac{\delta I}{I}$$

where W (called the *Weber ratio*) remains constant for a wide range of intensities. It means that the answer of a *Human Visual System*, when observing a homogeneous object on a homogeneous background, appears proportional to their intensities variation divided by the background intensity. Such a ratio is expressed as the differential function of a logarithmic function, showing that the human vision is of *logarithmic nature*.

Remark 4. The logarithmic modeling of the human eye is not (and will never be) rigorously demonstrated. It results from various observations and experiences. I think that a mathematician would say: "in a system Oxy, the curve representing the eye's answer (y axis) to intensity variations (x axis) presents a parabolic branch in Ox direction."

Let us remark that the same observation seems to be true for other human senses: concerning hearing, for example, nobody contests the logarithmic definition of decibels. More recently, some experiments resulted in a logarithmic modeling of smell to estimate the quality of wines or perfumes. For the moment, I have no information on the senses of touch and taste.

Now we want to express our gratefulness to Brailean, Sullivan, Chen, and Giger (1991), who early published a paper showing the consistency of the LIP model with human vision.

Consequences: This result considerably enlarges the application field of the LIP framework. In fact, when we have to process or analyze images acquired in reflected light, the goal generally consists of imitating (and, if possible, automating) what a human eye would do. In such cases, *we are perfectly justified in applying LIP tools and algorithms.*

In conclusion, LIP tools appear to be efficient for a wide typology of images. It is the reason why this book illustrates multiple examples concerning all areas.

3. FIRST CONSEQUENCES OF THE VECTOR SPACE STRUCTURE

We have seen that the space $I(D, [0, M[)$ of images is the positive cone of the vector space $F(D,]-\infty, M[)$. This implies that we can legitimately apply to $I(D, [0, M[)$ all the tools and properties associated with this kind of space, as long as we remain inside $I(D, [0, M[)$—ie, we do not need negative gray levels.

3.1 Logarithmic Interpolation

One of the first "tool transfers" (from vector spaces to images) I proposed was the use of a logarithmic interpolation. Such a concept is very simple: considering two images f and g, we can define the "segment" $[f, g]$ constituted of all the barycenters of f and g obtained with a scalar $\lambda \in [0, 1]$:

$$\forall (f,g) \in I^2\left(D, \left[0, M\right[\right), [f,g] = \{(\lambda \triangle f) \triangle ((1-\lambda) \triangle g)\}_{\lambda \in [0,1]} \tag{11}$$

Note that when λ moves from 1 to 0 in Eq. (11), we get intermediate images where the respective contributions of f and g are seen *in transparency*. This result is comparable to what is obtained in cinema when moving gradually from a scene to another (fade-out).

Let us now present two applications as examples of the logarithmic interpolation.

3.1.1 First Application: Three-Dimensional Reconstruction of T_4-Lymphocytes from Serial Cuts (See, for Example, Bron et al., 1990; Gremillet, Jourlin, Bron, Schüpbach, & Gautchi, 1991; Gremillet, Jourlin, & Pinoli, 1994)

In the Introduction, I evoked this research made in collaboration with the Swiss National Center for Retroviruses and the Electron Microscopy Laboratory (Zurich University). It took place during the years 1985-89 and aimed to visualize T_4 -lymphocytes removed from a culture in presence of HIV viruses in 3D. This removed sample was fixed with glutaraldehyde–acrolein, contrasted with uranyl acetate, dehydrated, and embedded in an Araldite or Epon block. Before the cutting procedure, it was necessary to mark the block with laser shots to facilitate the future image registration of successive cuts. Excimer lasers deliver an ultraviolet (UV) light of short wavelength with a high surface energy, which causes the disruption of intramolecular chemical bonds. Due to the small size of lymphocytes (diameter approximately 10–12 μm), it was necessary to realize an ultrathin serial cutting (40 nm) to expect a good quality of the 3D visualization. This step was performed with an Ultracut E (Reichert–Jung) and a Diatome cutting diamond. The sections were retrieved on carbon-reinforced, Formvar-coated, one-hole specimen grids (Bakers) according to a procedure of Galey and Nilsson. This procedure allowed the Zurich team to collect up to 500 serial cuts with a loss of less than 2%. The electron microscope (EM) images were acquired with a Philips EM 400T (cf. Fig. 5).

Our contribution to the project consisted of processing the images resulting of the previous procedure (around 250–300 images for a lymphocyte). The steps were as follows:

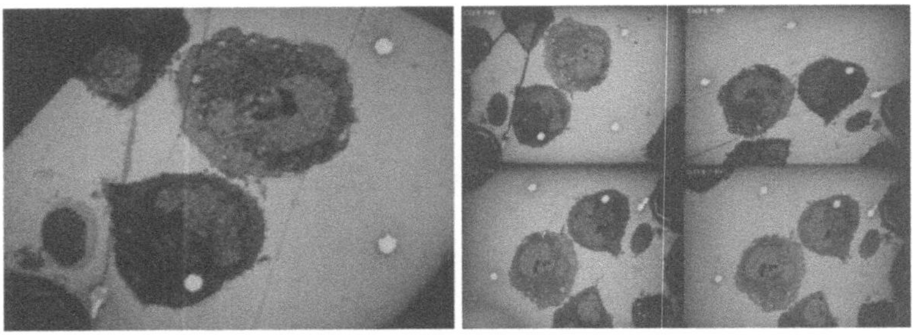

Fig. 5 Examples of images of T4-lymphocytes cuts. Note the laser marks (*white disks*). (See the color plate.)

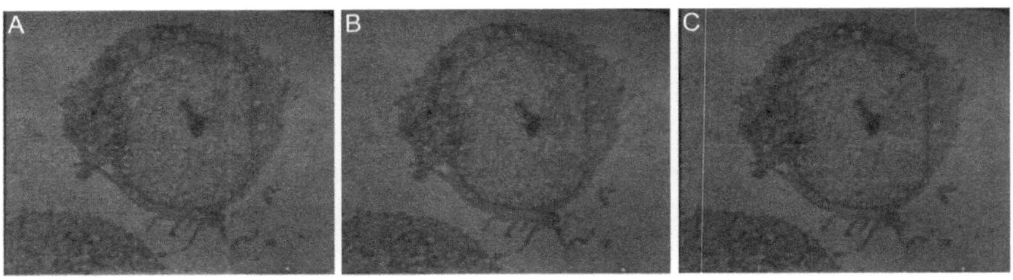

Fig. 6 Virtual changing of a cut thickness: here +10% (A) and +20% (B). (A) Initial image, (B) Image (A) ⩕1.1, (C) Image (A) ⩕1.2. (See the color plate.)

- Missing cuts: This problem was solved by a logarithmic interpolation between the cuts adjacent to the missing ones.
- Image enhancement: Even with perfect cutting techniques, two adjacent sections rarely have the same gray-level distribution. This is due to small variations in the thickness of the cuts, which result in different absorptions of the electrons. The LIP scalar multiplication allowed us to normalize the cuts' thickness and then the gray-level distribution in all imaged sections (cf. Fig. 6).
- Image restoration: The laser holes of successive images were used to counterbalance the deformations due to the cutting (cf. Fig. 7).
- Image analysis: The cutting procedures and the magnification chosen resulted in pixels of 40-nm length. The restoration of the third dimension thus yielded isotropic volume elements (voxels) of 40-nm side length. The restored 3D information of the whole reconstructed block is thus composed of 40 nm voxels. The reconstructed block is called a *virtual block* because it is mathematically reconstituted.
- Image reconstruction: This is easily done after the previous steps.

After the reconstruction step, it becomes possible to perform sections in the block. A section located along a laser hole permits observing the quality of deformations correction (cf. Fig. 7).

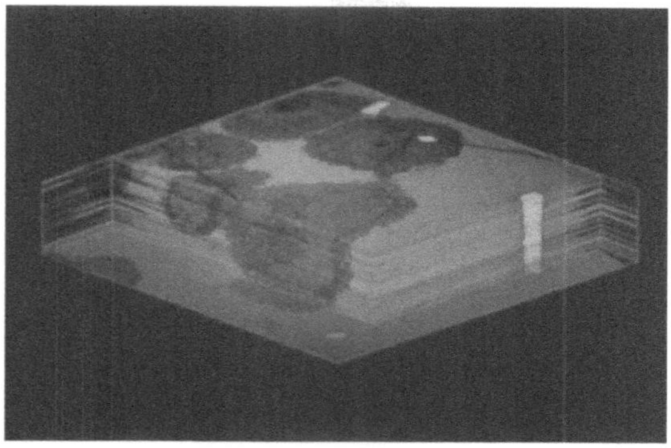

Fig. 7 Vertical cutting along a laser hole. (See the color plate.)

3.1.2 Second Application: Dental Surgery

The problem here concerns computer-aided surgery using 3D reconstruction of a jaw. The initial data consist of serial *optical* cuts of a jaw by means of an X-ray scanner (cf. Fig. 8). On such cuts, the opaque objects are bright (teeth) and the transparent ones black (air in the oral cavity, for example).

On this subject, we developed a long-term collaboration with the Faculté Dentaire (Lyon, France). The expectations of dentists were multiple and rather difficult to satisfy: in a first step, they wanted to visualize a 3D image of the set of scanner optical cuts in order to improve their understanding and interpretation. In a second one, they hoped it would be possible to perform an automated segmentation of this 3D image whose segments correspond to the anatomic elements like teeth, mandibular bone, sinuses, etc.

One initial problem was due to the anisotropy concerning the scanner spatial resolution (approximately 0.2 mm per pixel) and the thickness of each scanner cut (approximately 1 mm). With such a distortion, it was very difficult to produce a 3D image of good quality. Thus, we decided to compute four interpolated images between two successive cuts (cf. Fig. 9) in order to get cubic voxels.

Thanks to the interpolated cuts, it was possible to obtain an interpretable 3D visualization of the scanner serial cuts. One of our first results is presented in Fig. 10. Here the problem was to prepare a surgical operation consisting of removing a tooth grown in excess. It is necessary to precisely locate the concerned tooth; the difficulty is that it is under the gum and thus not visible. Classically, the dentist makes a sufficient incision to be sure to meet the tooth. A visual location becomes possible (extremity of the white arrow), through the 3D information in transparency, which permits an optimal

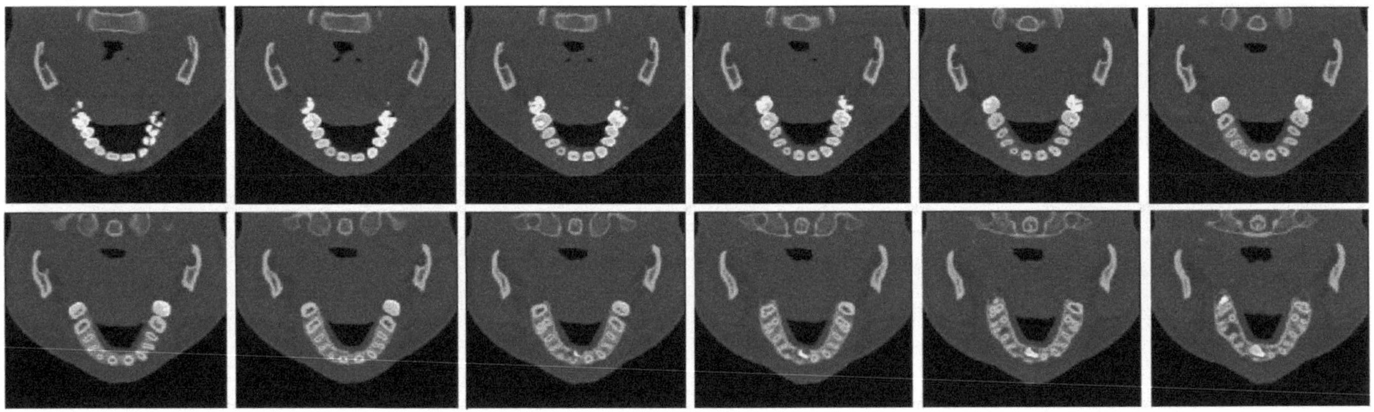

Fig. 8 Optical cuts of a jaw.

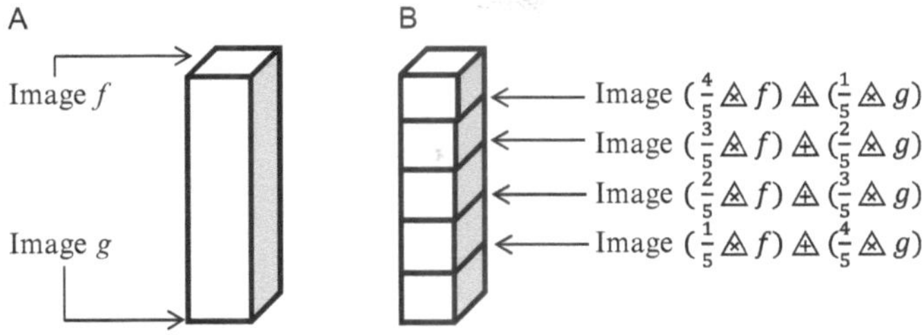

Fig. 9 Interpolation of two successive scanner cuts. (A) Initial voxel (height five times greater than the spatial resolution). (B) Initial voxel sliced into five cubic voxels based on four logarithmic interpolations.

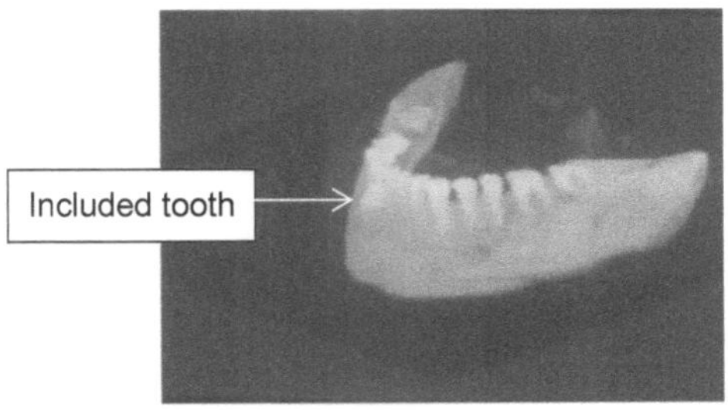

Fig. 10 Visualization of the tooth to be extracted. (See the color plate.)

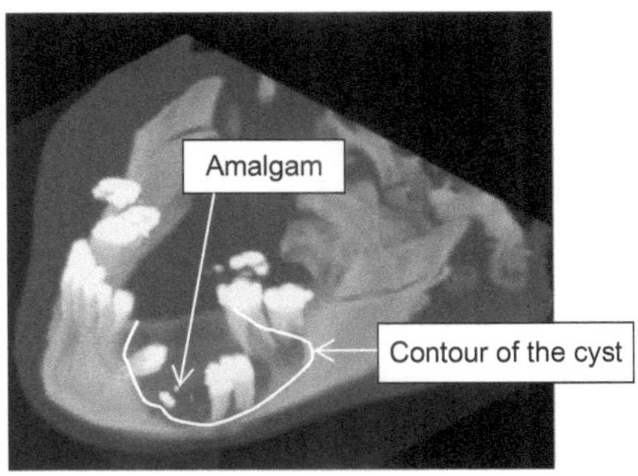

Fig. 11 Visualization of a cyst due to an amalgam piece.

surgical action. Among a large number of applications realized with the Faculté Dentaire (Lyon, France), I would present the amazing case of a young girl whose teeth did not grow normally. After her dentist asked a radiologist to realize X-ray scanner images, the cause of the problem appeared clearly on the 3D image: an enormous cyst due to the penetration in her gum of an amalgam element was invading the jaw. The 3D image presented in Fig. 11 gives a good visualization of the cyst and permits precisely locating the amalgam piece in order to remove it with a minimal incision at the gum surface. Thereafter, an antiinfectious treatment resulted in a progressive resorption of the cyst. Finally, an orthodontic step achieved a good positioning of the teeth (cf. Fortin et al., 1996).

Let us now examine the segmentation problem of these 3D images. Considering the dentists specifications, a first step was imagined before foreseeing a full automated segmentation. It consisted of performing a 3D *Region Growing* algorithm (see, for example, Gonzalez & Woods, 2002) starting from a seed chosen by the user inside a cut in order to mark the anatomic element of interest. In fact, we noticed that a standard *Region Growing* produced more or less acceptable results. Nevertheless, it appeared necessary to improve the algorithm, due in particular to the *Partial Volume Effect*: this problem is a consequence of the voxels located near the boundaries of two (or more) anatomical elements. In such a case, the voxel is the union of subsets of various anatomical elements and its gray level is not characteristic of a precise element. We can remark that this *Partial Volume Effect* produces blur in the boundaries neighborhoods (cf. Fig. 12, based on the visualization of the representative surface of a scanner cut region). In other words, the blurred parts are made up of "*Bridge Points*," mentioned by Asano and Yokoya (1981) as those permitting the *Region Growing* to jump over a boundary. The first efficient algorithm I proposed with Revol (Revol & Jourlin, 1997) is not, despite its title, strictly speaking a *Region Growing* algorithm: at each step of the algorithm, the points previously aggregated are reassessed and some of them may be removed of the current region if their suppression opens the way to affect more points to the region. This *systolic* approach permitted us to perform the first meaningful segmentations of teeth and sinuses.

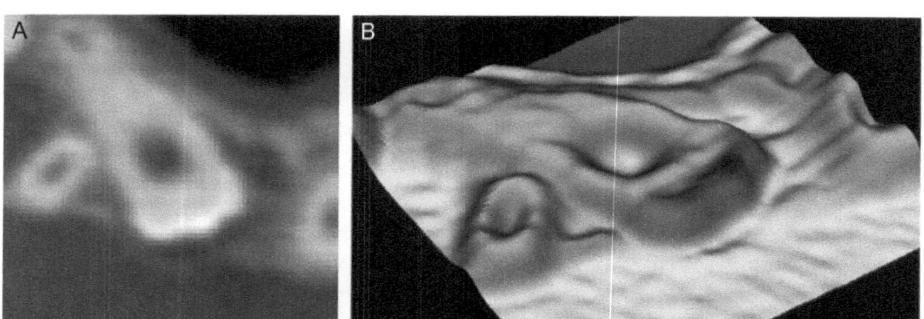

Fig. 12 Visualization of the blurred transitions due to the partial volume effect. (A) Gray-level image of a subset of an X-ray scanner cut. (B) Representative surface of image (A). (See the color plate.)

Comment on Fig. 12: The choice of false colors is only done to help in the visual interpretation of blur.

At the same time, another technique for the segmentation problem was studied by Darsonville in his PhD thesis (cf. Darsonville, 1998). Due to their comparable opacity to X-rays, the main challenge was to succeed in separating the teeth roots from the mandibular bone. The idea consisted of applying a preprocessing step to each X-ray optical cut, including two operations: a noise filtering (for example, a Median filter of size 5×5 to 9×9) in order to get a better homogeneity of regions and a strong deblurring (for example, using a Laplacian (cf. Jourlin, 1987) for a basic technique with a 3×3-neighborhood) in order to perform a boundary "crispening" (cf. Figs. 13A and B, and 14A and B).

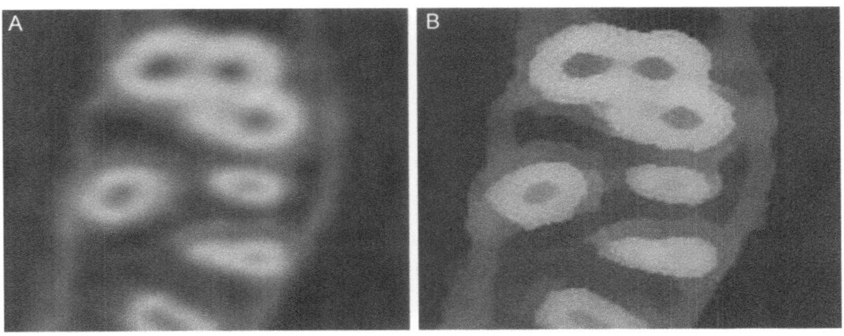

Fig. 13 Preprocessing of an X-ray scanner cut preparing a region growing. (A) Gray-level image of a subset of an X-ray scanner cut. (B) Result of a preprocessing of (A) including noise filtering and deblurring (edge crispening).

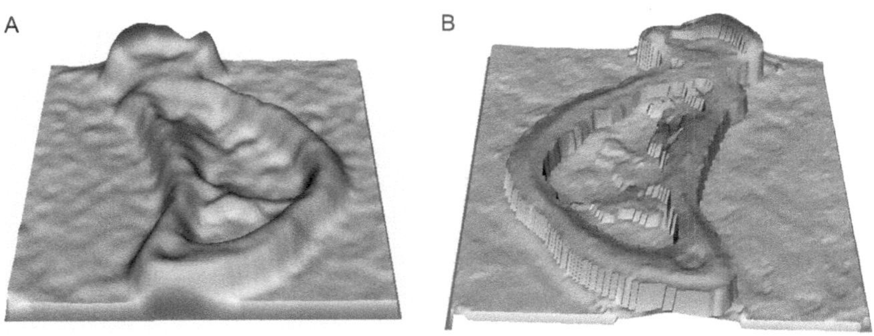

Fig. 14 Visualization in false colors of the deblurring preprocessing effect. (A) Representative surface of an X-ray scanner cut. (B) Effect of a preprocessing of (A) including noise filtering and deblurring (edge crispening). (See the color plate.)

Another manner of observing the preprocessing efficiency consists of applying any gradient operator to images of Fig. 13A and B. It becomes obvious to notice that the boundaries are thinner after edge crispening (cf. Fig. 15A and B). Such a result will make the interpretation of the gradient image, which is a gray-level image, easier and thus necessitates the extraction of binary contours.

In such a way, it becomes extremely difficult for the growing process to pass over the transitions between regions and we are ready to confront the challenging problem of separating teeth roots from mandibular bone. To be precise, it is not difficult to extract the regions corresponding to air and flesh (gum). Once these regions were removed from the 3D block, our systolic approach roughly succeeded in solving the problem. The first (unprocessed) results we obtained are visualized in Fig. 16 that shows the teeth and the sinuses after removing the mandibular bone. The classes' surfaces appear a little irregular before any smoothing: the main reason is certainly the low resolution of the initial image, in other words, the voxels large size.

Now, to take into account the requirement of dentists, we dedicated the next step to improve the aesthetic aspect of our 3D images. We will not look in depth in the algorithms used to achieve this goal (cf. Fig. 17A–C) because this question is out of the present book subject.

Comment on Fig. 17A: The quality of such an image permits observation of the proximity of molar roots with sinuses. Note the presence of a wisdom tooth (under the gum surface) to be extracted. It is possible to precisely

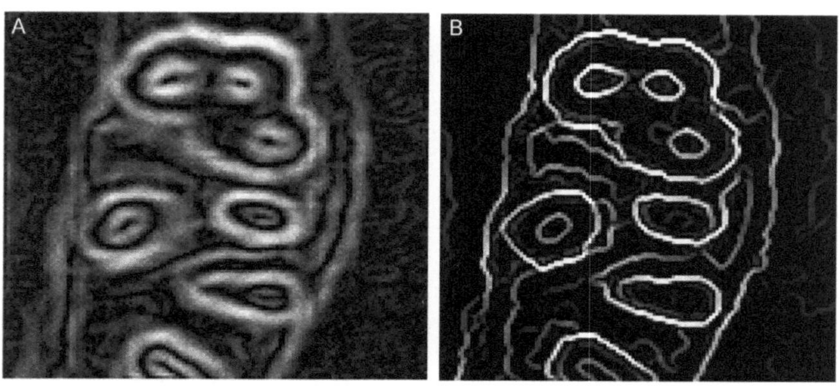

Fig. 15 Efficiency of the preprocessing in thinning boundaries. (A) Result of a Sobel gradient applied to image Fig. 13A). (B) Result of a Sobel gradient applied to image Fig. 13B).

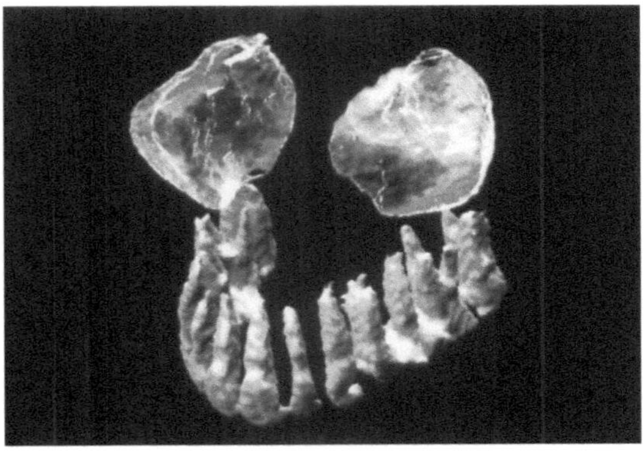

Fig. 16 3D visualization of two anatomic elements of the superior part of a jaw: the teeth roots and the sinuses. (See the color plate.)

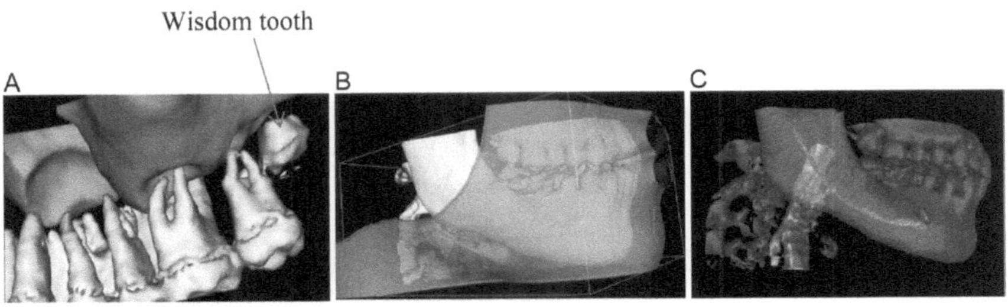

Fig. 17 Examples of 3D images improvements. (A) Interaction between teeth roots and sinuses. (B) Visualization of the jaw and the flesh (semitransparent rendering of the flesh). (C) Visualization of the jaw without the flesh. Semitransparent rendering of the jaw bone: in *red* the inferior dental nerve, in *green* the trachea. (See the color plate.)

evaluate its distance to the sinus in order to avoid a clumsy gesture with the scalpel.

General comment concerning this application: to temper the dentist's enthusiasm, it remained a fundamental question to answer: the realism and visual quality of the produced 3D segmentations are not sufficient. We expect the produced results are confident, which means that the 3D shapes extracted are the same as those of the true anatomical elements. Such a question is justified for intellectual satisfaction. Moreover, we observed that if some parameters were modified in our algorithms (for example, the aggregation tolerance driving the *Region Growing*), the shape of segmented regions was obviously influenced. Fig. 18 shows two examples of this phenomenon: in case (a) (resp. (b)), we display four (resp. two) images of a same tooth, depending on the chosen parameters.

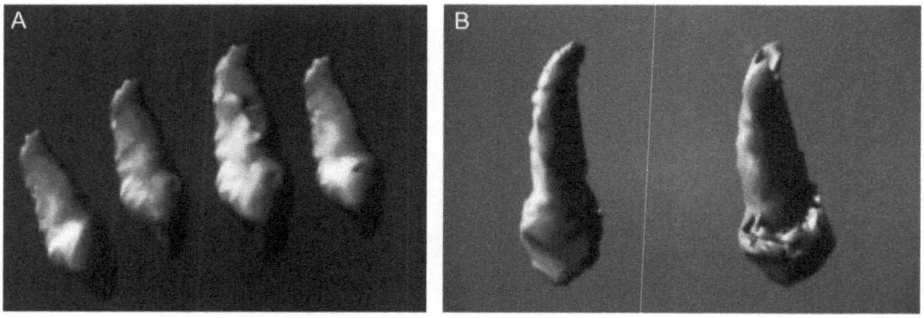

Fig. 18 (A) and (B) Examples of shapes modification of segmented regions due to the selected algorithm. (See the color plate.)

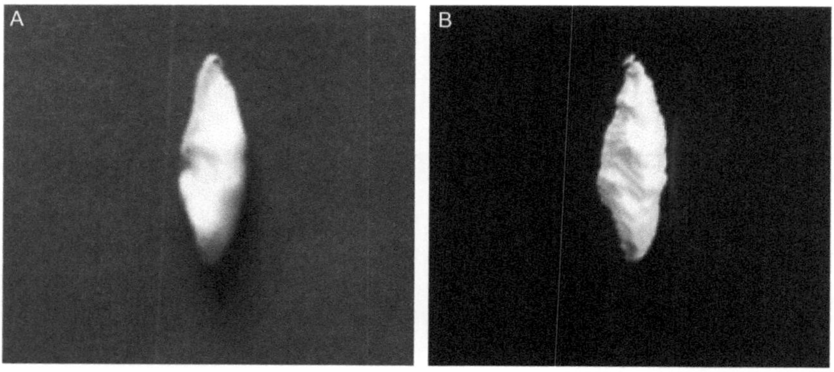

Fig. 19 Comparison of a real tooth to its 3D image before extraction. (A) Image of the real tooth after extraction. (B) Segmentation of the same tooth obtained from the 3D image before extraction. (See the color plate.)

The situations presented in Fig. 18 are rather extreme. Nevertheless, we decided to adjust the *Region Growing* parameters because of cases where we had the real model of an anatomic element, for example, an extracted tooth and a 3D image of the same tooth before extraction. It then becomes possible to fix the parameters values so that the shapes of the real tooth (Fig. 19A) and the segmented one (Fig. 19B) appear as similar as possible considering the poor resolution of X-ray scanner cuts in thickness. After applying this method on a significant number of cases, we were able to definitely determine the algorithm parameters.

Dental project extension: We initialized some other applications in my LISA laboratory (CPE Lyon), in particular concerning facial hemiatrophies, which are anomalies of the first branchial arch. The treatment of these dysmorphoses consists of bone distraction, resulting in an elongation of the atrophied mandibular bone. Using 3D segmented images, it was possible to precisely compute the displacement we must apply in the three directions to approach a symmetrization of the face (cf. Diemunch et al., 2007).

3.2 Other "Historical" Applications

Among the numerous possibilities made possible by the two logarithmic laws and their properties, I explored from the beginning two other consequences of the vector space structure:

a. Definition of a logarithmic additive contrast

b. Dynamic range expansion of an image

These two subjects will be studied in depth, respectively, in Chapter "Various Contrast Concepts" for (a) (sections 1.1, 1.2, 1.3) and in Chapter "Dynamic Range Expansion, Night Vision. Stabilization, Centering. Industrial and Biomedical Applications" for (b). Concerning this last point, the initial method was based on the scalar multiplication by the "optimal" real number λ_0 (Section 2.1.1).

4. ABOUT A GOOD USE OF THE LIP MODEL

4.1 Interest of the Isomorphism Between [0, M[and $\mathbb{R}^+$

Due to the fact that Ln is a strictly increasing function, J.-C. Pinoli observed that the application φ defined according to the equation:

$$\varphi(l) = -MLn\left(1 - \frac{l}{M}\right)$$

was obviously an isomorphism from $[0, M[$ onto $\mathbb{R}^+$, associating a positive real number to a given gray-level l. The same application is then able to transform a gray-level image $f \in I(D, [0, M[)$ in a function $\varphi(f) \in F(D, \mathbb{R}^+)$ or a function of the over-space $F(D,]-\infty, M[)$ in a function defined on D with values in $\mathbb{R}$ [ie, an element of $F(D, \mathbb{R})$].

In such conditions, when we work with 8-bits images, the corresponding gray scale $[0, \ldots, 255]$ is distributed inside $\mathbb{R}^+$ according to the values: $\varphi(0), \varphi(1), \ldots, \varphi(254), \varphi(255)$, ie, $0, 1.002, \ldots, 1242.120$, 1419.565 (cf. Fig. 20).

The continuous logarithmic representative curve of φ is shown in Fig. 21, where the x axis represents the initial gray scale [0,256[and the y axis represents $\mathbb{R}^+$.

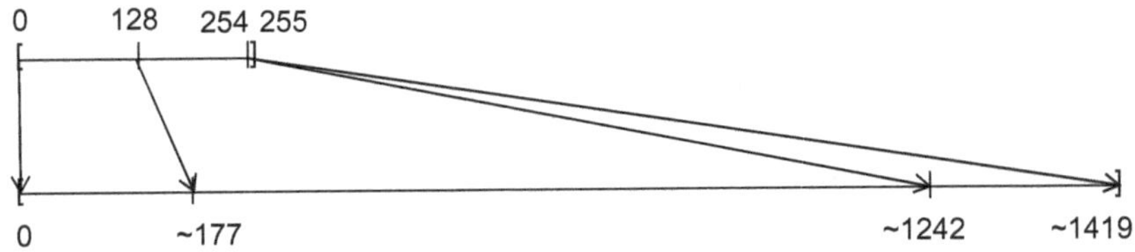

Fig. 20 Correspondence between the gray scale [0,255] and $\varphi([0,255])$.

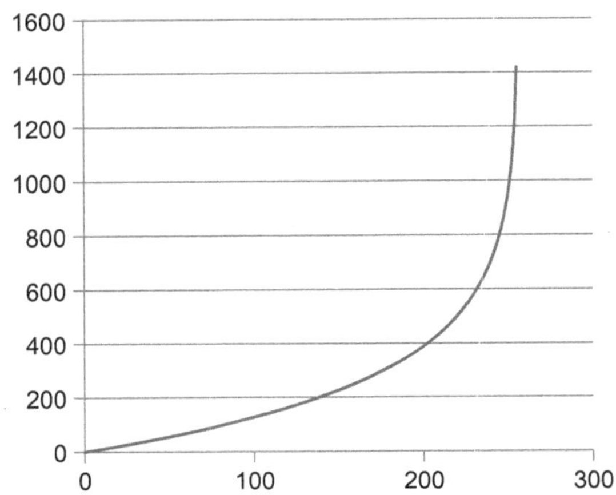

Fig. 21 Logarithmic representative curve of isomorphism φ.

Remark 5. It is important to note that the part $[255, 256[$ of the gray scale corresponds to the infinite interval $]\sim 1419, +\infty[$.

Consequence: We will demonstrate throughout the next chapters that the LIP tools are strongly sensitive in the dark-gray levels (for us the great values of the scale, cf. Warning 1), but the previous remark shows that all the very dark values are confused at the value 255, which constitutes an obstacle for applying our algorithms to very low-light images and, for example, in the case of night vision. Nevertheless, the LIP model is developed at a theoretical level, which makes it able to adapt to future sensors more sensitive in the dark-gray levels, with a digitization onto 10 bits, 12 bits, or more. Such a perspective appears very exciting, but I want to reassure the reader: even with this current limitation, we will demonstrate the practical interest of the LIP tools through numerous applications.

The advantage of the previous isomorphism is that it transforms the linear initial gray scale into a logarithmic one where the LIP operators $\triangle\!\!\!\!+$, $\triangle\!\!\!\!\times$, $\triangle\!\!\!\!-$ are replaced by ordinary operations $+$, $\times$, $-$.

In practice, it is not so simple. For example, to compute the LIP addition of two gray levels l and m, the steps are as follows:

- Compute $\varphi(l)$ and $\varphi(m)$
- Compute the classical sum $s = \varphi(l) + \varphi(m)$
- Compute the inverse isomorphism $\psi(s) = \varphi^{-1}(s) = \varphi^{-1}(\varphi(l) + \varphi(m)) = l \triangle\!\!\!\!+ m$

Such successive operations increase significantly the computation time, particularly in the case of addition. For the scalar multiplication, the two approaches present similar costs due to the exponential function present in $\lambda \triangle\!\!\!\!\times$.

In any case, if the goal is to speed up the algorithms, the execution times of the two methods are strongly shortened by the use of precomputed LUTs.

So what? In my opinion, the existence of the isomorphism φ is intellectually satisfactory. Practically, it can avoid writing programs corresponding to LIP operators. Nevertheless, its systematic use presents two drawbacks:

1. It does not simplify the notations. For example, if we want to make the discrimination between gray-level images considered in the initial gray scale and their transformed functions by φ. Moreover, it sometimes occurs that we need to perform both linear and logarithmic operations between images inside the same algorithm. In this case, the use of notations $\triangle\!\!\!\!+$, $\triangle\!\!\!\!\times$, $\triangle\!\!\!\!-$ on one hand and $+$, $\times$, $-$ on the other easily permits the distinction.
2. As already apparent, the isomorphism φ is not *onto* $\mathbb{R}^+$ in the digital case: it only offers a logarithmic distribution of the initial linear gray scale. Furthermore, for digital images, other logarithmic distributions would play the same role if φ is replaced by any function $-kLn\left(1 - \frac{l}{M}\right)$ where k represents any real number strictly positive.

At this step, the definition of an isomorphism like φ appears equivalent to the question developed in the next section.

4.2 Is It Possible to Define a Logarithmic Gray Scale?

I asked myself this question years ago and formulated it as follows:

"Considering the Transmittance Law as founding the LIP laws, is it possible to define inside the gray scale [0, 255] a logarithmic scale of 255 levels respecting the following property: each gray level represents the same transmittance information?"

Another way to express that is:

"Is it possible to define a gray level unit u with which a logarithmic gray scale is defined according to: $u, 2 \triangle\!\!\!\!\times u, \ldots, 255 \triangle\!\!\!\!\times u$?"

The answer I suggest includes two aspects:

1. If we try $\in$ to build such a logarithmic scale inside the (analogic) interval $[0, M[$ without more specification, every value $u \in]0, M[$ is suitable: $255 \triangle u$ is always $< M$ and is thus an element of the interval $]0, M[$.
2. To determine u, we need to fix a superior bound to $255 \triangle u$ and the natural solution is to consider the darkest level of the linear gray scale $[0, 255]$, ie, 255. This implies [based on Eq. (10)]:

$$255 \triangle u = 255 \Leftrightarrow 256 - 256\left(1 - \frac{u}{M}\right)^{255} = 255 \Rightarrow u \sim 5.5$$

This means that the *logarithmic* gray-level unit u is equivalent to more than five *linear* gray-level units, which confirms that in the LIP framework, we will be more sensitive in the dark levels and less sensitive in the bright ones, compared with the linear processing. This leads us to the next section.

4.3 Is LIP Interest Limited to the Use of a Logarithmic Look-Up-Table?

Before we answer this question, we must define a Look-Up-Table (LUT) in image processing. A LUT is a correspondence table whose goal may be "*to transform the input data into a more desirable output format*" (Wikipedia). The term "more desirable" is sufficiently imprecise to cover various interpretations: in image processing, it often means "more desirable for human vision": in fact, LUTs are generally based on mathematical transformations (logarithmic, exponential, power, or gamma functions) that favor the visual interpretation of dark regions, for example, (or bright ones) by affecting a larger dynamic range than in the initial image. Another use of LUTs is the precomputation of equations or data we will apply later in order to speed-up the calculus. Here, we are concerned by the first aspect and the isomorphism φ is nothing but a LUT.

Nevertheless, the interest of LIP operators is not limited to that role, due to the numerous notions we will define (interpolation, contrasts, metrics, etc.) with efficient applications in various domains: dental surgery, 3D reconstruction of cells, interpretation of cars crash tests, confocal microscopy, automated interpretation of drone's images. Furthermore, the physical (optical) justification of LIP laws permit simulation of exposure time changing, source intensity variations, thickness modification of the obstacle, etc., so that, even when we exploit the LIP laws as logarithmic LUTs, *we know exactly why and under what hypotheses it is justified,* contrary to other transformations.

5. TRANSITION TO NEXT CHAPTERS

When I discussed the first consequences of the vector space structure in Section 3, I also planned to discuss three subjects, according to the chronology of the LIP development: the logarithmic interpolation (widely done through biomedical examples), but also the definition of a novel contrast notion and the use of the logarithmic multiplication to enhance, stabilize, and center an image. In fact, these three applications were the first I imagined.

Regarding the definition of a logarithmic contrast, let me mention the following reference (Jourlin, Pinoli, & Zeboudj, 1989) and for image enhancement and centering, I suggest the interested reader look to Jourlin and Pinoli (1995). Nevertheless, these subjects have been so expanded during these past years that I decided to dedicate a specific chapter to each of them: Chapter "Various Contrast Concepts" for *Various Contrasts Concepts* and Chapter "Dynamic Range Expansion, Night Vision. Stabilization, Centering. Industrial and Biomedical applications" for *Stabilization, Centering, Dynamic Range Expansion, and Night Vision.*

ACKNOWLEDGMENT

To conclude the second application "Dental surgery" dedicated to computer- (and image-) assisted dental surgery, I must thank all the teams that participated with us in the European project MINOSQUARE (Maxillary Imaging Network on Optimization Strategies and Quality Assurance of Radiographs in Europe) headed by Reinhilde Jacobs, Professor at the University of Leuven and concluded by a Congress in Leuven (2000).

REFERENCES

Asano, T., & Yokoya, N. (1981). Image segmentation schema for low-level computer vision. *Pattern Recognition, 14*(1–6), 267–273.

Brailean, J., Sullivan, B., Chen, C., & Giger, M. (1991). Evaluating the EM algorithm using a human visual fidelity criterion. In *International conference on acoustics, speech and signal processing* (pp. 2957–2960).

Bron, C., Gremillet, P., Launay, D., Jourlin, M., Gautchi, H. P., Bächi, T., et al. (1990). Three-dimensional electron microscopy of entire cells. *Journal of Microscopy, 157*(1), 115–126.

Darsonville, F. (1998). *Segmentation and visualization of 3D images: Application of region growing to dental images.* PhD, Saint-Etienne, France: University Jean Monnet.

Diemunch, C., Faure, F., Trunde, F., Morgon, L., Bossard, D., Jourlin, M., et al. (2007). Three-dimensional modeling system for unilateral mandibular bone distraction: A clinical case. *Computer Aided Surgery, 12*(5), 262–269.

Fortin, T., Coudert, J. L., Francois, B., Huet, A., Niogret, F., Jourlin, M., et al. (1996). Marsupialization of dentigerous cyst associated with foreign body using 3D CT images: A case report. *The Journal of Clinical Pediatric Dentistry*, *22*(1), 29–33.

Gonzalez, R. C., & Woods, R. E. (2002). *Digital image processing* (2nd ed.). New Jersey: Prentice Hall.

Gremillet, P., Jourlin, M., Bron, C., Schüpbach, J., & Gautchi, H. P. (1991). Dedicated image analysis techniques for three-dimensional reconstruction from serial sections in electron microscopy. *Machine Vision and Applications*, *4*(4), 263–270.

Gremillet, P., Jourlin, M., & Pinoli, J. C. (1994). LIP-model based three-dimensional reconstruction of HIV-infected entire cells. *Journal of Microscopy*, *174*(1), 31–38.

Jourlin, M. (1987). Role of the underlying Laplacian in the edge improvement. *Journal de microscopie et de spectroscopie électroniques*, *12*(1), 89–93.

Jourlin, M., & Pinoli, J.-C. (1985). *A model for logarithmic image processing*. Département de Mathématiques, no. 3, Université de Saint-Etienne.

Jourlin, M., & Pinoli, J.-C. (1987). Logarithmic image processing. *Acta Stereologica*, *6*(3), 651–656.

Jourlin, M., & Pinoli, J.-C. (1988). A model for logarithmic image processing. *Journal of Microscopy*, *149*(1), 21–35.

Jourlin, M., & Pinoli, J.-C. (1995). Image dynamic range enhancement and stabilization in the context of the logarithmic image processing model. *Signal Processing*, *41*(2), 225–237.

Jourlin, M., & Pinoli, J.-C. (2001). Logarithmic image processing: The mathematical and physical framework for the representation and processing of transmitted images. *Advances in Imaging and Electron Physics*, *115*, 129–196.

Jourlin, M., Pinoli, J.-C., & Zeboudj, R. (1989). Contrast definition and contour extraction for logarithmic images. *Journal of Microscopy*, *156*(1), 33–40.

Lang, S. (2002). *Algebra, Graduate texts in mathematics*: *Vol. 211* (Revised 3rd ed.). New York: Springer-Verlag.

Revol, C., & Jourlin, M. (1997). A new minimum variance region growing algorithm for image segmentation. *Pattern Recognition Letters*, *18*(3), 249–258.

Schaefer, H. (1999). *Topological vector spaces* (2nd ed.). New York: Springer-Verlag.

CHAPTER TWO

Various Contrast Concepts

Contents

In this chapter, we will return to the notion of logarithmic additive contrast (LAC), which was one of the first consequences I foresaw from the additive law. By definition, this contrast will have a precise physical (optical) interpretation. Moreover, an explicit link between the LAC and the classical Michelson contrast will be demonstrated.

In image processing, there exists a wide range of applications where the concept of contrast plays a fundamental role. We will insist on the benefit, in such applications, to replace the classical contrast notions by the LAC, due to its double advantage of being consistent with human vision and to possess a physical meaning. Nevertheless, the LAC interest is not limited to this replacement and this chapter will propose novel applications. Moreover, this notion will be essential in other sections of the book, for example, in the

Advances in Imaging and Electron Physics, Volume 195
ISSN 1076-5670
http://dx.doi.org/10.1016/bs.aiep.2016.04.002

possible definition of novel metrics (see Chapter "Metrics Based on Logarithmic Laws").

Finally, the concept of logarithmic multiplicative contrast (LMC) will be studied.

Definitions of contrast

- General definition (dictionary): *Opposition between two things, one being emphasized by the other*
- In the field of vision (Wikipedia): *The difference in color and light between parts of an image*

Regarding this last definition, we will limit our study to the situation of gray-level images. Two cases must be distinguished: the general case where the goal is to estimate the global contrast inside a region $R \subset D$ and the local case which concerns a pair of pixels x and y, generally neighbors.

If f is a gray-level image and R a subset of D, the global contrast of f inside R (Michelson contrast) is often estimated as:

$$C_R^{\mathrm{Mi}}(f) = \frac{\mathrm{Max}_R(f) - \mathrm{Min}_R(f)}{\mathrm{Max}_R(f) + \mathrm{Min}_R(f)} \tag{1}$$

where $\mathrm{Max}_R(f)\,[\text{resp.}\,\mathrm{Min}_R(f)] = \mathrm{Max}_{x\in R}(f(x))\,[\text{resp.} = \mathrm{Min}_{x\in R}(f(x))]$

In the particular case of a region uniformly null, ie, $\mathrm{Max}_R(f) = \mathrm{Min}_R(f) = 0$, we admit the equality $C_R^{\mathrm{Mi}}(f) = 0$.

Remark 1. Be careful with this definition which takes place in a classical context where the value 0 in the gray scale $[0, M[$ represents the black level, which means that $\mathrm{Max}_R(f)$ corresponds to the brightest point of f on R, contrary to our conventions in the logarithmic image processing (LIP) context.

Let us now consider the local case: if the region R is restricted to a pair of neighboring points (x, y), Eq. (1) becomes:

$$\begin{aligned} C_{(x,y)}^{\mathrm{Mi}}(f) &= \frac{|f(x) - f(y)|}{f(x) + f(y)} && \text{if } f(x) \neq 0 \text{ or } f(y) \neq 0 \\ \text{and } C_{(x,y)}^{\mathrm{Mi}}(f) &= 0 && \text{if } f(x) = f(y) = 0 \end{aligned} \tag{2}$$

Remark 2. The quotient $\frac{|f(x) - f(y)|}{f(x) + f(y)}$ in Eq. (2) may be interpreted as the standard deviation $\frac{|f(x) - f(y)|}{2}$ of the pair $(f(x), f(y))$ divided by the mean value $\frac{f(x) + f(y)}{2}$. Such a quotient must be compared with the Weber ratio

$W = \frac{\delta I}{I}$ (cf. section 2.4 of Chapter "Gray-Level LIP Model. Notations, Recalls, and First Applications").

Remark 3. Let us note that a very simple contrast notion is also used, consisting of evaluating the difference $|f(x) - f(y)|$ between the gray levels of x and y, in absolute value.

1. LOGARITHMIC ADDITIVE CONTRAST

1.1 Definition

This section is devoted to LAC, which was published for the first time in Jourlin, Pinoli, and Zeboudj (1989). Considering a gray-level function f and a pair (x, y) of neighboring pixels in the spatial support D, we start with the simplest notion of contrast $|f(x) - f(y)|$ in which we project to replace the classical subtraction by a logarithmic one. Let us recall that in the Chapter "Gray Level LIP Model. Notations, Recalls, and First Applications," we noticed that $|f \triangle g| \neq |g \triangle f|$. Finally, to remove any ambiguity to such a logarithmic difference, it is possible to define:

$$C^{\triangle}_{(x,\,y)}(f) = \mathrm{LAC}_{(x,\,y)}(f) = \mathrm{Max}(f(x), f(y)) \triangle \mathrm{Min}(f(x), f(y)) \quad (3)$$

In such a way, $\mathrm{Max}(f(x), f(y))$ is darker than $\mathrm{Min}(f(x), f(y))$. The subtraction in Eq. (3) *always results in a gray level* and using the definition of the logarithmic subtraction yields:

$$C^{\triangle}_{(x,\,y)}(f) = \mathrm{LAC}_{(x,\,y)}(f) = \frac{|f(x) - f(y)|}{1 - \dfrac{\mathrm{Min}(f(x), f(y))}{M}} \quad (4)$$

Remark 4. If $f(x)$ is noted f and $f(y)$ becomes $f + \delta f$, the definition of Eq. (4) shows that the LAC is expressed as $M\frac{\delta f}{M - f} = MW$ according to the equality established in Jourlin and Pinoli (1988) between the Weber constant and the quotient $\frac{\delta f}{M - f}$. Thus we can assert that the LAC follows the Weber–Fechner law.

1.2 First Properties of the LAC

In Section 1 the applicative importance of the LAC will appear in various situations. Let us begin by providing some properties:

(1) $\forall f \in I(D,[O,M[),\forall(x,y)\in D^2: \qquad C^{\triangle\!\!\!\!+}_{(x,y)}(f)\in[0,M[$

(2) $\forall f \in I(D,[O,M[),\forall(x,y)\in D^2,\forall\lambda\in\mathbb{R}: C^{\triangle\!\!\!\!+}_{(x,y)}(\lambda \triangle\!\!\!\!\times f)=\lambda \triangle\!\!\!\!\times C^{\triangle\!\!\!\!+}_{(x,y)}(f)$

(3) $\forall(f,g)\in I^2(D,[O,M[),\forall(x,y)\in D^2: \quad C^{\triangle\!\!\!\!+}_{(x,y)}(f \triangle\!\!\!\!+ g)\le C^{\triangle\!\!\!\!+}_{(x,y)}(f) \triangle\!\!\!\!+ C^{\triangle\!\!\!\!+}_{(x,y)}(f)$

Comments on Eq. (4) and on These Properties

- To define the LAC, we started with a logarithmic difference (Eq. 3), but the resulting expression (Eq. 4) is roughly similar to that of Michelson's contrast. In fact, the two expressions possess the same numerator and we could interpret their denominators as terms dedicated to *weight* the numerator according to the location of the gray levels $f(x)$ and $f(y)$ in the gray scale. To be more precise, we will clearly establish that the LAC and the Michelson contrast have a common property: for a same difference $|f(x)-f(y)|$, both of them overestimate the contrast of dark points compared with the contrast of bright ones. Such a common property incited us to examine if these two notions were explicitly linked to each other. For more precision, see Section 1.3.
- The formula expressed in Property (2) shows the homogeneity of the LAC related to λ, ie, it does not change when the thickness of the obstacle varies. Property (3) shows the LAC is subadditive. With these properties (2) and (3), the LAC becomes for mathematicians a positively subLinear form.

Proof of Property (2).

$$\forall f \in I\big(D,[O,M[\big),\forall(x,y)\in D^2,\forall\lambda\in\mathbb{R}:$$
$$C^{\triangle\!\!\!\!+}_{(x,y)}(\lambda \triangle\!\!\!\!\times f)=\mathrm{Max}(\lambda \triangle\!\!\!\!\times f(x),\lambda \triangle\!\!\!\!\times f(y)) \triangle\!\!\!\!- \mathrm{Min}(\lambda \triangle\!\!\!\!\times f(x),\lambda \triangle\!\!\!\!\times f(y))$$

Knowing that the logarithmic scalar multiplication is expressed as an exponential function, it is therefore increasing and we have:

$$\begin{aligned} C^{\triangle\!\!\!\!+}_{(x,y)}(\lambda \triangle\!\!\!\!\times f) &= \lambda \triangle\!\!\!\!\times [\mathrm{Max}(f(x),f(y)) \triangle\!\!\!\!- \mathrm{Min}(f(x),f(y))] \\ &= \lambda \triangle\!\!\!\!\times C^{\triangle\!\!\!\!+}_{(x,y)}(f) \end{aligned}$$

□

Property (3) is obvious.

1.3 Physical Meaning of the LAC

We estimate this section is very important, considering that the most common contrast notions are not optically justified.

Fundamental remark: Eq. (3) is very easy to interpret. The LAC between two gray levels appears as *the gray level which must be added* (in transmitted light, superposed) *to the brightest in order to obtain the darkest one.*

Such an interpretation incited us in a practical verification (Mayet, Pinoli, & Jourlin, 1996): 15 semitransparent sheets of plastic were stacked between the source and the camera with a small gap separating two successive sheets (cf. Fig. 1). From this experience, we verify the linearity, *in a logarithmic scale*, of gray levels acquired by the camera related to the corresponding sheets number.

Another interesting observation is that the LACs computed between the gray levels of two successive sheets are perfectly identical. This result is not surprising, due to the definition of the LAC, but it shows that the detection of an object observed over an inhomogeneous background becomes possible. Let us recall that in the situation of a background with lighting drift, a classical solution in mathematical morphology is the use of the "Top-Hat operator," based on a gray-level opening, which extracts the peaks and crest lines (see, for example, Serra, 1982). Nevertheless, the chosen height of significant peaks is the same whatever the background while it is obvious that in transmitted light, an observed object will generate a small gap when the background is dark and a larger one otherwise. The LAC is the correct answer to this problem and this consideration incited us to define a logarithmic "Top-Hat" (Jourlin & Montard, 1997).

1.4 Explicit Link Between Michelson's Contrast and the Logarithmic Additive One

In the comments on Eq. (4), we evoked the common behavior of these two different notions consisting of overweighting the contrasts in the dark parts of the gray scale.

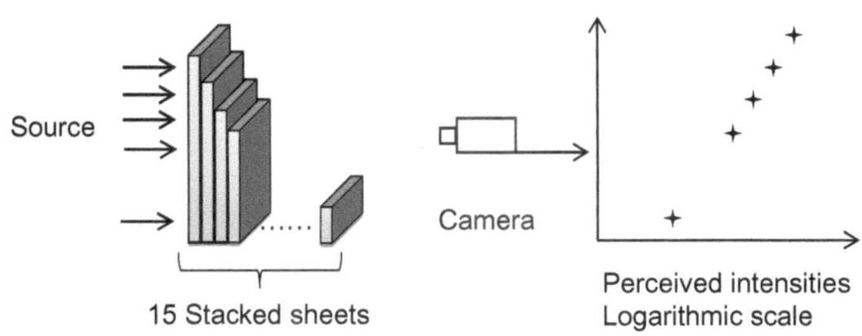

Fig. 1 Plastic sheets stacked between the source and the sensor.

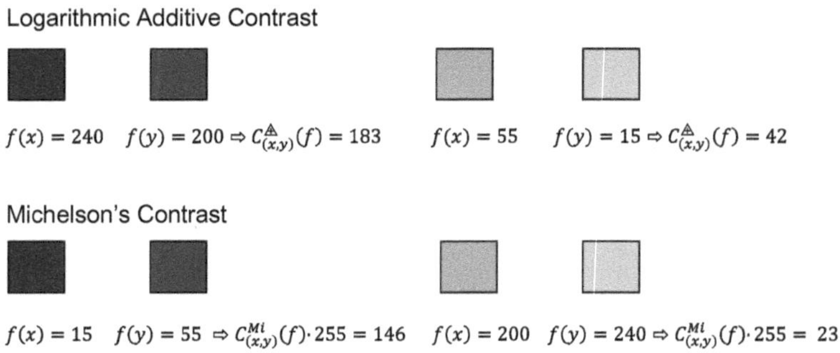

Fig. 2 Behavior comparisons of logarithmic additive and Michelson contrasts.

To illustrate that point, let us consider two pairs of points having, respectively, the gray levels 240 and 200 in the LIP context (0 = white) and the gray levels 15 and 55 in the standard gray scale (inverse of the LIP one, ie, 0 = black) corresponding to Michelson and let us compute their contrasts (Fig. 2).

Comment of Fig. 2: The Michelson contrast lies in the interval $[0, 1]$ and we can exclude the value 1, which corresponds to $f(x) = f(y) = 0$ (black pixels not considered in the scale $[0, M[$ within the LIP context. To obtain comparable values, the Michelson contrast has been multiplied by 255. For the same reason, the chosen gray level values take into account the gray-scale inversion.

The observed contrasts values confirm that the two notions overweight the dark part of the gray scale. Moreover this overweighting appears stronger in the LAC case than in the Michelson one.

Let us now extend the previous observation to the general case of two pixels x and y whose gray levels differ from a constant k, ie, $f(y) = f(x) + k$, with $f(y)$ darker than $f(x)$. It is possible to compare $C^{\triangle}_{(x,y)}(f)$ and $C^{\mathrm{Mi}}_{(x,y)}(M - 1 - f)$. We first decided to compute the curves of these two contrasts for some values of the constant k (cf. Fig. 3).

As observed in the example (Fig. 2), these curves confirm that the LAC is always greater than Michelson's contrast. Moreover, the similitude of the curves shapes suggests the existence of a link between the two contrasts. Precisely, the following result holds:

The Michelson's contrast is explicitly linked to the LAC according to the Eq.:

$$M \cdot C^{\mathrm{Mi}}(f(x), f(x) + 2k) = C^{\triangle}(M - f(x), M - f(x) - k) \tag{5}$$

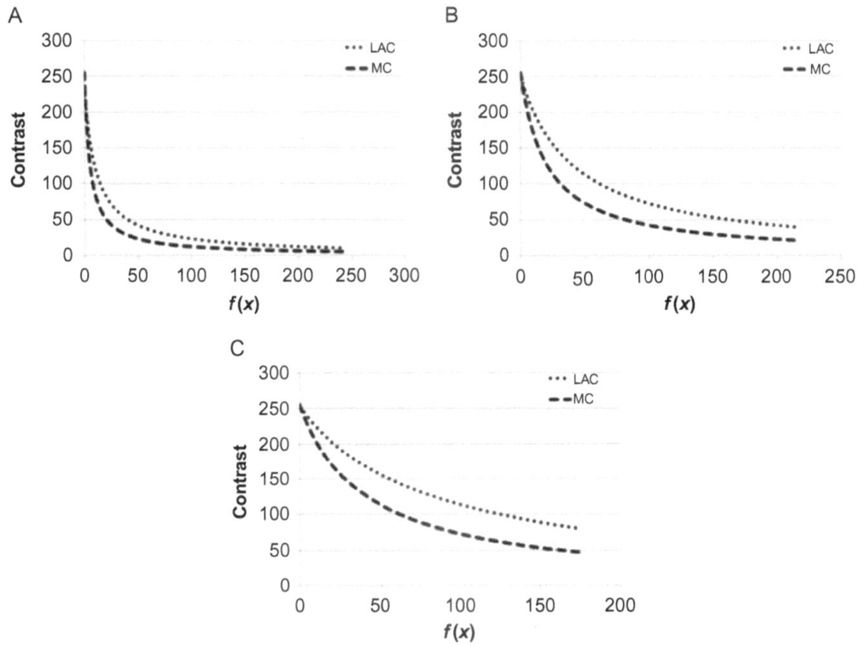

Fig. 3 Comparison of LAC and Michelson contrast for various values of k. $k=10$ (resp. 40, resp. 80) in (A) (resp. (B), resp. (C)).

Proof. The values $f(x)$ and $f(y)=f(x+2k)$ are represented in a classical gray scale, while the values $M-f(x)=X$ and $M-f(x)-k=Y$ are represented in a LIP gray scale (cf. Fig. 4). □

We can write:

- In the classical gray scale: $C^{\mathrm{Mi}}_{(x,y)}(f)=\dfrac{f(y)-f(x)}{f(y)+f(x)}=\dfrac{2k}{2f(x)+2k}=\dfrac{k}{f(x)+k}$
- In the LIP gray scale: $C^{\triangle}_{(X,Y)}(f)=\dfrac{|X-Y|}{1-\dfrac{\mathrm{Min}(X,Y)}{M}}=\dfrac{k}{1-\dfrac{M-f(x)-k}{M}}=\dfrac{Mk}{f(x)+k}$

Thus Eq. (5) is established.

We attach some importance to this result, which assigns a physical meaning to the Michelson's contrast. Note however that the difference between the two gray levels must be doubled for Michelson, which explains why the LAC seemed more sensitive than Michelson in our first investigations.

1.5 Other Contrast Concepts in Link with the LAC

Inner Contrast associated to a region (cf. Jourlin et al., 1989): Given a gray level f and a region R, we are going to evaluate, for each point x of R an average

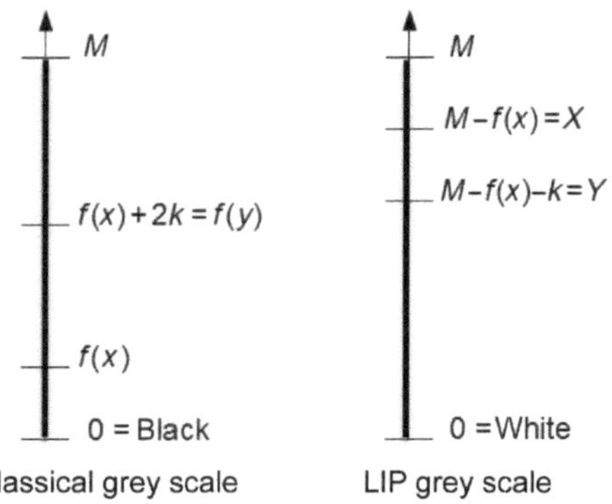

Fig. 4 Representation of the pairs of gray levels $f(x)$ and $f(x)+2k$ (resp. $M-f(x)-k$ and $M-f(x)$ in a classical gray scale (resp. a LIP gray scale).

contrast between x and the elements x_i of R lying in a neighborhood $N(x)$ of x. According to the location of x inside R, the cardinal (number of elements) of $N(x)\cap R$ is variable and will be denoted $\mathrm{Card}(N(x)\cap R)$. Thus we define the inner logarithmic contrast of R as follows:

$$\mathrm{IC}_R^{\triangle}(f)=\frac{1}{\mathrm{Card}(R)}\triangle\sum\nolimits_{x\in R}^{\triangle}\left\{\frac{1}{\mathrm{Card}(N(x)\cap R)}\triangle\sum\nolimits_{y\in N(x)\cap R}^{\triangle}C_{(x,y)}^{\triangle}(f)\right\}$$

where $\sum^{\triangle}$ designs the logarithmic sum of designed terms.

Note that the term $\left\{\frac{1}{\mathrm{Card}(N(x)\cap R)}\triangle\sum\nolimits_{y\in N(x)\cap R}^{\triangle}C_{(x,y)}^{\triangle}(f)\right\}$ represents the average additive contrast $\mathrm{AC}_x^{\triangle}(f)$ of the pixel x with its neighboring points. This notion will be introduced later (Eq. 11) and used as a contour detector. Here, the expression $\mathrm{IC}_R^{\triangle}(f)$ is nothing but the mean value, over a region R, of the local average contrasts computed at each point $x\in R$.

The interest to do this is to globalize in a unique number how these local average contrasts vary inside R.

An example is proposed in Fig. 5: to the initial image f "Pétanque Ball" (A), we associate the map of local average contrasts (B) and we compute the inner contrasts $\mathrm{IC}_R^{\triangle}(f)$ for each of the three regions delimited in (C).

The obtained values show that the inner contrast is low in the region included in the ball (11,31 gray levels) and high in the region located in the bright background (38,82 gray levels). The third tested zone has been chosen inside the ball's shadow where the contrasts are apparently

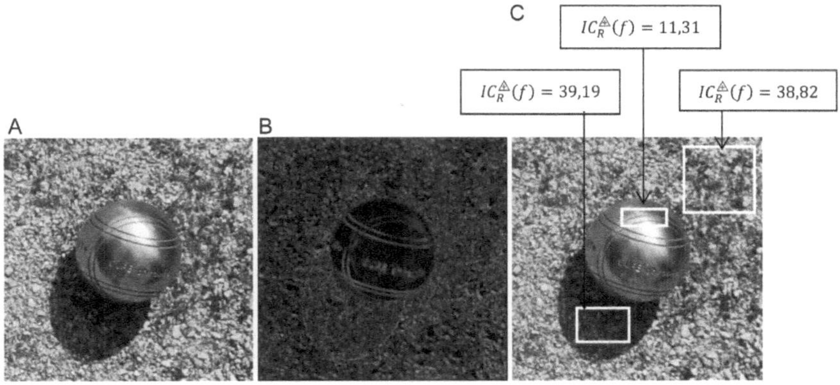

Fig. 5 Examples of inner contrast associated with a region. (A) Initial image (B) Map of local average contrasts (C) inner contrasts $IC_R^{\triangle}(f)$ associated to three regions.

attenuated. Nevertheless, based on the LIP additive contrast, the result (39,19) is similar to that of the bright background. Such a property will be used later to perform the detection of the ball (Fig. 12).

Contrast associated with a boundary: In Jourlin et al. (1989), a simple approach was proposed. This point simultaneously concerns automated thresholding and contour detection; in fact, when various thresholds are compared to select the best one, this selection is performed based on the optimization of some criterion and the threshold generating the boundaries of maximal contrast (in a sense to be defined) is generally significant. The problem of contour detection is of same nature because it aims to extract the most relevant boundaries. These two points will be detailed in the next section.

1.6 Applications of the LAC

In this section, our only aim is to show the efficiency of the LAC on the basis of examples.

1.6.1 Automated Thresholding and Multithresholding

Thresholding may be considered as the simplest approach of the difficult problem of image segmentation. In fact, it allows associating a gray-level image with a binary one where the two classes represent generally the studied object and the background.

In order to confer this binarization a strong reliability level, it is useful to drive it by some parameter optimization that results in an automated thresholding. Such techniques were mainly studied more than 30 years ago. Let us recall some of them:

- Interclass variance maximization by Otsu (1979)
- Entropy maximization by Pun (1981)
- Moment preserving by Tsaï (1985)

and for multithresholding:

- Nuées dynamiques by Diday (1971) written in French followed by various improvements, such as:
- "*k*-means" clustering by Jain and Dubess (1988), or Kanungo et al. (2002)

Remark 5. All these methods are based on the histogram of the considered image or notions of barycenters (means) and thus do not take into account the spatial distribution of the gray levels. For this reason, we will focus in this section on the method proposed by Köhler (1981), based on a contrast concept associated with the boundary generated by a given threshold. Moreover, the simple contrast used by Köhler will be replaced with benefit by the LAC.

1.6.1.1 Notations and Recalls on Köhler's Method

We start with a gray-level image f defined on the spatial support D. In Köhler's paper, f takes its values in the gray scale $[0, M[$, where 0 represents classically the black extremity and M the white one.

For each pixel $x \in D$ we define the neighborhoods: $N_4(x)$ and $N_8(x)$, respectively, constituted of the four and eight nearest neighbors of x.

To a given threshold t lying in $[0, \ldots, 255]$, Köhler associates the two classes:

$$C_0^t(f) = \{x \in D, f(x) < t\} \text{ and } C_1^t(f) = \{x \in D, f(x) \geq t\}$$

The boundary $B(t)$ generated by the threshold t, ie, the boundary separating the classes $C_0^t(f)$ and $C_1^t(f)$ is defined as follows:

$$B(t) = \{(x, y) \in D^2, \text{such that } x \in C_0^t(f) \text{ and } y \in C_1^t(f) \cap N_4(x)\} \quad (6)$$

For each pair of pixels (x, y) in $B(t)$, a "local" contrast $C^{K,t}(x, y)$ (in the sense of Köhler) is defined:

$$\begin{aligned} C^{K,t}(x, y) &= \text{Min}(|f(x) - t|, |f(y) - t|) \\ &= \text{Min}(t - f(x), f(y) - t) \end{aligned} \quad (7)$$

as well as a "global" contrast noted $C^K(B(t))$ in order to evaluate the significance of the boundary $B(t)$, which means the significance of t:

$$C^K(B(t)) = \left[\frac{1}{\mathrm{Card}(B(t))}\sum\nolimits_{(x,\,y)\in B(t)} C^{K,t}(x,y)\right] \tag{8}$$

where Card($B(t)$) designs the "cardinal" (number of elements) of $B(t)$. Thus $C^K(B(t))$ represents the average contrast of $B(t)$.

At this step, Köhler proposes the "optimal" threshold t_0 producing the boundary of maximal contrast, which means:

$$C^K(B(t_0)) = \mathrm{Max}_{t\in[0,\,\ldots,\,255]}\left\{C^K(B(t))\right\} \tag{9}$$

Remark 6. Concerning the local contrast $C^{K,t}(x,y)$ defined by Köhler in Eq. (7), let us recall that $f(x) < t \le f(y)$ and that each difference $f(y) - t$ and $t - f(x)$ may be arbitrarily small, which means $C^{K,t}(x,y)$ may tend to zero even when the difference $f(y) - f(x)$ is large. In a first analysis it appears surprising, but the expression of $C^K(B(t_0))$ as a maximum (Eq. 9) makes Köhler's approach a "Max Min" one. In such conditions, if the initial image f is already binary with gray levels n_1 and n_2, Köhler's algorithm produces $\frac{n_1 + n_2}{2}$ as optimal threshold, while other techniques would consider all the thresholds included in the interval $[n_1, n_2]$ equivalent.

1.6.1.2 New Method: LIP-Köhler Automated Thresholding

In this section, we propose first to replace the contrast used by Köhler by the LAC, in order to produce a Köhler-type automated thresholding perfectly adapted to transmitted images (microscopy in transmission, for example).

The LAC of Köhler is noted $C^{K,t,\triangle\!\!\!\!\!\triangle}$ and is defined as follows:

$$\begin{aligned} C^{K,t,\triangle\!\!\!\!\!\triangle}(x,y) &= \mathrm{Min}\left[C^{\triangle\!\!\!\!\!\triangle}(f(x),t),\, C^{\triangle\!\!\!\!\!\triangle}(f(y),t)\right] \\ &= \mathrm{Min}\left[t \triangle\!\!\!\!\!\triangle f(x), f(y) \triangle\!\!\!\!\!\triangle t\right] \end{aligned}$$

Then an averaging of such contrasts is computed as in Eq. (8) through logarithmic operations:

$$C^{K,\triangle\!\!\!\!\!\triangle}(B(t)) = \left[\frac{1}{\mathrm{Card}(B(t))} \triangle\!\!\!\!\!\times \sum\nolimits^{\triangle\!\!\!\!\!\triangle}_{(x,\,y)\in B(t)} C^{K,t,\triangle\!\!\!\!\!\triangle}(x,y)\right] \tag{10}$$

where $\sum^{\triangle\!\!\!\!\!\triangle}$ represents the sum in the LIP sense.

Note: Another variant of Köhler automated thresholding will be obtained (Section 2.2.1) by replacing the LAC with the LMC.

Fig. 6 permits to compare the results of Köhler and LIP-Köhler thresholdings (Fig. 6C and E, respectively). The use of the LAC produces a threshold which favors the detection of the bright part of the image.

In fact, the contrast values of $C^K(B(t))$ (Köhler, Eq. 8) are represented as a curve when t varies from 0 to 255 (Fig. 6B) and the contrast values of $C^{K,\triangle\!\!\!\!\triangle}(B(t))$ (LIP-Köhler, Eq. 10) as a curve (Fig. 6D). Note that the gray scale is here represented in the classical way (0 is the black extremity). The best threshold, ie, the maximal value of the contrast curve, is roughly located at the middle of the gray scale for Köhler and at the bottom of the gray scale for LIP-Köhler. The resulting binary images are shown, respectively, in Fig. 6C and E.

Comment on Fig. 6: the curve of contrasts (B), for example) shows the presence of five maxima. This remark obviously opens the way to multithresholding: the selection, as thresholds, of the five corresponding gray levels will generate six classes whose boundaries are *significant*. In such a way, the corresponding multithresholded image (possessing six gray levels) will appear very similar to the initial image (256 gray levels). Before developing this point, let us give other comparisons of Köhler and LIP-Köhler (Figs. 7 and 8).

Comment on Fig. 7: We observe the presence of some bright points on the initial image which generate a strong contrast at their boundaries, in the Köhler sense. In such a situation, the threshold produced by Köhler selects these points (Fig. 7C), while the LIP-Köhler threshold proposes here a resulting image (Fig. 7E) probably more significant. However, we must avoid hasty conclusions: we observe that the two contrast curves are bimodal

Fig. 6 Comparison between Köhler and LIP-Köhler thresholdings. (A) initial image, (B) and (D) curves representing the Köhler and LIP-Köhler contrasts, (C) and (E) thresholding of image (A) at the maximal value of (B) and (D). (See the color plate.)

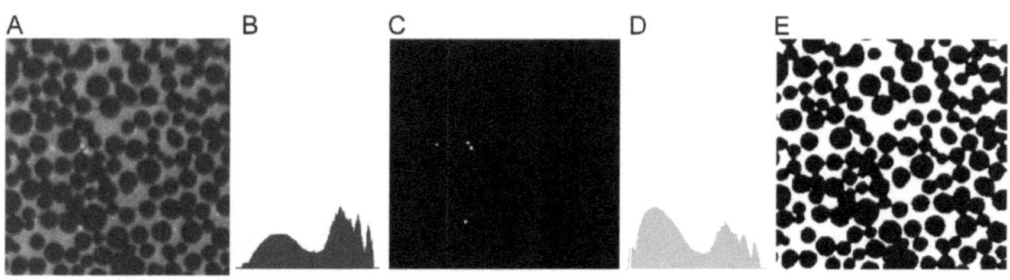

Fig. 7 Comparison between Köhler and LIP-Köhler thresholdings. (A) Initial image. (B) (resp. (D)) Curves representing the Köhler (resp. LIP-Köhler) contrasts. (C) (resp. (E)) Thresholding of image (A) at the maximal value of (B) (resp. (D)). (See the color plate.)

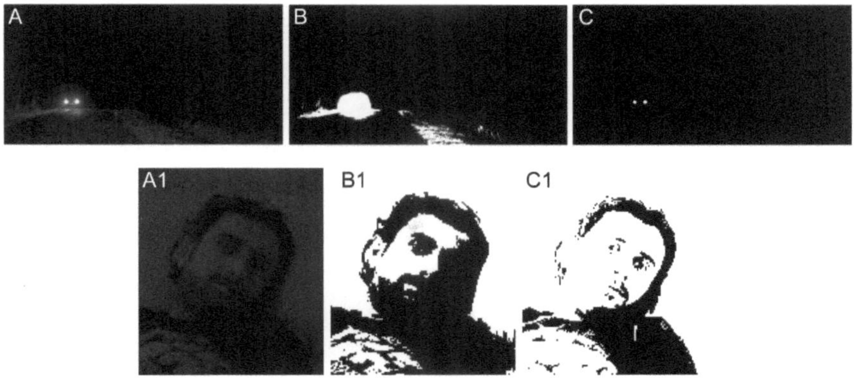

Fig. 8 Comparison between Otsu and LIP-Köhler thresholdings. (A) and (A1) Initial images. (B) and (B1) Otsu thresholding applied to (A) and (A1). (C) and (C1) LIP-Köhler thresholding of (A) and (A1).

in both cases and that the main peak is not in the same mode, which explains the results. In fact, a reasonable behavior in such a situation would be to detect a peak in each mode, producing two thresholds and three classes.

In Fig. 8, we show two cases of globally dark images, one presenting bright points with a lighting drift around them. The comparison between the classical Otsu thresholding (interclass variance maximization) and Köhler-LIP is here in favor of this last method.

1.6.1.3 Extension of Köhler's Method to Multithresholding

Most of thresholding methods are not adaptable to multithresholding. In the Comment on Fig. 6, we highlighted the ability of Köhler's method to give access to the local maxima of the contrast curve, which correspond to the most significant thresholds. Here we will not detail the various techniques permitting the extraction of these local maxima and will provide two examples (cf. Figs. 9 and 10).

Fig. 9A and B: We start with the image of Fig. 7A and its associated contrast curve on which the five detected local maxima are represented (Fig. 9B). In Fig. 9C, we display the resulting multithresholded image on which each class is affected with the corresponding gray-level value computed on the initial image. It is encouraging to observe that this image, which contains only six gray levels, is visually rather similar to the initial one with 256 gray levels.

Fig. 10A–H: This figure presents multithresholding results obtained with Köhler and LIP-Köhler contrasts. We again note the better sensitivity of the LIP version to detect classes in the dark parts of the image.

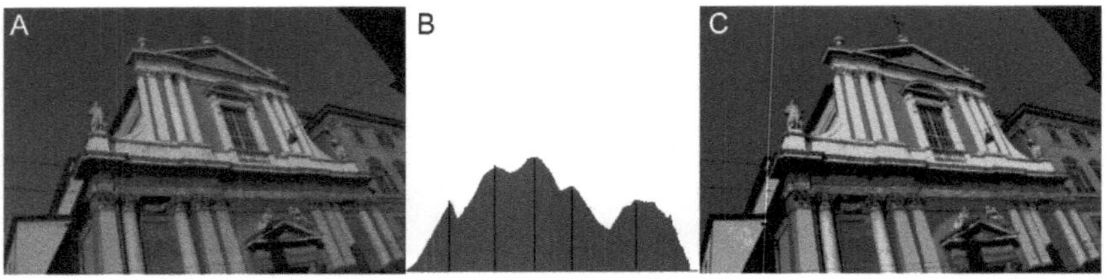

Fig. 9 Example of multithresholding based on Köhler's method. (A) Initial image. (B) Curve of Köhler's contrasts with local maxima (values 39, 79, 112, 145, 202). (C) Multithresholding of the initial image based on the local maxima of (B). Example of multithresholding based on Köhler's method. (See the color plate.)

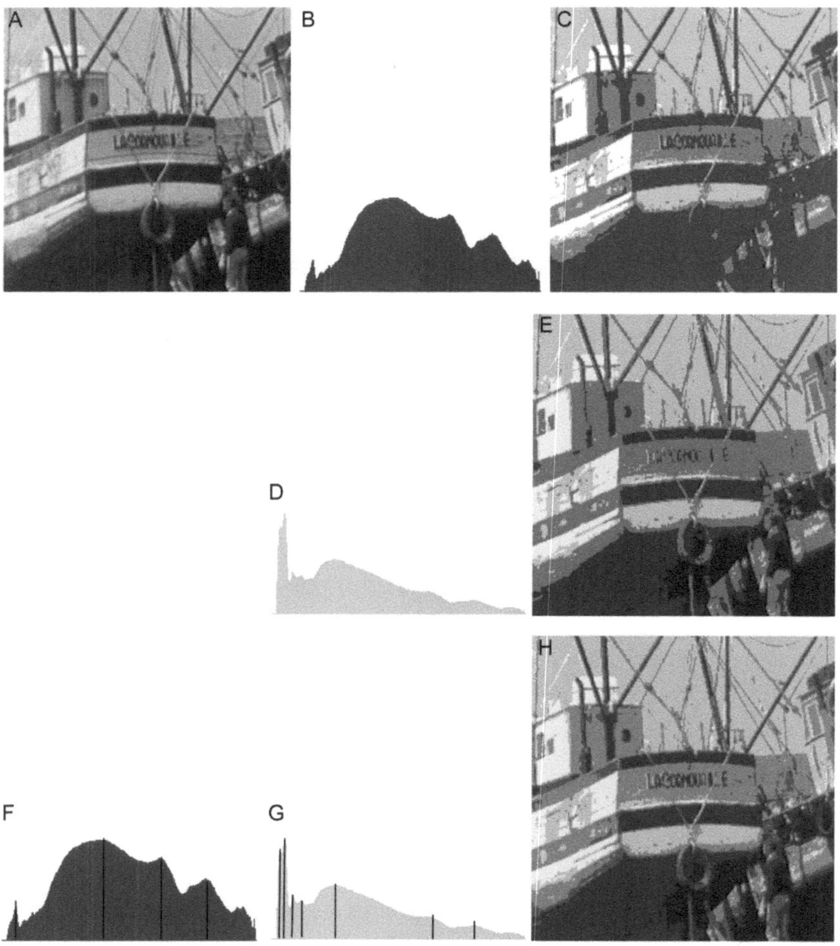

Fig. 10 Multithresholding gathering Köhler and LIP-Köhler multithresholdings. (A)–(C) Initial image, (resp. Köhler contrast curve), (resp. result: five classes generated by the four thresholds values 13, 100, 160, and 205, displayed in (F)). (D) and (E) LIP-Köhler Multithresholding of (A) (local maxima of (D)): 6, 13, 21, 30, 64, 160, and 205. (F) and (G) Köhler and LIP-Köhler thresholds on the corresponding contrast curves. (H) Multithresholding obtained by gathering the local maxima of Köhler and LIP-Köhler: 6, 13, 21, 30, 64, 100, 160, 205. (See the color plate.)

For example, the person on the bottom right is better segmented with the LIP contrast (Fig. 10E) than with the Köhler's one (Fig. 10C). On the opposite side, the boat's name is less visible in the LIP case, due to its location in a bright region.

In such conditions, it seems interesting to use the union of thresholds detected by each method (Fig. 10H).

1.6.2 Contour Detection

It is possible to apply LIP tools in order to perform contour detection operations. One way to do that consists of replacing in existing contour detection filters the classical laws of addition, subtraction, and scalar multiplication with their LIP version.

An example of this approach concerns the changing of Sobel filter in LIP-Sobel. For more details, the interested reader can refer to Deng and Pinoli (1998) and Palomares, Gonzalez, and Ros (2005). In the latter paper, the authors proposed a speeding up of the LIP-Sobel and a *Laplacian of Gaussian* edge detector. Typically, the efficiency of such techniques is due to the nonlinearity of LIP operators, which behave as a look-up table (LUT). Moreover, it would be easy to define LIP-Prewitt, LIP-Kirsh, LIP-Compass, and so on. Such concepts may present some applicative interest, even if they have no theoretical interest.

Next, we come back to the use of the LAC as an edge detector. Two definitions are possible, each of them comparing a given pixel x to its neighbors in terms of contrast:

Average Contrast: Given a gray-level image f, let us note $(v_i(x))_{i=1,\ldots,8}$ the eight nearest neighbors of x. It is possible to compute the average value of the LACs of x with each of its eight neighbors:

$$\mathrm{AC}_x^{\triangle\!\!\!\triangle}(f) = \frac{1}{8} \triangle\!\!\!\!\times \left(\sum\nolimits_{i=1,\ldots,8}^{\triangle\!\!\!\triangle} C_{(x,v_i(x))}^{\triangle\!\!\!\triangle}(f) \right) \tag{11}$$

Maximal Contrast: With the same notations, we compute the maximal LAC noted $\mathrm{MC}_{(x)}^{\triangle\!\!\!\triangle}(f)$:

$$\mathrm{MC}_{(x)}^{\triangle\!\!\!\triangle}(f) = \mathrm{Max}_{i=1,\ldots,8}\, C_{(x,v_i(x))}^{\triangle\!\!\!\triangle}(f) \tag{12}$$

Starting with an image f, each of these two notions associates a contrast image if we replace the gray level $f(x)$ of every pixel x by $\mathrm{AC}_x^{\triangle\!\!\!\triangle}(f)$ or

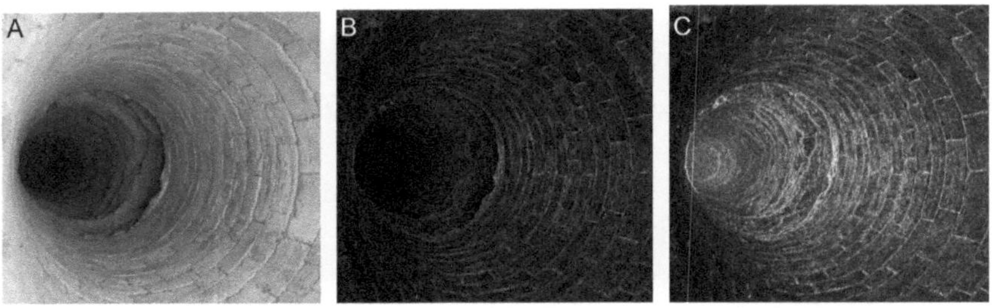

Fig. 11 Contour detection by means of the LMC. (A) Initial gray level image of a well. (B) Contour detection by means of a classical Sobel operator. (C) Contour detection by means of the maximal contrast $MC_x^{\triangle}(f)$.

$MC_x^{\triangle}(f)$. The generated results highlight the efficiency of the method in presence of lighting variations, compared with classical methods (cf. Fig. 11).

Comment on Fig. 11: we observe that the maximal contrast detects the contours in the dark regions as well as in the bright ones. The Sobel operator fails in extracting the information in dark zones. This difference is explained by the use, in our case, of a notion that adapts to lighting variations, while Sobel is based on a simple difference between gray levels.

Note that a "true" contour detection would require a binarization of images 11B and C in order to get binary boundaries. It is not the purpose here, but we note that this step is not so easy: a simple automated thresholding will result in a serious loss of information and it would be preferable to orient toward the use of watersheds (Beucher & Lantuejoul, 1979; Beucher & Meyer, 1993).

Remark 7. To perform the extraction of a region of interest from an image, two main approaches are possible: the edge detection algorithms and the segmentation algorithms. When it is possible, we prefer the latter for various reasons: in fact, the notion of "gradient" refers to the mathematical notion of *derivative* and it is clear that such a notion is not rigorously defined for digitized images. It is the reason why, in image processing, the lack of a mathematical guide allowed for the introduction of plenty of empirical gradients: Prewitt, Sobel, Roberts, Wallis, De Riche, Canny, compass gradient, morphological gradient, and many others, no one being demonstrated as always better than the others. Moreover, we have seen that, after applying a gradient, it remains the binarization step in order to get boundaries. Due to the general weak quality of gradients, this binarization step generally produces incomplete boundaries, whose thickness is variable. On the other hand, when it is possible to perform a satisfying segmentation, the required object is obtained as a filled binary shape, and it is quite easy to associate it with a continuous and closed (internal or external) boundary possessing a thickness of one pixel. For an example in the biomedical context, see Jourlin et al. (1989).

Concerning the choice between $AC_x^{\triangle\!\!\!\!\triangle}(f)$ or $MC_x^{\triangle\!\!\!\!\triangle}(f)$, the second one is obviously more sensitive (a single neighbor of x can determine a high value of $MC_x^{\triangle\!\!\!\!\triangle}(f)$), with the risk of sensitivity to noise.

Another situation where $MC_x^{\triangle\!\!\!\!\triangle}(f)$ proves its efficiency as a contour detector is given in Fig. 12 where we aim to detect an object here a "*pétanque ball*" on a textured background in presence of shadow (Fig. 12A).

The idea is to work on the gray-level luminance image. If we apply a classical contour detector like Sobel (Fig. 12C) to it, we observe that the textured background and the object produce very different contours, except in the shaded zone, which has the same behavior as the object and will be wrongly attributed to the same region.

The maximal contrast $MC_x^{\triangle\!\!\!\!\triangle}(f)$ is nearly insensitive to the presence of shadow (Fig. 12B), which considerably facilitates the extraction of the ball from the background.

Before introducing the LMC, let us highlight another interesting application of the LAC. For example, when we look at a cellular preparation in microscopy, one cell is composed of a cytoplasm, a nucleus, and possibly of one (or more) nucleolus. If we are interested in the nucleolus, for example, and we observe it in the microscope or by means of a camera, the object we

Fig. 12 Robustness to light condition of texture detection using the LIP additive maximal contrast. On the *left*: original image, at the *center*: contour detection using $MC_x^{\triangle\!\!\!\!\triangle}(f)$, on the *right*: contour detection by Sobel operator. (See the color plate.)

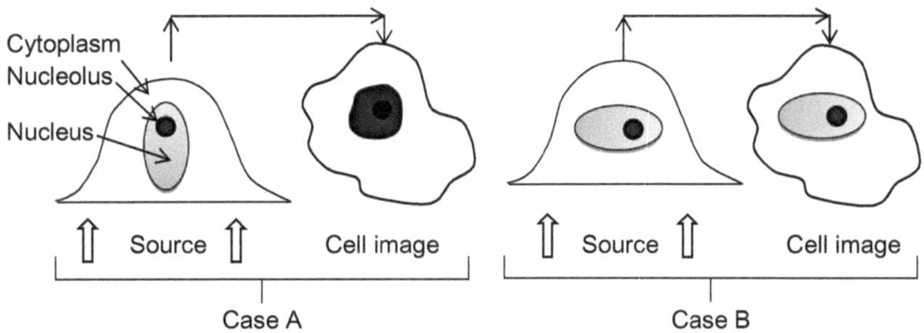

Fig. 13 Nucleolus appearance varies according to the nucleus orientation. In case (A), the nucleolus is seen through a dark background, due to the nucleus vertical orientation. In case (B), the nucleolus appears brighter (observed through a thinner thickness of nucleus).

look at is not properly the nucleolus, but an addition (in transmission, thus in the LIP context) of various components (cf. Fig. 13A and B). According to the 3D orientation of the nucleus inside the cellular preparation, the nucleolus appearance (in terms of gray levels) will change, even if we consider it as a homogeneous sphere. It becomes possible to estimate the *real* (true gray level) of the nucleolus thanks to the LAC between neighboring points of the nucleolus and the nucleus.

Now let us present an example of image segmentation in regions: Section 1.6.3.

1.6.3 Region Growing

Region growing is a very popular segmentation method. It consists of the following steps:

(1) Selection of a *seed s*, ie, an initial pixel. This step can be done:
 (a) Interactively (the user chooses the seed) or
 (b) Automatically: the seed is randomly selected
(2) Choose a similarity criterion between two pixels
(3) Aggregate to s its neighboring *similar* pixels, resulting in a first region R_1
(4) Choose a *similarity* criterion between a pixel and a region
(5) Aggregate to R_1 its neighboring (ie, belonging to its external boundary) *similar* pixels, resulting in a region $R_2 \supset R_1$
(6) Carry on the process until no more pixels can be aggregated

Remark 8. The region growing method is easily generalized to three-dimensional images (cf., for example, Sekiguchi, Sano, & Yokoyama, 1994; Sivewright & Elliott, 1994)

Remark 9. For a gray-level image f, the similarity criterion between two pixels x and y is often expressed as a tolerance ε on their gray level's difference in absolute value:

$$|f(x) - f(y)| \leq \varepsilon$$

Nevertheless, the similarity criterion between a pixel and a region can be estimated in various manners (cf., for example, Baraldi & Parmiggiani, 1996; Copty, Ranka, Fox, & Shankar, 1994; Wu, 1993).

Remark 10. The similarity criterion $|f(x) - f(y)| \leq \varepsilon$ corresponds to an upper bound on a basic contrast notion. One can write this condition with a LAC:

$$C^{\triangle\!\!\!\!\triangle}_{(x,\,y)}(f) = \mathrm{Max}(f(x), f(y)) \triangle\!\!\!\!\triangle \mathrm{Min}(f(x), f(y)) \leq \varepsilon$$

Note that this point will be considered in Chapter "Transfer of Classical Tools to LIP Context" and it will be shown through examples that this LIP approach presents some advantages compared with the classical one.

2. LOGARITHMIC MULTIPLICATIVE CONTRAST

2.1 Definitions, Notations

The LAC is based on the logarithmic additive law. It seemed interesting to study a possible definition of a contrast notion based on the scalar multiplicative law of the LIP. It resulted in a contrast notion completely different from existing ones. Nevertheless, we took care to give it a physical meaning, once more based on the transmittance law.

This Section 2 is widely inspired by a previous paper (cf. Jourlin, Carré, Breugnot, & Bouabdellah, 2012).

The LMC we are going to introduce is a real number instead of a gray level and it will present the advantage of being more sensitive near the white extremity of the gray scale than near the black one, contrary to the LAC.

If f represents a gray-level image and (x, y) a pair of points of the spatial support D, we define the LMC, noted $C^{\triangle\!\!\!\!\triangle}_{(x,\,y)}(f)$ or $\mathrm{LMC}_{(x,y)}(f)$ as the logarithmic ratio of $\mathrm{Max}(f(x), f(y))$ by $\mathrm{Min}(f(x), f(y))$. In other words, it represents the number by which the brightest gray level must be multiplied, in the LIP sense, to obtain the darkest gray level:

$$C^{\triangle\!\!\!\!\triangle}_{(x,\,y)}(f) \triangle\!\!\!\!\times \mathrm{Min}(f(x), f(y)) = \mathrm{Max}(f(x), f(y)) \tag{13}$$

The same reasoning permits to define the LMC $C_x^{\triangle}(f,g)$ for a pair of gray-level functions (f,g) at each point x of D:

$$C_x^{\triangle}(f,g) \triangle \text{Min}(f(x),g(x)) = \text{Max}(f(x),g(x)) \tag{14}$$

From Eq. (13) (resp. (14)), it is possible to explicitly define the value of the LMC $C_{(x,y)}^{\triangle}(f)$ (resp. $C_x^{\triangle}(f,g)$) through the LIP scalar multiplication (Eq. 10, chapter "Grey level LIP model. Notations, recalls and first applications," Section 2.2):

$$\begin{aligned}
\text{Eq.}(13) &\Leftrightarrow M - M\left(1 - \frac{\text{Min}(f(x),f(y))}{M}\right)^{C_{(x,y)}^{\triangle}(f)} = \text{Max}(f(x),f(y)) \\
&\Leftrightarrow M - \text{Max}(f(x),f(y)) = M\left(1 - \frac{\text{Min}(f(x),f(y))}{M}\right)^{C_{(x,y)}^{\triangle}(f)} \\
&\Leftrightarrow 1 - \frac{\text{Max}(f(x),f(y))}{M} = \left(1 - \frac{\text{Min}(f(x),f(y))}{M}\right)^{C_{(x,y)}^{\triangle}(f)} \\
&\Leftrightarrow Ln\left(1 - \frac{\text{Max}(f(x),f(y))}{M}\right) = C_{(x,y)}^{\triangle}(f) x Ln\left(1 - \frac{\text{Min}(f(x),f(y))}{M}\right) \\
&\Leftrightarrow C_{(x,y)}^{\triangle}(f) = \frac{Ln\left(1 - \frac{\text{Max}(f(x),f(y))}{M}\right)}{Ln\left(1 - \frac{\text{Min}(f(x),f(y))}{M}\right)}
\end{aligned} \tag{15}$$

The same result holds for Eq. (14):

$$\text{Eq.}(14) \Leftrightarrow C_x^{\triangle}(f,g) = \frac{Ln\left(1 - \frac{\text{Max}(f(x),g(x))}{M}\right)}{Ln\left(1 - \frac{\text{Min}(g(x),f(x))}{M}\right)} \tag{16}$$

Physical interpretation: In each of the previous situations (Eqs. 13 and 14), the multiplicative contrast clearly corresponds to the number of times we must stack (physically superpose) the Min gray level between the source and the sensor to obtain an attenuation equivalent to the Max gray level.

Remark 11. The LAC $C_{(x,y)}^{\triangle}(f)$ was by definition a gray level. The multiplicative one is a real number, which reaches $+\infty$ if the pair of gray levels $f(x)$ and $f(y)$ satisfy at the same time: $\text{Min}(f(x),f(y)) = 0$ and $\text{Max}(f(x),f(y)) \neq 0$. In the particular case where $f(x) = f(y) = 0$, we can decide to estimate the multiplicative contrast to the unit value, similar to the general case of two identical gray levels.

In order to overcome the problem of the possible infinite value, we can propose two solutions:

- For digital images, a very simple and practical approach consists of replacing each null gray level by the unit value, which does not affect significantly either the aspect or the interpretation of the studied image. In such conditions, all contrast values are finite and may be visualized as gray levels, after a normalization step, if necessary
- A more rigorous approach based on a classical mathematical technique: when a metric d_1 is susceptible to reach infinite values, it is common to associate it with a novel metric d according to:

$$d(x,y)=\frac{d_1(x,y)}{1+d_1(x,y)} \quad \text{if } d_1(x,y) \text{ is finite}$$
$$d(x,y)=1 \quad \text{if } d_1(x,y)=\infty$$

This method produces a metric d lying in the interval $[0,1]$ and respecting the nearness order between points, because the function $u \to \frac{u}{1+u}$ is increasing.

In such conditions, all contrast values are finite and may be visualized as gray levels, after a normalization step, if necessary.

Remark 12. When working in the classical situation of 8-bit images, if we chose the solution of replacing null gray-level values with 1, we dispose of a gray scale of 256 (limited to 255) gray levels (from 1 to 255) with a nearly linear distribution. Thus the largest possible value $C_{\text{Max}}^{\triangle\!\!\!\times}$ for the multiplicative contrast is reached for the pair of gray levels (1, 255) and satisfies the following equation:

$$C_{\text{Max}}^{\triangle\!\!\!\times} \triangle\!\!\!\times 1 = 255$$

from which is deducted: $C_{\text{Max}}^{\triangle\!\!\!\times} \sim 1416.79$. At this step, we dispose of various possibilities:

First solution: We define the normalized LMC (lying in the interval [0, 255]) noted $C_{(x,y)}^{\triangle\!\!\!\times N}(f)$:

$$C_{(x,y)}^{\triangle\!\!\!\times N}(f) = 255x\frac{1}{C_{\text{Max}}^{\triangle\!\!\!\times}} C_{(x,y)}^{\triangle\!\!\!\times}(f) \tag{17}$$

sometimes denoted $C^{\triangle N}_{(f(x),t)}(f)$ in the following. This notation $C^{\triangle N}_{(f(x),t)}(f)$ will be used when one of the considered gray levels is not defined by f but corresponds to a threshold t, for example.

In the same way, the contrast of a pair (f,g) of images at a point $x \in D$ is given by:

$$C_x^{\triangle N}(f,g) = 255x \frac{1}{C^{\triangle}_{\text{Max}}} C_x^{\triangle}(f,g) \tag{18}$$

Weakness of this normalization: The LMC appears concentrated on the black side of the gray scale and then furnishes limited information.

Second solution: As is commonly done for gradients, for example, we can truncate the LMC values: Each value greater than 255 is limited to 255. We reject this method for its arbitrary aspects.

Third solution: For each situation, we can choose the *best* display by transforming the interval $[\text{Min}(\text{LMC}), \text{Max}(\text{LMC})]$ in $[0,255]$ (for an example, see Fig. 16C).

Fourth solution: We keep the initial values of the LMC and process them in each application: contour detection, thresholding, metrics, etc., in order to preserve all the information throughout the computation. At the end, the third approach is finally applied when a display is needed.

To illustrate the better sensitivity of the LMC in the bright part of the gray scale, let us consider, for example, a pair of gray levels differing from a constant, namely, $f(x)$ and $f(x) + k$, and let us compute the LMC between them when $f(x)$ varies in the interval $[0, 255 - k]$ in a classical gray scale where 0 is black. The curves representing such contrasts for various values of k $(k = 30, 50, 100, 150, \text{and } 200)$ are displayed (Fig. 14). The nonlinearity of the LMC appears clearly, as well as its attenuation toward the black side of the scale and its high sensitivity in the white part. Such a behavior must be compared with the LAC presented in Fig. 3.

As we have done for the LAC, let us now present some applications examples, like automated thresholding and contour detection.

2.2 Applications of the LMC

2.2.1 Application to Automated Thresholding and Multithresholding

Let us return to Köhler's automated thresholding method. As we have done for the LAC, it is possible to replace the LIP-Köhler's additive contrast $C^{K,\triangle}(B(t))$ defined in Eq. (10) with its corresponding LMC version noted $C^{K,\triangle}(B(t))$ and defined as follows:

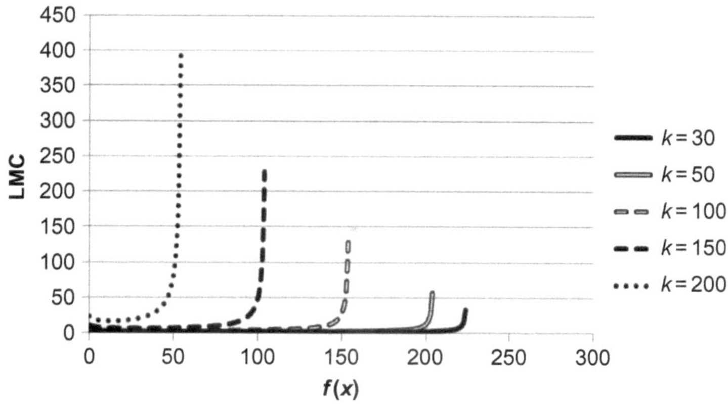

Fig. 14 Curves representing the LMC between $f(x)$ and $f(x)+k$, for $k=30, 50, 100, 150, 200$.

$$C^{K,\triangle}(B(t)) = \left[\frac{1}{\mathrm{Card}(B(t))} \triangle \sum\nolimits^{\triangle}_{(x,y)\in B(t)} C^{K,t,\triangle}(x,y)\right] \tag{19}$$

where $C^{K,t,\triangle}(x,y) = \mathrm{Min}\left[C^{\triangle}_{(f(y),t)}, C^{\triangle}_{(t,f(x))}\right]$

Remember that the values of the average contrast $C^{K,\triangle}(B(t))$ when the threshold t varies in [0,255] permit to determine the "optimal" threshold t_0 producing a boundary $B(t_0)$ whose average contrast is maximal.

It is quite easy to extend this method to multithresholding. An example is given Fig. 15: The initial infrared image (Fig. 15A) represents the harbor of "La Rochelle".

The dark (cold) zones are constituted mainly of the sea and the clouds. On its histogram (Fig. 15B), two classes appear: the cold one on the left, very homogeneous, and the hot one corresponding to the earth, rather scattered around its average. The optimal threshold given by the *Interclass Variance Maximization* (cf. Otsu, 1979, with the Mahalanobis correction) corresponds with the gray level 79. This threshold produces a binary image (Fig. 15C). A possible way to exhibit a hot (bright) class inside the earth (not visible on the histogram) consists of computing the LMC map of (A) whose histogram is presented in (D) with a significant peak. Thresholding the initial image (A) at this peak value generates a binary image (Fig. 15E) whose contours are superimposed on the initial image (Fig. 15F).

We previously noted as important that the LMC emphasizes contrasts in the white part of the gray scale. This point is illustrated by Fig. 15D–F.

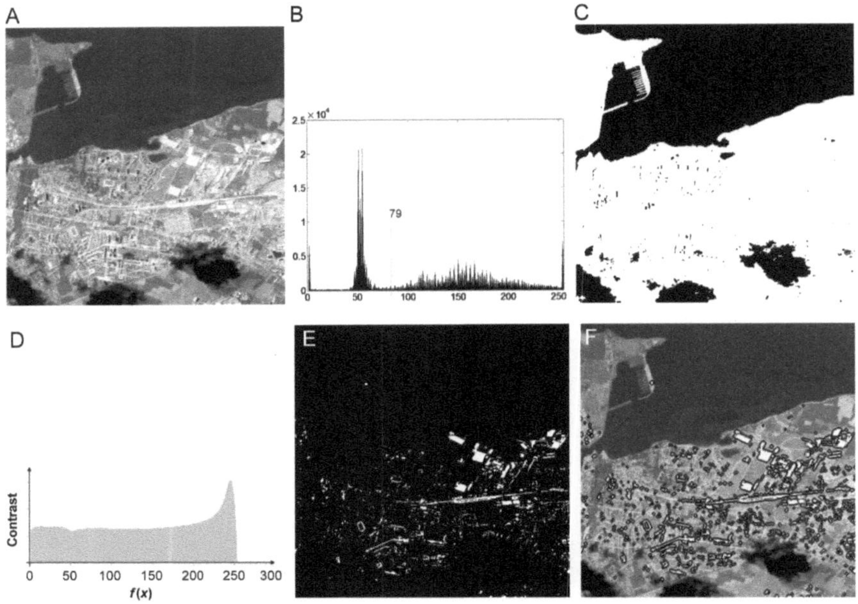

Fig. 15 Comparison of thresholdings based on interclass variance maximization and on LMC map. (A) Initial image; (B) Histogram of (A) with highlighted value 79 corresponding to the threshold based on interclass variance maximization; (C) Binary image after thresholding (A) at gray level 79; (D) Histogram of the LMC map of (A) with a unique peak; (E) Binary image after thresholding (A) at the peak of (D); (F) Thresholded area highlighted on (A). (See the color plate.)

2.2.2 Application to Contour Detection

As we have done for the LAC (Eqs. 11 and 12), it is possible to define an average and a maximal multiplicative contrasts:

Average LMC: Considering a gray-level image f and x a point of D, we denote as usually $(\nu_i(x))_{i=1,\ldots,8}$ the eight nearest neighbors of x. It is then possible to compute the average value of the LMCs of x with each of its eight neighbors:

$$AC_x^{\triangle\!\!\!\triangle} \cdot (f) = \frac{1}{8} \cdot \left(\sum\nolimits_{i=1,\ldots,8} C^{\triangle\!\!\!\triangle}_{(x,\nu_i(x))}(f) \right) \tag{20}$$

Contrary to the additive case, note that the sum $\Sigma_{i=1,\ldots,8}$ and the scalar product by $\frac{1}{8}$ are intended in the classical sense and not the logarithmic one: in fact, as we have already remarked, the contrasts $C^{\triangle\!\!\!\triangle}_{(x,\nu_i(x))}(f)$ are real numbers and not gray levels.

Maximal LMC: With the same notations, we compute the maximal LMC noted $MC_x^{\triangle}(f)$:

$$MC_{(x)}^{\triangle}(f) = \text{Max}_{i=1,\ldots,8} C_{(x,\nu_i(x))}^{\triangle}(f) \tag{21}$$

Starting with an image *f*, each of these two notions associates a contrast image if we replace each gray level *f*(*x*) of a pixel *x* by $AC_{(x)}^{\triangle}(f)$ or $MC_{(x)}^{\triangle}(f)$, these two values being gray levels under normalization condition (cf. Remark 12).

In such conditions, both Eqs. (20) and (21) allow the computation of a contrast map for a given image *f*. Considering as evident that the contour points of *f* correspond to those presenting a significant contrast, it seems natural that the contrast maps computed from the average multiplicative contrast $AC_{(x)}^{\triangle}(f)$ or the maximal multiplicative contrast $MC_{(x)}^{\triangle}(f)$ have the effect of enhancing boundary points. Fig. 16 corroborates this hypothesis and has been chosen to emphasize the efficiency of the LMC on over-lighted images.

In the situation of over-lighted images or regions, no doubt exists concerning the preeminence of the multiplicative logarithmic approach compared with a classical one: the initial image possesses a saturated zone around the sun, whose intensity decreases when the distance to the sun grows. Compared with classical gradients operators, only the maximal

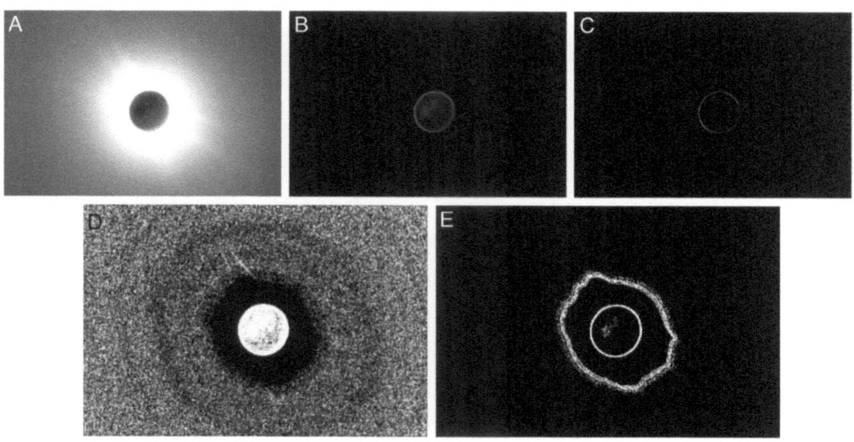

Fig. 16 Contour detection based on LMC. (A) Initial image "Sun eclipse 2006" (with authorization of the author L. Ferrero); (B) Classical gradient of (A); (C) Contrast map of (A) corresponding to $MC_x^{\triangle}$; (D) Thresholded gradient; (E) Thresholded contrast map. (See the color plate.)

multiplicative contrast $MC^{\triangle\!\!\!\!\cdot}_{(x)}(f)$ is able to detect a boundary in the bright part of the image (cf. Fig. 16E). More precisely, the gradient map (Fig. 16B) and the map of multiplicative contrasts (Fig. 16C) have been thresholded at a very low level in order to put in evidence a large number of potential boundary pixels. A classical gradient then detects points everywhere in the image (sun texture and noise of the sensor in the background) except in the saturation zone (Fig. 16D) while the LMC contour detector extracts only the transitions inside the region with strong brightness (Fig. 16E).

2.3 Another Contrast Concept in Link with the LMC

Inner contrast associated to a region: As we have proposed for the LAC, it is possible, for a gray-level image f and a region R, to evaluate, for each point x of R an average contrast between x and the elements x_i of R lying in a neighborhood $N(x)$ of x. According to the location of x inside R, the cardinal (number of elements) of $N(x) \cap R$ is variable and will be denoted $\text{Card}(N(x) \cap R)$. Thus we define the inner LMC of R as:

$$IC^{\triangle\!\!\!\!\cdot}_{(R)}(f) = \frac{1}{\text{Card}(R)} \triangle\!\!\!\!\cdot \sum\nolimits_{x \in R} \left\{ \frac{1}{\text{Card}(N(x) \cap R)} \sum\nolimits_{y \in N(x) \cap R} C^{\triangle\!\!\!\!\cdot}_{(x,y)}(f) \right\}$$

3. WHAT ABOUT COLOR IMAGES?

We have already said that the present book focuses mainly on gray-level images, but from time to time, we explore some ideas in direction of color image processing. Here we propose two ways concerning the definition of a color contrast notion.

3.1 Logarithmic Image Processing for Color Images

The interested reader will find a detailed presentation of the logarithmic image processing for color images (LIPC) model in Jourlin, Breugnot, Itthirad, Bouabdellah, and Closs (2011). In this paper, we have already defined a notion of color contrast and a norm associated with the color contrast. Here we will only give an illustration of these concepts to show their low sensitivity to lighting drift (cf. Fig. 17). In fact, the results of color contrast or associated norm are very similar when applied on the image "Peppers" represented on Fig. 17A and on the image "Peppers with lighting drift" (Fig. 17D).

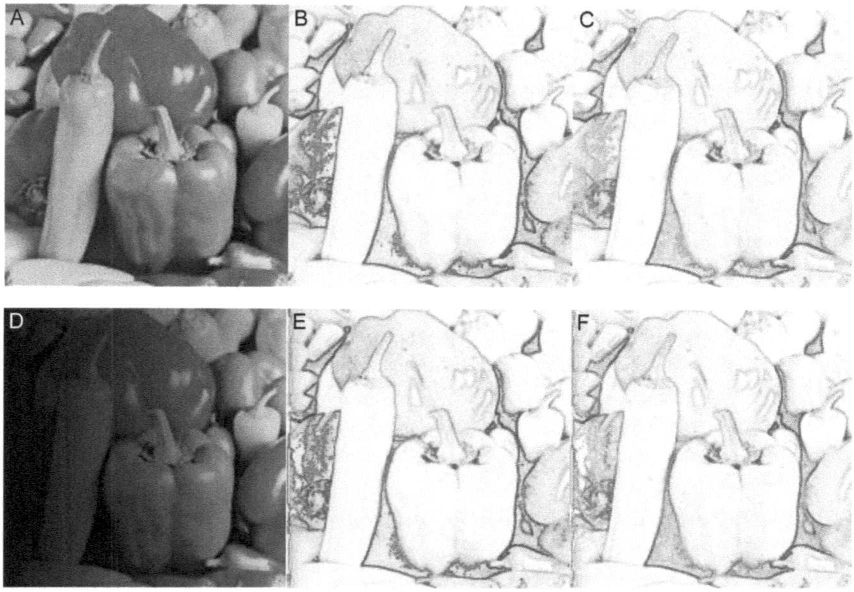

Fig. 17 Insensitivity to lighting drift of LIPC color contrast and norm. (A) Initial image "Peppers"; (D) Image "Peppers" with lighting drift; (B) resp. (E)) Color contrast image of (A) (resp. (D)); (C) (resp. (F)) Norm image of (A) (resp. (D)). (See the color plate.)

Note that the color contrast highlights the transition pixels, ie, located at the boundaries separating different colors. In fact, it appears as a contour detector and produces a color image. The image resulting from the norm cumulates the information of the three channels and is then displayed in gray levels.

3.2 Extension of the LAC to Color Images

Considering two colors $C_1 = (R_1, G_1, B_1)$ and $C_2 = (R_2, G_2, B_2)$ in the R, G, B space, we propose to introduce the LAC in each channel. For example, in the red channel, $C_R^{\triangle\!\!\!\!\triangle}(C_1, C_2)$ or $\text{LAC}_R(C_1, C_2)$ or LAC_R when no confusion is possible, is defined according to:

$$C_R^{\triangle\!\!\!\!\triangle}(C_1, C_2) = \text{LAC}_R(C_1, C_2) = \text{Max}(R_1, R_2) \triangle\!\!\!\!\triangle \text{Min}(R_1, R_2) = \frac{|R_1 - R_2|}{1 - \dfrac{\text{Min}(R_1, R_2)}{M}}$$

The question is: how can we gather the information carried by each of the three contrasts $\text{LAC}_R(C_1, C_2), \text{LAC}_G(C_1, C_2), \text{LAC}_B(C_1, C_2)$? Several answers can be proposed:

First solution: If we remember that the LAC computed for two gray levels is itself a gray level, it means that $LAC_R(C_1, C_2)$ is interpretable in the red scale as a red level. In the same way, $LAC_G(C_1, C_2)$ and $LAC_B(C_1, C_2)$ represent, respectively, a green level and a blue one.

In such conditions, it seems natural to introduce in each channel an average contrast and a maximal contrast as we have done in Eqs. (11) and (12).

Red Average Contrast: Given a color image f and a pixel $x \in D$, let $(v_i(x))_{i=1,\ldots,8}$ denote the eight nearest neighbors of x. It is possible to compute the red LAC $LAC_R[f(x), f(v_i(x))]$ of x with each of its eight neighbors. Then the average of these eight values, denoted $ALAC_R(f(x))$ is defined according to:

$$ALAC_R(f(x)) = \frac{1}{8} \triangle \left(\sum\nolimits^{\triangle}_{i=1,\ldots,8} LAC_R[f(x), f(v_i(x))] \right) \tag{22}$$

Red Maximal Contrast: With the same notations, we compute the red maximal LAC noted $MLAC_R(f(x))$:

$$MLAC_R(f(x)) = Max_{i=1,\ldots,8} LAC_R[f(x), f(v_i(x))] \tag{23}$$

For the green and blue channels, we dispose of the same equations. In other words, it is possible to associate to each point x of an image f the vector:

$$\{ALAC_R(f(x)), ALAC_G(f(x)), ALAC_B(f(x))\} \tag{24}$$

and the vector:

$$\{MLAC_R(f(x)), MLAC_G(f(x)), MLAC_B(f(x))\} \tag{25}$$

Note that *each of these vectors is nothing but a color*. Moreover, such colors take high values for transition pixels. In such conditions, if we affect $x \in D$ to one of these colors, we get a contour detector (Fig. 18).

Remark 13. For the moment, this approach remains at an exploratory level. Two points could be explored in depth:

(1) These color contour detectors likely permit detection of an object of a given color seen on a background of another color. It seems that in image (A) the transitions *red/green* produce similar colors on images (B) or (C).

(2) These contour detectors are rather insensitive to the brightness of the image (cf. Fig. 19). In fact, the initial version of the image "Peppers" (Fig. 19A) and the brighter version (Fig. 19C) produce very similar images (Fig. 19B and D) when they are processed with the maximal LAC.

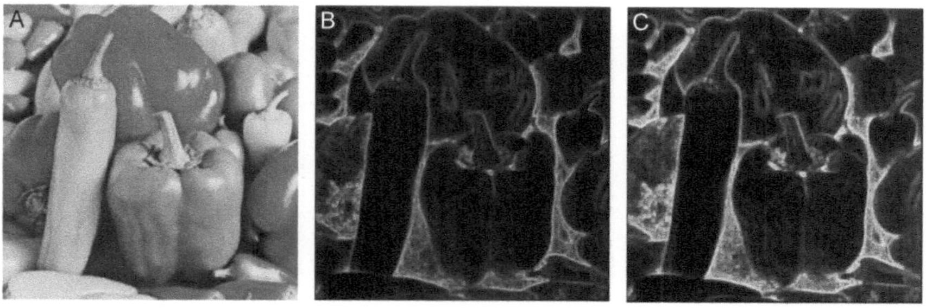

Fig. 18 Average and maximal LAC as color vectors. (A) Initial image "Peppers". (B) Image (A) processed with the average LAC: each pixel of "Peppers" has been replaced by the vector given in Eq. (24). (C) Image (A) processed with the maximal LAC: each pixel replaced by the vector given in Eq. (25). (See the color plate.)

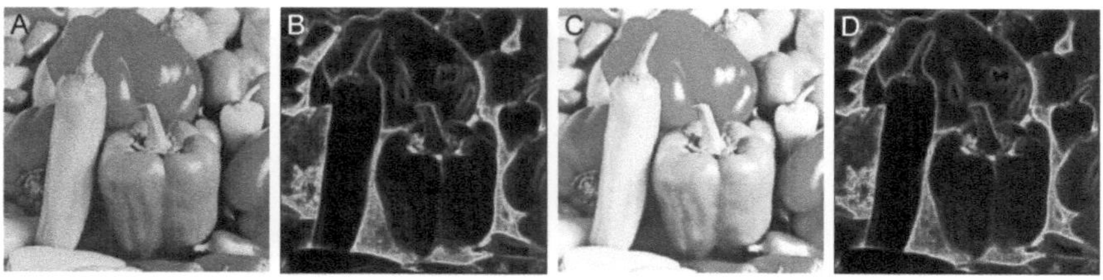

Fig. 19 Insensitivity of the MLAC vector to brightness variations. (A) Initial "Peppers", (C) Image (A) with higher brightness, (B) and (D) Images (A) and (B) processed by the MLAC vector. (See the color plate.)

Another recurrent property of logarithmic contrasts is their insensitivity to a significant lighting drift: see, for example, Fig. 20. In 20A, we observe the initial image "Peppers" and in 20B a strong lighting drift. Applied to 20A, this drift produces Image 20C. In Fig. 20D (resp. 20E), we display the MLAC vector applied to (A) (resp. (C)). These two images are quite similar, except in the very dark part of (C), where false colors appear (arrows in Fig. 20E). Nevertheless, for a lightly lower drift (Fig. 20F), the presence of false colors decreases significantly (Fig. 20G).

Second solution: From the three contrasts $\mathrm{LAC}_R(C_1, C_2), \mathrm{LAC}_G(C_1, C_2)$, $\mathrm{LAC}_B(C_1, C_2)$ we can deduct various color metrics, for example:

$$d(C_1, C_2) = \frac{1}{\sqrt{3}}\sqrt{[\mathrm{LAC}_R(C_1, C_2)]^2 + [\mathrm{LAC}_G(C_1, C_2)]^2 + [\mathrm{LAC}_B(C_1, C_2)]^2} \tag{26}$$

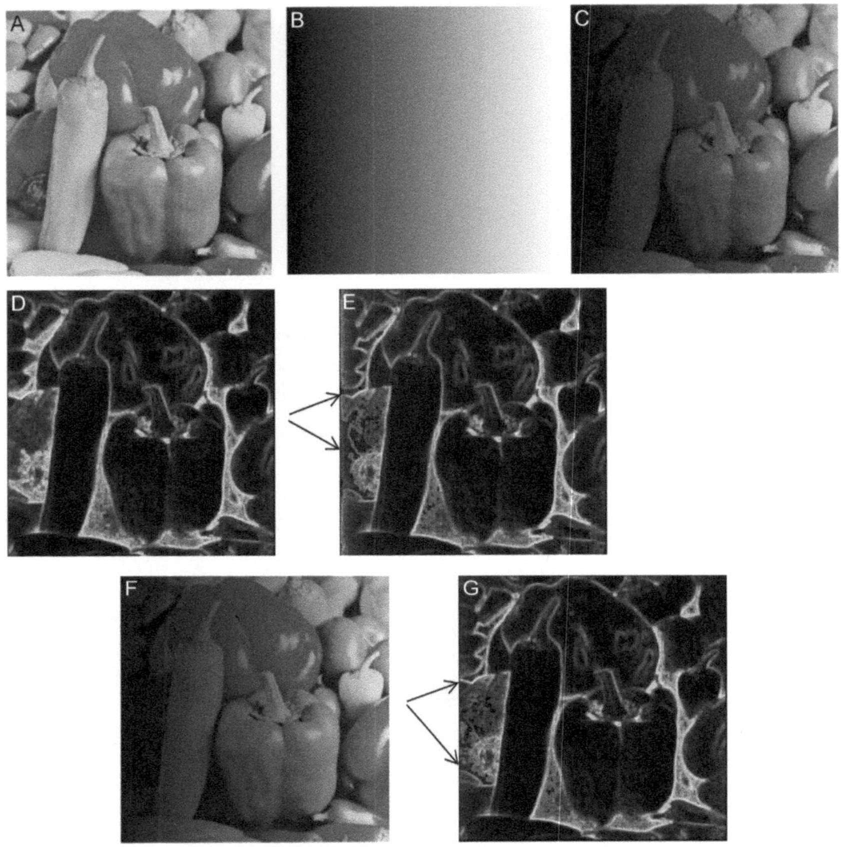

Fig. 20 Behavior of the MLAC vector in presence of lighting drift. (A) Initial image "Peppers". (B) An example of strong lighting drift. (C) "Peppers" with lighting drift (B). (D) (resp. (E)), MLAC vector applied to (A) (resp. (C)). (F) "Peppers" with a lightly lower lighting drift. (G) MLAC vector applied to (F). (See the color plate.)

Or

$$d'(C_1, C_2) = \frac{1}{3} \overset{\triangle}{\times} [\mathrm{LAC}_R(C_1, C_2) \overset{\triangle}{+} \mathrm{LAC}_G(C_1, C_2) \overset{\triangle}{+} \mathrm{LAC}_B(C_1, C_2)] \quad (27)$$

Note that these two metrics take their values in $[0, M[$ without any normalization: the distance between two colors is simply a gray level.

In Fig. 21, we propose to visualize the gray levels corresponding to each of these metrics for some pairs of colors.

Comment on Fig. 21: For a number of colors pairs chosen at random, we observe that the gray levels corresponding to $d(C_1, C_2)$ and $d'(C_1, C_2)$ are very similar (for (B)–(F)). We also notice a significant difference in the case (A) corresponding to a same tone.

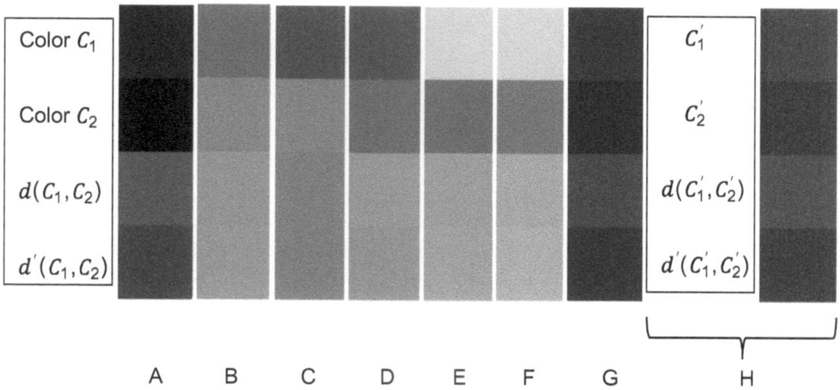

Fig. 21 Representation, in gray levels, of the distances $d(C_1, C_2)$ and $d'(C_1, C_2)$ for various pairs of colors (C_1, C_2). (See the color plate.)

This finding incited us to select a pair of colors (C_1, C_2) of the same blue tone (Fig. 21G) and to acquire them with a longer exposure time in order to brighten them, resulting in a pair (C_1', C_2') (Fig. 21H). The distances $d(C_1, C_2)$ and $d\left(C_1', C_2'\right)$ on one hand and $d'(C_1, C_2)$ and $d'\left(C_1', C_2'\right)$ on the other present exactly the same gray levels: this shows that d *and* d' are insensitive to lighting variations due to exposure time changing.

Third solution: Now we propose a particular focus on a novel family of metrics: the Asplünd's one, which previously existed for binary shapes. We have extended this initial metric to gray-level images and then to color and multispectral ones. They will be presented in depth in Chapter "Metrics Based on Logarithmic Laws." Such metrics are particularly well adapted to the LIP framework, there are very few sensitive to lighting variations, to exposure time changing, and to diaphragm aperture variations.

To conclude this section dedicated to metrics issued from color contrast, we would just evoke the "multiplicative Asplünd's distance" between two colors $C_1 = (R_1, G_1, B_1)$ and $C_2 = (R_2, G_2, B_2)$, denoted $d_{As}^{\triangle}(C_1, C_2)$ and defined according to:

$$d_{As}^{\triangle}(C_1, C_2) = \ln\left(\frac{\alpha}{\beta}\right)$$

where $\alpha = \mathrm{Inf}_{k \in \mathbb{R}^+}\{k, k\triangle R_1 \geq R_2, k\triangle G_1 \geq G_2, k\triangle B_1 \geq B_2\}$ and $\beta = \mathrm{Sup}_{k \in \mathbb{R}^+}\{k, k\triangle R_1 \leq R_2, k\triangle G_1 \leq G_2, k\triangle B_1 \leq B_2\}$

Remark 14. Throughout Section 3.2, it appeared that the introduced contrast notions are naturally exploitable to define associated metrics. This subject is presented with more detail in Chapter "Metrics Based on Logarithmic Laws."

3.3 Link with Mac-Adam Experiment

Remark 15. The main interest of the color metrics proposed arises from the consistency of the LIP framework with human vision. This point allows us to expect a link between the shapes of color neighborhoods computed for each metric in the (x,y)-space and the elliptic shapes obtained by Mac Adam (1942).

We only took a quick look on this subject, which justifies a deeper study.

In fact, we decided to display the neighborhoods associated to our metrics in the (x,y) space, hoping some similarity with Mac-Adam ellipses. However, we failed in creating colors of constant luminosity as the Mac-Adam ones because of our R, G, B representation. For some selected colors, we display in Fig. 22B their neighborhoods generated by the metric d_1 for a given tolerance ε. It means that for a reference color C, all the colors C' satisfying $d_1(C_1, C_2) < \varepsilon$ are considered as neighboring C. The same reasoning has been done in Fig. 22A for the Euclidean metric.

The neighborhoods are limited to R, G, B colors because our metrics are based on the LIP model. Euclidian metric leads to circular-shaped neighborhoods, with a small distortion in the monochromatic parts of the diagram

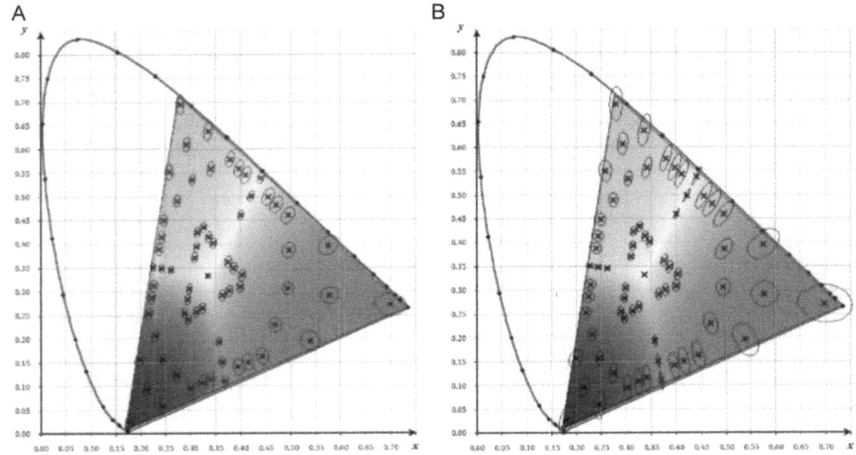

Fig. 22 Comparison of colors neighborhoods shapes in (*x*,*y*)-space. (A) Euclidean neighborhoods, (B) Neighborhoods associated with metric d_1. (See the color plate.)

(Fig. 22A). However, LIP-based neighborhoods are elliptically shaped with a small size in the blue part and a larger one in the red and green ones (Fig. 22B) as in the Mac-Adam experiment.

4. CONCLUSION

In this chapter, we have presented and studied novel contrast concepts. In particular, we have established a link between the LAC and the classical Michelson contrast commonly used by physicists, permitting us to produce a physical interpretation of this contrast. We proposed some examples of LAC usefulness, such as automated thresholding or contour detection.

The LMC seems a notion less natural and we think it could be an interesting subject of research, due, for example, to its sensitivity to small variations in the "white" part of the gray scale. This remains to be done, as well as its interest for automated thresholding, contour detection, and its possible extension to color images.

We think it would be interesting to deepen the link with Mac-Adam experiment, which was only suggested.

REFERENCES

Baraldi, A., & Parmiggiani, F. (1996). Single linkage region growing algorithms based on the vector degree of match. *IEEE Transactions on Geoscience and Remote Sensing*, *34*(1), 137–148.

Beucher, S., & Lantuejoul, C. (1979). *Use of watersheds in contour detection. In Workshop on image processing, real-time edge and motion detection.* http://cmm.ensmp.fr/~beucher/publi/watershed.pdf.

Beucher, S., & Meyer, F. (1993). The morphological approach to segmentation: The watershed transformation. In E. R. Dougherty (Ed.), *Mathematical morphology in image processing* (pp. 433–481).

Copty, N., Ranka, S., Fox, G., & Shankar, R. V. (1994). A data parallel algorithm for solving the region growing problem on the connection machine. *Journal of Parallel and Distributed Computing*, *21*, 160–168.

Deng, G., & Pinoli, J.-C. (1998). Differentiation-based edge detection using the logarithmic image processing model. *Journal of Mathematical Imaging and Vision*, *8*, 161–180.

Diday, E. (1971). Une nouvelle mèthode en classification automatique et reconnaissance des formes. La méthode des nuées dynamiques. *Revue de Statistiques Appliquées*, *19*(2), 19–33.

Jain, A. K., & Dubes, R. C. (1988). *Algorithms for clustering data.* Los Altos: Prentice Hall.

Jourlin, M., Breugnot, J., Itthirad, F., Bouabdellah, M., & Closs, B. (2011). Logarithmic image processing for color images. *Advances in Imaging and Electron Physics*, *168*(2), 65–107.

Jourlin, M., Carré, M., Breugnot, J., & Bouabdellah, M. (2012). Logarithmic image processing: Additive contrast, multiplicative contrast, and associated metrics. *Advances in Imaging and Electron Physics*, *171*, 357–406.

Jourlin, M., & Montard, N. (1997). A logarithmic version of the top-hat transform in connection with the Asplund distance. *Acta Stereologica, 16*(3), 201–208.

Jourlin, M., & Pinoli, J. C. (1988). A model for logarithmic image processing. *Journal of Microscopy, 149*, 21–35.

Jourlin, M., Pinoli, J. C., & Zeboudj, R. (1989). Contrast definition and contour detection for logarithmic images. *Journal of Microscopy, 156*(1), 33–40.

Kanungo, T., Mount, D., Netanyahu, N., Piatko, C., Silverman, R., & Wu, A. (2002). An efficient k-means clustering algorithm: Analysis and implementation. *IEEE Transactions on Pattern Analysis and Machine Intelligence, 24*, 881–892.

Köhler, R. (1981). A segmentation system based on thresholding. *Computer Vision, Graphics, and Image Processing, 15*, 319–338.

Mac Adam, D. L. (1942). Visual sensitivities to color differences in daylight. *Journal of the Optical Society of America, 32*, 247–274.

Mayet, F., Pinoli, J.-C., & Jourlin, M. (1996). Physical justifications and applications of the LIP model for the processing of transmitted light images. *Traitement du Signal, 13*(3), 251–262.

Otsu, N. (1979). A threshold selection method from gray level histograms. *IEEE Transactions on Systems, Man and Cybernetics, SMC, 9*, 62–66.

Palomares, J., Gonzalez, J., & Ros, E. (2005). Designing a fast convolution under the LIP paradigm applied to edge detection. In *Third international conference on advances in pattern recognition* (pp. 560–569).

Pun, T. (1981). Entropic thresholding: A new approach. *Computer Graphics and Image Processing, 16*, 210–239.

Sekiguchi, H., Sano, K., & Yokoyama, T. (1994). Interactive 3-dimensional segmentation method based on region growing method. *Systems and Computers in Japan, 25*(1), 88–97.

Serra, J. (1982). In *Image analysis and mathematical morphology: Vol. 1*. London: Academic Press. 610 pp.

Sivewright, J. G., & Elliott, J. (1994). Interactive region and volume growing for segmenting volumes in MR and CT images. *Medical Informatics, 19*(1), 71–80.

Tsaï, W. H. (1985). Moment preserving thresholding: A new approach. *Computer Vision, Graphics, and Image Processing, 29*, 377–393.

Wu, Z. (1993). Homogeneity testing for unlabeled data: A performance evaluation. *CVGIP, Graphical Models and Image Processing, 55*(5), 370–380.

CHAPTER THREE

Metrics Based on Logarithmic Laws

Contents

Advances in Imaging and Electron Physics, Volume 195
ISSN 1076-5670
http://dx.doi.org/10.1016/bs.aiep.2016.04.003

1. INTRODUCTION

The interest of image processing and image understanding considerably developed during the past three decades for various reasons: the power of computers exponentially increased while new tools appeared, making possible the common use of Computer Vision for biomedical applications (automated detection of cancerous cells, automated analysis of eye fundus images, 3-D reconstruction from serial or optical cuts issued, for example, of X-ray scanner), industrial applications (robotics, real-time quality control on conveyors), safety and surveillance (sensitive sites, highways, streets, subway stations), images acquired by a drone (inspection of dams, bridges, nuclear power plant cooler), and many other situations.

For all these reasons, the access to efficient tools in image analysis became a key need in order to make automated reliable decisions. In this aim, we often have to compare an image with a reference one in order to detect what is wrong and/or what has changed. The most classical approaches consist of detecting matching points or applying correlation tools.

Among all the possible techniques, we can consider that metrics answer the previous requests well. Nevertheless, there remain two properties that are generally not taken into account:

- The consistency with human vision, on which image interpretation was previously founded.
- The insensitivity to lighting variations.

When defining metrics in the logarithmic image processing (LIP) framework, the first property is a result of LIPs compatibility with the human visual system.

It is much more difficult to satisfy the second property, but in this chapter we will propose an efficient solution based on a novel kind of metrics that we have extended to functions (gray level, color, or multispectral images): the family of Asplünd's metrics, which were initially restricted to binary shapes.

2. RECALLS ON SOME EXISTING METRICS

2.1 Definition of a Metric

E is a set and d an application of ExE in the space $\mathbb{R}$ of real numbers.

We say that d is a *metric* defined on E if it satisfies the four following properties:

$$
\begin{array}{lll}
\forall (x, y) \in E^2 & d(x, y) \geq 0 & \\
\forall (x, y) \in E^2 & d(x, y) = d(y, x) & \text{(symmetry property)} \\
\forall (x, y) \in E^2 & d(x, y) = 0 \Leftrightarrow x = y & \text{(separation property)} \\
\forall (x, y, z) \in E^3 & d(x, z) \leq d(x, y) + d(y, z) & \text{(triangular inequality)}
\end{array}
$$

The notion of metric spaces is a particular case of topologic spaces, ie, those spaces based on the concept of neighborhoods. In fact, given a metric d on E, we can define specific neighborhoods of a point $x \in E$, called "balls" and constituted of elements of E whose distance to x is less than a given tolerance $\varepsilon > 0$. In spaces like $\mathbb{R}$, $\mathbb{R}^2$, and $\mathbb{R}^3$, balls become segments, disks, and spheres. Nevertheless, in general metric spaces as functional ones (spaces of images, for example), it is possible to observe neighborhoods with various shapes (cf. next section).

2.2 Examples of Functional Metrics

2.2.1 The Most Classical

We will limit our discussion to the most classical examples of functional metrics, noted d_1 and d_∞. In the case of gray-level images f and g defined on the same spatial support D or on a region R of D, we have:

$$d_1(f, g) = \iint_{D \text{ or } R} |f(x, y) - g(x, y)| dxdy$$

$$d_\infty(f, g) = \operatorname{Sup}_{(x, y) \in D \text{ or } R} |f(x, y) - g(x, y)|$$

Such metrics are extremely different.

Properties and usefulness of d_1

- d_1 is a "global" metric in the sense that it takes into account all the points lying in the considered region
- The value $d_1(f,g)$ represents the volume delimited by the representative surfaces of f and g
- In such conditions, it is possible to evaluate if two images are roughly similar, but it is unable to extract small-sized differences
- In the digital version, it is transformed into the double sum of the differences between pixels' gray levels according to the rows and columns, multiplied by the area of one pixel:

$$d_{1, D \text{ or } R}(f, g) = \Sigma\Sigma_{(i, j) \in D \text{ or } R} |f(i, j) - g(i, j)| \times (\text{area of a pixel})$$

Properties and usefulness of d_∞

- d_∞ is an "atomic" metric in the sense that its value is computed on a single point
- In such conditions, it is able to detect very small-sized differences between f and g, possibly limited to a single pixel in digital version. This is why we have introduced an intermediate metric between d_1 and d_∞in Jourlin, Carre, Breugnot, and Bouabdellah (2012).

Now let us return to the ability of metrics to define neighborhoods and examine their shapes.

Neighborhoods generated by the previous metrics

The neighborhoods shapes are totally different for d_1 and d_∞. In fact, given a function f, each function g verifying $d_\infty(f,g) \leq \varepsilon$ satisfies $|f(x) - g(x)| \leq \varepsilon$ for every point x lying in the considered region. It means that g belongs to a "tolerance tube" around f (cf. Fig. 1A). This remark explains why d_∞ is called "uniform convergence metric." This result holds for images, the tolerance tube becoming the volume located between the representative surfaces of $f + \varepsilon$ and $f - \varepsilon$.

When considering the "global" metric d_1, a ε-neighborhood of a given function f is different from a tube: it is an *unbounded set.* In fact, a function g belonging to the ε-neighbor of f may present at some point x an arbitrary large difference $|f(x) - g(x)|$ and a very small area located between f and g (cf. Fig. 1B)

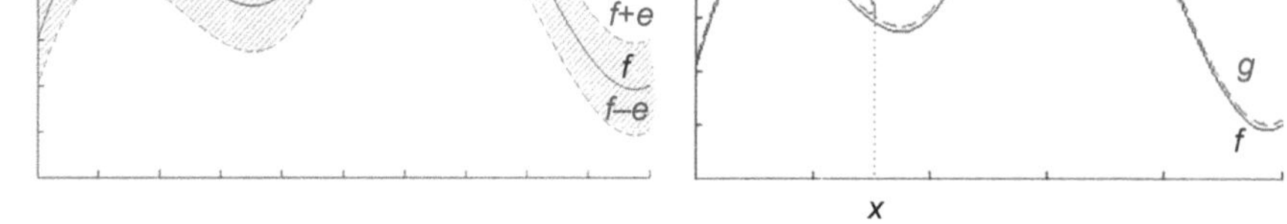

Fig. 1 Neighborhoods shapes for d_1 and d_∞. (A) The tolerance tube of f is represented by the hatched area. (B) The local difference between a function f and a function g lying in the ε-neighbor of f may be arbitrarily large. (See the color plate.)

2.2.2 *Examples of Other Metrics*

2.2.2.1 Intermediate Metric Between "Global" and "Atomic"

For some applications of industrial control, for which we aim to detect defects, the metric d_1 can be too global and thus insensitive to the purchased small defects while the atomic metric d_∞ will extract acceptable defects reduced to a single pixel. This is why we proposed an intermediate solution.

The intermediate solution consists of making a compromise between the size of an unacceptable defect and its intensity (contrast) in relation to the background. Let us explain this concept in one dimension, knowing that there is no difficulty to write it in two (or three) dimensions. In order to do this, we define an interval I_ε of length ε, which is moved along the interval $[a, b]$. We calculate the following integral:

$$d_\varepsilon(f,g) = \mathrm{Sup}_{x \in [a,\, b-\varepsilon]} \int_x^{x+\varepsilon} |f(t) - g(t)|dt$$

An illustration of this metric is given Fig. 2

In two dimensions, I_ε is simply replaced by a region of the spatial support, which is moved across it while computing the Sup:

$$d_{1,\mathrm{Sup}R}(f,g) = \mathrm{Sup}_{R \subset D} \iint_R |f(x,y) - g(x,y)|dxdy$$

For this approach, the region R is sized at the defect's desired dimension. In order to apply it to digital images, the double integral $\iint_R$ is replaced by the double summation $\sum\sum$ according to rows and columns.

2.2.2.2 A Bounded Metric Associating Binary and Gray-Level Approaches

Let us consider two gray-level functions f and g. The support notion $S(\cdot)$ of each of them is the subset of D where they take strictly positive values:

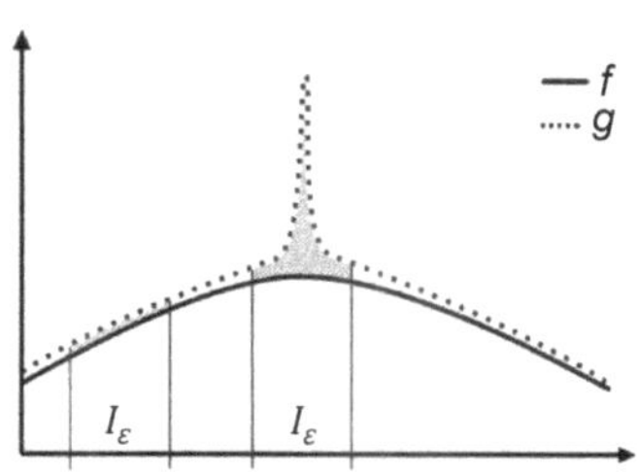

Fig. 2 Value of $d_\varepsilon(f,g) =$ largest hatched zone.

$$S(f) = \{(x, y) \in D, f(x, y) \neq 0\}$$

The spatial domain D is then separated into three disjointed subsets:

$$\begin{aligned} D_1 &= \{(x, y) \in D, f(x, y) \neq 0 \text{ and } g(x, y) \neq 0\} \\ &= S(f) \cap S(g) \end{aligned}$$

$$D_2 = \{(x, y) \in D, f(x, y) = 0 \text{ and } g(x, y) \neq 0 \textit{ or } f(x, y) \neq 0 \text{ and } g(x, y) = 0\}$$

$$= S(f) \Delta S(g)$$

where Δ represents the symmetric difference between two sets, ie, $A \Delta B = A \cup B \;\; A \cap B$

$$\begin{aligned} D_3 &= \{(x, y) \in D, f(x, y) = 0 \text{ and } g(x, y) = 0\} \\ &= S(f)^c \cap S(g)^c = (S(f) \cup S(g))^c \end{aligned}$$

At this step, the multiplicative contrast $C^{\triangle}_{(x,y)}(f, g)$ is computed on each of these subsets:

- If $(x, y) \in D_1$, $C^{\triangle}_{(x,y)}(f, g)$ is classically defined:

$$C^{\triangle}_{(x,y)}(f, g) \triangle \text{Min}(f(x, y), g(x, y)) = \text{Max}(f(x, y), g(x, y))$$

- If $(x, y) \in D_2$:

$$C^{\triangle}_{(x,y)}(f, g) = \infty$$

- If $(x, y) \in D_3$:

$$C^{\triangle}_{(x,y)}(f, g) = 0 \text{ by convention}$$

In order to replace the infinite value (case of $(x, y) \in D_2$), we apply the method proposed in remark 12 of Chapter "Various Contrast Concepts." Thus, a novel punctual distance d is obtained:

$$\begin{aligned} d_{(x,y)}(f, g) &= \frac{C^{\triangle}_{(x,y)}(f, g)}{1 + C^{\triangle}_{(x,y)}(f, g)} \text{ if } (x, y) \in D_1 \\ d_{(x,y)}(f, g) &= 1 \text{ if } (x, y) \in D_2 \\ d_{(x,y)}(f, g) &= 0 \text{ if } (x, y) \in D_3 \end{aligned}$$

Now, it is possible to perform a summation of these elementary distances when (x, y) lies in D (or a region of D), resulting in a metric $d_{S,GL}$, which takes into account the supports' shapes as well as the distance between gray levels on the supports intersection:

$$d_{S,GL}(f,g) = \sum_{(x,y)\in D} d_{(x,y)}(f,g)$$

$$= \sum_{(x,y)\in D_1} \left[\frac{C^{\triangle}_{(x,y)}(f,g)}{1 + C^{\triangle}_{(x,y)}(f,g)} \right] + \sum_{(x,y)\in D_2} [1]$$

However, $\sum_{(x,y)\in D_2}[1] = \mathrm{Card}(D_2)$ which we can assimilate to the area of D_2 or, by definition of D_2 to the area of $S(f)\,\Delta\, S(g)$.

Comment: The first sum of $d_{S,GL}(f,g)$ represents a distance in terms of gray levels and the second one a distance in terms of shapes.

The main interest we imagine for this metric is to quantify the evolution in shape and in gray level (concentration) of a polluting cloud (fumes, radioactive emissions, etc.).

We successfully tested the metric in such cases, but we have not the authorization to publish images.

2.3 Examples of Metrics Defined on Binary Shapes

Symmetric difference metric

Consider two binary shapes A and B (in $\mathbb{R}^2$ or $\mathbb{R}^3$). Their symmetric difference $A\triangle B$ (cf. Fig. 3A) is given by the following formula:

$$A\triangle B = (A\cup B)\ \ (A\cap B) = (A\ \ B)\cup(B\ \ A)$$

The symmetric difference distance between A and B is then defined according to:

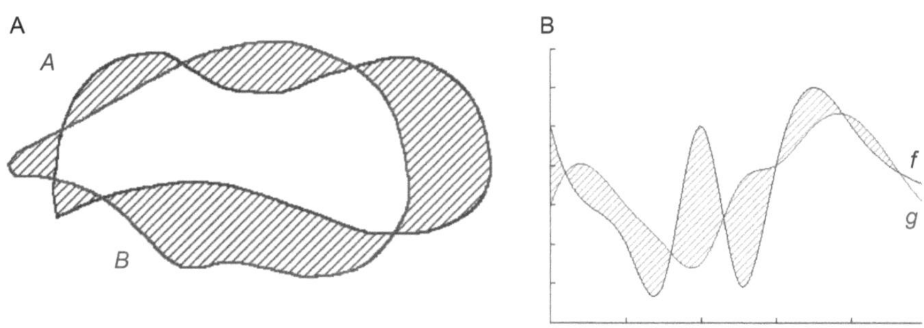

Fig. 3 Illustration of symmetric difference in the case of binary shapes and in the case of functions. (A) Symmetric difference between A and B (hatched area). (B) Symmetric difference between the subgraphs of functions f and g (hatched area). (See the color plate.)

$$d_\Delta(A, B) = \text{area}(A \,\Delta\, B)$$

Note that this metric is very easy to program and very fast in computer time. Its principle must be compared with the functional metric d_1 defined in previous Section 2.2 (cf. Fig. 3B). In fact:

$$\begin{aligned} d_1(f,g) &= \iint_{D\,\text{or}\,R} |f(x,y) - g(x,y)|\,dxdy \\ &= \iint_{D\,\text{or}\,R} |SG(f(x,y) - SG(g(x,y)|\,dxdy \end{aligned}$$

where $SG(f)$ designs the subgraph of f (set of points located under the representative surface of f).

Hausdorff's Metric

Given two binary shapes A and B (in $\mathbb{R}^2$ or $\mathbb{R}^3$), we define the Hausdorff's metric d_H according to the formula:

$$\begin{aligned} d_H(A, B) &= \text{Max}\left[\text{Sup}_{a \in A}\, \text{Inf}_{b \in B}\, d(a, b),\ \text{Sup}_{b \in B}\, \text{Inf}_{a \in A}\, d(a, b)\right] \\ &= \text{Max}\left[\text{Sup}_{a \in A}\, d(a, B), \text{Sup}_{b \in B}\, d(b, A)\right] \end{aligned}$$

This metric takes into account the point of A (or B) whose distance from the other shape is maximal (cf. Fig. 4A and B).

As we have done between d_1 and d_Δ, we can observe the similitude between d_∞ and d_H: a single point determines these two metrics. Thus, Hausdorff's metric generates neighborhoods representable as "tolerance tubes." More precisely, if A is a binary shape and ε a strictly positive real number, the neighborhood of A generated by the error tolerance ε is the set of all binary shapes B, satisfying the double condition: B contains the eroded set of A by a ball of radius ε and B is included in the dilated set of A by the same ball. In such conditions, the "tolerance tube" of A is the dilated of the boundary of A by a ball of radius ε.

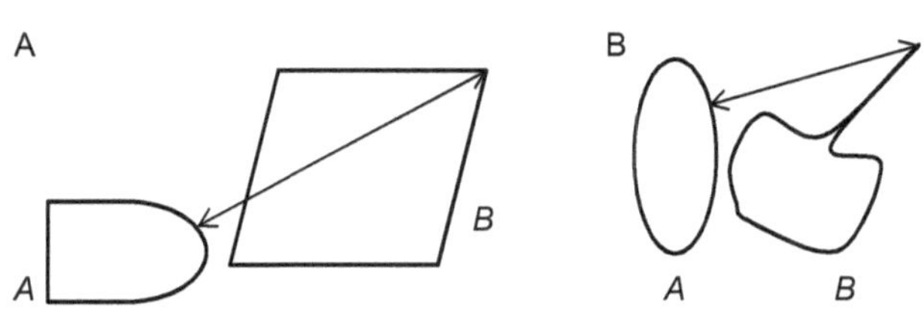

Fig. 4 Hausdorff's distance between two binary shapes: length of the double arrow.

3. TWO WAYS TO INTRODUCE NOVEL METRICS IN THE LIP FRAMEWORK

3.1 Extension of d_1 and d_∞ in the LIP Sense

This point has been studied in a previous paper Jourlin et al., 2012 and is briefly recalled here: it consists of introducing the metrics $d_1^{\triangle\!\!\!-}$ and $d_\infty^{\triangle\!\!\!-}$ on the basis of the logarithmic addition law (more precisely on the subtraction law $\triangle\!\!\!-$, due to the definition of the two classical metrics d_1 and d_∞):

$$d_1^{\triangle\!\!\!-}(f,g) = \iint_{D \text{ or } R} |f(x,y) \triangle\!\!\!- g(x,y)| dxdy$$

$$d_\infty^{\triangle\!\!\!-}(f,g) = \text{Sup}_{(x,y)\in D \text{ or } R} |f(x,y) \triangle\!\!\!- g(x,y)|$$

Remark 1. *What is the interest of such an extension?* The logarithmic subtraction between two gray levels represent, by definition (cf. Chapter "Various Contrast Concepts"), the logarithmic additive contrast (LAC) between them and it has been seen that such a contrast possesses a rigorous physical interpretation based on the transmittance law. This point results in a strong property: the LAC takes perfectly into account the intensity of the background on which an object is observed in transmission. In Jourlin et al. (2012), various theoretical and applicative interests of that property have been presented: automated thresholding and multithresholding (cf. Fig. 5), contour detection (cf. Fig. 6), target tracking (cf. Fig. 7), processing of images acquired under variable lighting, insensitivity of covariograms to lighting drift and lighting variations (cf. Fig 8), etc. For more details concerning thresholding and multithresholding, see Chapter "Transfer of Classical Tools to LIP Context."

Fig. 5 Automated multithresholding done by Köhler's method in which the initial contrast is replaced by the logarithmic additive contrast. (A) Initial image with 256 gray levels. (B) Multithresholded image with six gray levels.

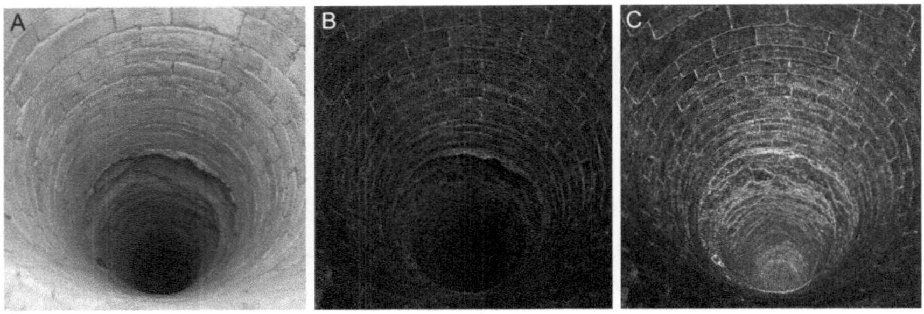

Fig. 6 Comparison of Sobel gradient and $MC_{(x)}$ (f) as contour detectors. (A) Initial gray-level image (a "well" with a very dark part). (B) Classical contour detection (Sobel gradient) applied on (A). (C) Visualization of the maximal contrast $MC_{(x)}$ (f) (notion defined in Chapter "Various Contrast Concepts") at each point of (A).

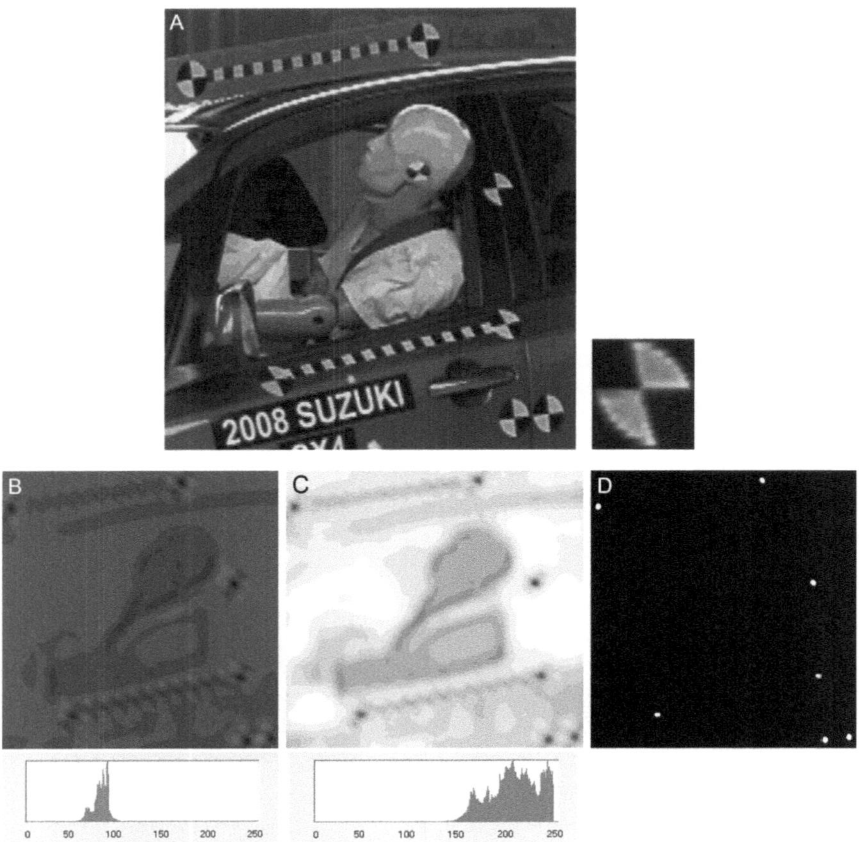

Fig.7 Example of target tracking based on the logarithmic metric $d_1^{\triangle}$. (A) Image of car crash test (with the authorization of "Insurance Institute for Highway Safety") with reference target (magnified). (B) Represents the distance map associated to the classical metric d_1. (C) Represents the distance map associated to the logarithmic metric $d_1^{\triangle}$. (D) Location of the targets obtained by an automated thresholding of (C). (See the color plate.)

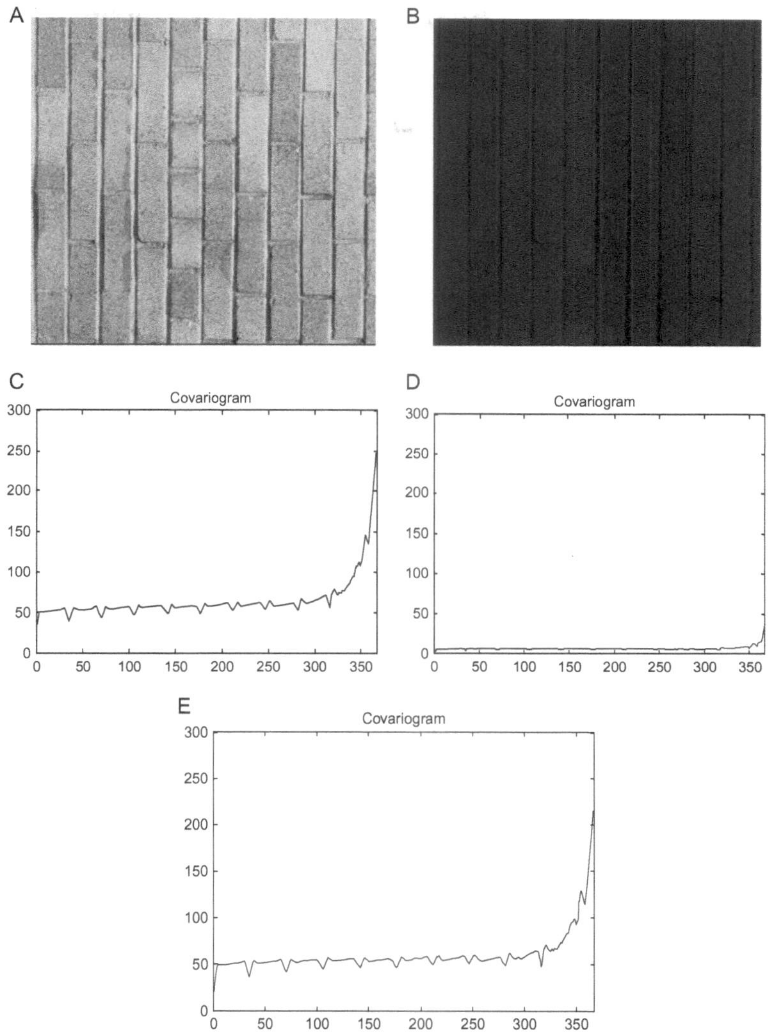

Fig. 8 Insensitivity to lighting variations of covariograms based on the logarithmic additive metric $d_1^{\triangle}$. (A) and (B) Same brick wall under variable lighting. (C) Horizontal covariogram of (A) computed based on the metric d_1 (similar result with $d_1^{\triangle}$). (D) Horizontal covariogram of (B) based on to the metric d_1. (E) Horizontal covariogram of (B) based on to the metric $d_1^{\triangle}$.

Comment on Fig. 5: we refer here to the automated thresholding method proposed by Köhler (1981), which has the advantage of accounting for the spatial information (contrasts of boundaries associated to a threshold), contrary to methods based only on histogram information. Moreover, to better take into account the human visual system, we have replaced the initial

Köhler's contrast, based on a simple difference between gray levels, with the LAC.

Comment on Fig. 6: The preeminence of the logarithmic approach is obvious when comparing images (B) and (C). Nevertheless, to achieve the binarization of the contours proposed in (C), which is not the purpose here, it is possible to apply a thresholding (not very efficient for complex images) or more accurately a watershed algorithm (see Beucher, 1991; Beucher & Lantuejoul, 1979).

Comments on Fig. 7: The dynamic range of the distance map obtained with $d_1^{\triangle\!\!\!\triangle}$ (Fig. 7C) is obviously better than that obtained with d_1 (Fig. 7B) as it is observable on the associated histograms. The automated thresholding applied on image (C) to get the final result (D) is from Otsu (1979), which is based on the classical technique of interclass variance maximization.

Let us now study two images of the same brick wall acquired under variable lighting (cf. Fig. 8). These images are clearly "pseudoperiodic" and the goal is to estimate the period. The chosen tool to reach this goal is the covariogram concept, applied here to compare an image with its horizontal translated. The comparison is made based on a metric and the covariogram represents the translation vector (horizontal axis) and the distance value (vertical axis).

On the covariogram curve, we observe local minima corresponding to the translation values for which the studied image and its translated are most similar. The distance between two successive minima produces an estimation of the most probable period observed on the initial image in the translation direction.

Comments on Fig. 8: The covariograms previously presented are nothing but an autocorrelation of the studied image where correlation is evaluated based on a distance, here d_1 and $d_1^{\triangle\!\!\!\triangle}$. It is interesting to note that these two metrics produce similar results for an image with standard lighting (image (A)). When the illumination decreases significantly, the metric $d_1^{\triangle\!\!\!\triangle}$ preserves the covariogram interpretability while d_1 fails.

3.2 Extension of the Intermediate Metric $d_{1,\mathrm{Sup}R}(f,g)$

This metric has been introduced in Section 2.2.2.1. Its LIP version is defined according to the following expression:

$$d^{\triangle\!\!\!\triangle}_{1,\,\mathrm{Sup}R}(f,g) = \mathrm{Sup}_{R\subset D} \iint_R \left[\mathrm{Max}(f(x,y),g(x,y)) \triangle\!\!\!\triangle \mathrm{Min}(f(x,y),g(x,y))\right] dxdy$$

whose digital expression, corrected by Card$R$, is:

$$d^{\triangle}_{1,\,\mathrm{Sup}R}(f,g)$$
$$=\mathrm{Sup}_{R\subset D}\frac{1}{\mathrm{Card}R}\left\{\sum\nolimits^{\triangle}\sum\nolimits^{\triangle}_{(i,j)\in R}[\mathrm{Max}(f(i,j),g(i,j))\triangle\mathrm{Min}(f(i,j),g(i,j))]\right\}$$

We have already explained the advantage of dividing the previous expression by Card$R$.

Remark 2. For a good use of the metric, the region R must be sized at the dimension of maximal purchased defects so the result will not be polluted by very small ones, while preserving the extraction of larger defects.

Remark 3. This intermediate distance is defined according to a superior bound. In such conditions, it detects only one defect: that corresponding to the greatest value. In order to avoid this weakness, it is possible to come back to the version of $d_{1,R}$ and to compute it for all the positions of R in D. It remains to retain as defects all the R locations where the computed distance is greater than a chosen threshold (Fig. 9).

3.3 Extension to Functions of the Binary Asplünd's Metric

Here we continue with the parallel previously established between functional and binary metrics. A few years ago, we focused on a little known metric defined in Asplund (1960) and Grünbaum (1963) for binary shapes: the Asplünd's metric.

Recall on the binary Asplünd's metric: In the initial definition of Asplünd's metric for a pair (A, B) of binary shapes, one shape, B, for example, is chosen to perform the double-sided probing of A by means of *homothetic* shapes of B: we compute the smallest number λ_0 such that $\lambda_0 B$ contains A and the greatest number μ_0 such that A contains $\mu_0 B$ (cf. Fig. 10). The Asplünd's distance $d_{\mathrm{As}}(A, B)$ between A and B is then defined according to:

$$d_{\mathrm{As}}(A,B)=Ln\left(\frac{\lambda_0}{\mu_0}\right)$$

Remark 4. Such a definition implies that d_{As} remains unchanged when one shape is magnified or reduced in any ratio k.

Comment on Fig. 10: After constructing the homothetic sets $\lambda_0 B$ and $\mu_0 B$, we can remark that for every shape A_1, we have the equivalence:

$$\mu_0 B\subset A_1\subset\lambda_0 B\Leftrightarrow d_{\mathrm{As}}(A_1,B)\leq d_{\mathrm{As}}(A,B)$$

This means that the set $\alpha B\backslash\beta B$ is interpretable as a tolerance tube.

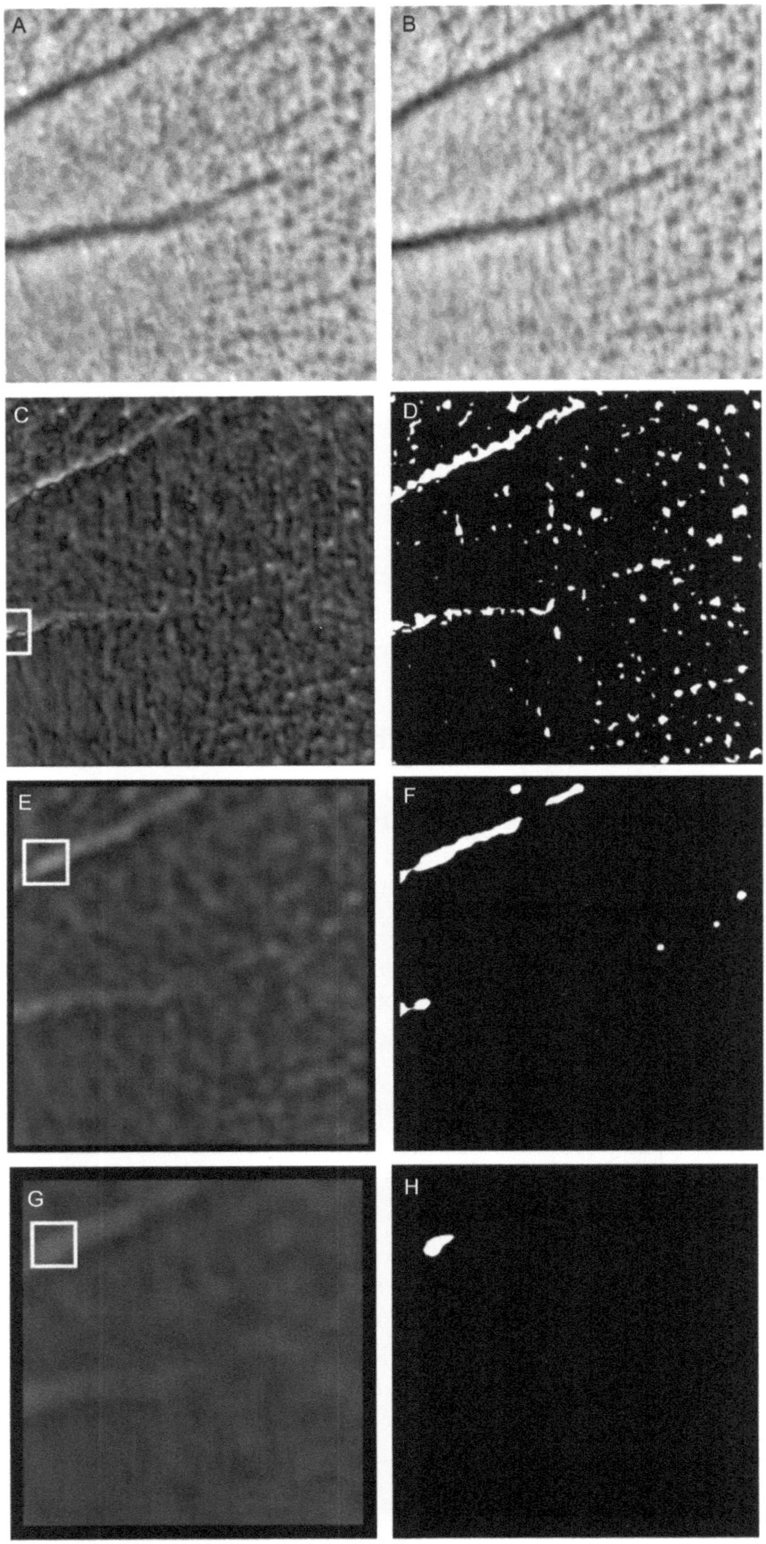

Fig. 9 See figure legend on opposite page.

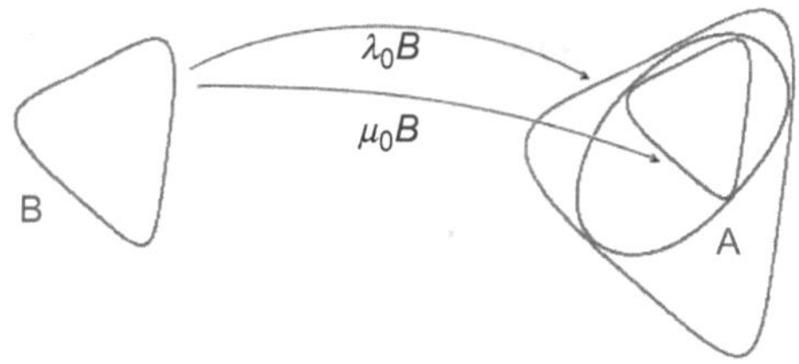

Fig. 10 Asplünd's metric for binary shapes. (See the color plate.)

Furthermore, given a reference shape A and a tolerance number $\varepsilon > 0$, it is easy to characterize the set $N_{\mathrm{As},\varepsilon}(A)$ of shapes B neighboring A in the sense of the following inequality:

$$d_{\mathrm{As}}(A,B) \le \varepsilon$$

Starting from the couple A and ε, we choose an arbitrary homothetic shape of A, say βA, and compute the (unique) number α such that $Ln\left(\frac{\alpha}{\beta}\right) = \varepsilon$. Thus, the two homothetic sets αA and βA delimit a tolerance tube constituted of their difference set $\alpha A \setminus \beta A$.

In fact, each element B satisfying the double inclusion $\beta A \subset \lambda B \subset \alpha A$ for some real number λ, is an element of the neighborhood $N_{\mathrm{As},\varepsilon}(A)$:

$$\beta A \subset \lambda B \subset \alpha A \Rightarrow B \in N_{\mathrm{As},\,\varepsilon}(A)$$

The difference with previous situations associated with atomic metrics is that the pair of initially considered shapes generates the tolerance tube and it is not of constant thickness. Nevertheless, the similarity of Asplünd's metric with atomic metrics means that we will have to consider the sensitivity to noise of Asplünd-like metrics.

First papers combining functional Asplünd's metrics, probing, and tolerance tubes: The idea of extending Asplünd's metric to functions and consequently to images was one I have long wanted to explore. In a paper coauthored by Nathalie Montard (Jourlin & Montard, 1998) dedicated to the notion of logarithmic Top-Hat, we made a link with Asplünd's metric. A few years

Fig. 9 Defects detection based on the logarithmic intermediate metric $d^{\triangle}_{1,\mathrm{SupR}}$. (A) Initial image f of skin wrinkles by fringes projection at time T_0. (B) Initial image g at $T_0 + 28$ days. (C), (E), and (G): $d^{\triangle}_{1,\mathrm{Sup}R}(f,g)$ for different sizes of R (1x1, 21x21, 41x41): the detected region is the white square. (D), (F), and (H) represent the images (C), (E), and (G) thresholded at gray level 90.

later, I proposed to Cécile Barat that she work on this subject during her PhD. She proposed two variants:

- The first, called "pattern matching using morphological probing" (Barat, Ducottet, & Jourlin, 2003) consisted of transferring to image analysis the principle of mechanical probing: it resulted in two morphological pattern-matching transforms: the single object matching using probing (SOMP) transform and its extension to multiple object detection, the multiple object matching using probing (MOMP) transform. It permits defining a priori tolerance tubes limited by two functions (a superior bound and an inferior one) in order to extract from an image all the patterns lying in the tolerance tube.
- In the second (Barat, Ducottet, & Jourlin, 2010), called virtual double-sided image probing (VDIP): "Virtual Double-sided Image Probing: A unifying framework for nonlinear grayscale pattern matching," Barat proposed a unified approach of seemingly different methods by placing them under a topological formulation. Let us note that Barat highlights a property which will be at the center of functional Asplünd's metric properties: "*Moreover, the VDIP transform compensates automatically an additive illumination effect or a multiplicative illumination effect when working on log intensities. This allows detecting a pattern independently of the illumination*" page 3438 (in section 4.2 Properties).

Definition of a Functional Multiplicative Asplünd's metric: Here we refer to previous papers where this subject has been addressed (Jourlin et al., 2012; Jourlin, Couka, Abdallah, Corvo, & Breugnot, 2014, for example).

The novelty of what we proposed in Jourlin et al. (2012) consists of using logarithmic homothetics $k \triangle\!\!\!\!\times f$ as probing patterns. Based on the transmittance law, this logarithmic scalar multiplication always remains in the grayscale. This property makes it possible to define a logarithmic multiplicative Asplünd metric as:

Given two images f and g defined on D, we choose, as for binary shapes, a probing element, say g, for example (cf. Fig. 11), and define the two numbers:

$$\alpha = \text{Inf}\{k, f \leq k \triangle\!\!\!\!\times g\} \text{ and } \beta = \text{Sup}\{k, k \triangle\!\!\!\!\times g \leq f\}$$

The corresponding multiplicative Asplünd's metric is noted $d_{\text{As}}^{\triangle\!\!\!\!\times}$ and is defined according to:

$$d_{\text{As}}^{\triangle\!\!\!\!\times}(f, g) = Ln\left(\frac{\alpha}{\beta}\right) \tag{1}$$

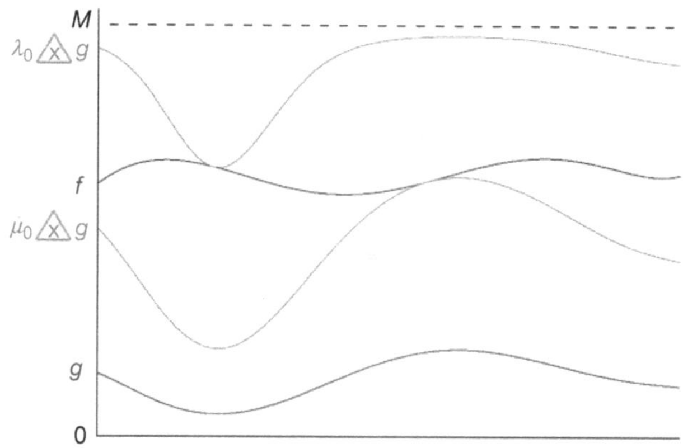

Fig. 11 Computation of the double-sided probing metric $d^{\triangle}_{As}(f,g)$, when g is the probing function. (See the color plate.)

Fundamental property: As noted for the binary Asplünd's metric (cf. Remark 4), let us focus on a strong advantage of the multiplicative Asplünd's metric—it remains unchanged when one function (f, for example) is replaced by any homothetic $k \triangle f$ for $k > 0$. Remember that $k \triangle f$ is brighter than f when $k \in]0, 1]$ and darker than f when $k \in [1, +\infty[$. Such a property shows the insensitivity of $d^{\triangle}_{As}$ to strong illumination variations modeled by the logarithmic multiplicative law.

Mathematicians Section: From the definition of $d^{\triangle}_{As}$ and the previous fundamental property, it would be more rigorous to present the multiplicative Asplünd's metric in the following way:

- First, we can define an *equivalence relation* R on the space of images:

  Given a pair of images f and g, we say they are "in relation" if they satisfy:

$$(fRg) \Leftrightarrow \exists k > 0, f = k \triangle g$$

  Note that the previous relation R is clearly an equivalence relation, because it satisfies the three required properties:
  - * Reflexivity: $\forall f \in I(D, [0, M[), fRf$
  - * Symmetry: $\forall (f, g) \in I^2(D, [0, M[), fRg \Leftrightarrow gRf$
  - * Transitivity: $\forall (f, g, h) \in I^3(D, [0, M[), (fRg \text{ and } gRh) \Rightarrow fRh)$
- Second, we associate each image $f \in I(D, [0, M[)$ with its equivalence class $f^{\triangle}$

$$f^{\triangle} = \{g, gRf\}$$

– Third, we are now able to give a rigorous version of the multiplicative Asplünd's metric defined on the space of equivalence classes $I^{\triangle}(D, [0, M[)$:

$$\forall (f^{\triangle}, g^{\triangle}) \in I^{\triangle 2}(D, [0, M[), d_{\text{As}}^{\triangle}(f^{\triangle}, g^{\triangle}) = d_{\text{As}}^{\triangle}(f_1, g_1)$$

where $d_{\text{As}}^{\triangle}(f_1, g_1)$ is the distance, according to Eq. (1) between two elements f_1 and g_1 of the equivalence classes $f^{\triangle}$ and $g^{\triangle}$.

The following section is devoted to the multiplicative Asplünd's metric $d_{\text{As}}^{\triangle}$, its properties, drawbacks, and main applications.

4. THE MULTIPLICATIVE ASPLÜND'S METRIC $d_{\text{AS}}^{\triangle}$

Remark 5. From the definition of $d_{\text{As}}^{\triangle}$ given in Eq. (1), we observe a situation previously met for the multiplicative contrast introduced in Chapter "Various Contrast Concepts:" this metric can produce infinite values when there exists at least a point x in D for which f or g is null and the other strictly positive. We propose to report to the solutions proposed in Chapter "Various Contrast Concepts" to answer this problem.

4.1 Local Processing and Application to Target Tracking

Let us remark that the metric $d_{\text{As}}^{\triangle}$ is adaptable to local processing, in particular to detect on an image f the place where a given pattern or target model is probably located. In such a case, the target corresponds to an image t defined on a spatial support D_t, smaller than the initial one D. For each location of D_t included in D, the distance $d_{\text{As}}^{\triangle}\left(t, f|_{D_t}\right)$ is computed, where the notation $f|_{D_t}$ represents the restriction of image f to D_t.

In the following example (cf. Fig. 12, previously presented in Jourlin et al., 2012), we consider a brick wall from which we extract two bricks, one bright and the other dark. For each of them, we produce the corresponding Asplünd's map: for each location of the target t inside the initial image f, we compute the distance $d_{\text{As}}^{\triangle}\left(t, f|_{D_t}\right)$ and we affect this number (normalized as a level in the standard grayscale $[0, 255[$) at the center of the target. The smallest distance values are the dark points located at the bricks centers.

Comments on Fig. 12

– Let us note the strong similarity between the Asplünd's maps (C) and (E). This shows that the multiplicative Asplünd's metric is roughly insensitive

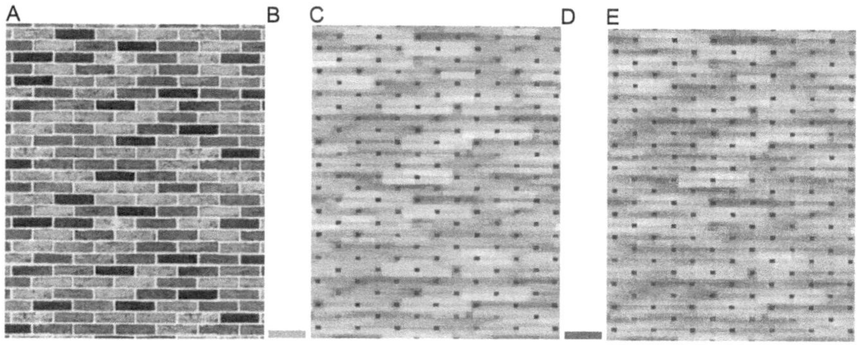

Fig. 12 Example of target detection by means of multiplicative Asplünd's metric. (A) Initial image. (B) Bright target t_1. (C) Corresponding Asplünd's map (values of $d_{\mathrm{As}}^{\triangle}\left(f|_{D_{t_1}}, t_1\right)$. (D) Dark target t_2. (E) Corresponding Asplünd's map (values of $d_{\mathrm{As}}^{\triangle}\left(f|_{D_{t_2}}, t_2\right)$.

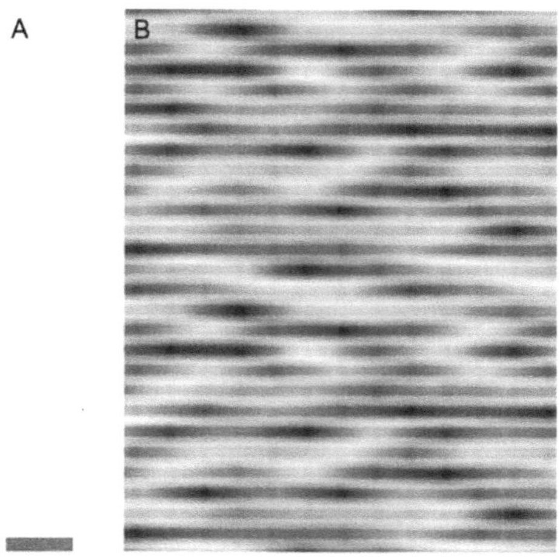

Fig. 13 Example of target detection based on correlation. (A) Chosen target. (B) Correlation map.

to "lighting" variations as long as these variations can be modeled by the logarithmic multiplicative law.

- The targets have been chosen lightly smaller than the bricks themselves in order to facilitate the probing of (A) by the targets. This explains the thickness of the dark points of (C) and (D).
- The final extraction (from (C) and (E)) of the bricks locations is easily realized by means of an automated thresholding (Otsu, 1979, for example).
- A classical approach by correlation (cf. Fig. 13) produces an image less easy to interpret. In the best case, we will extract the bricks presenting gray levels similar to the target ones.

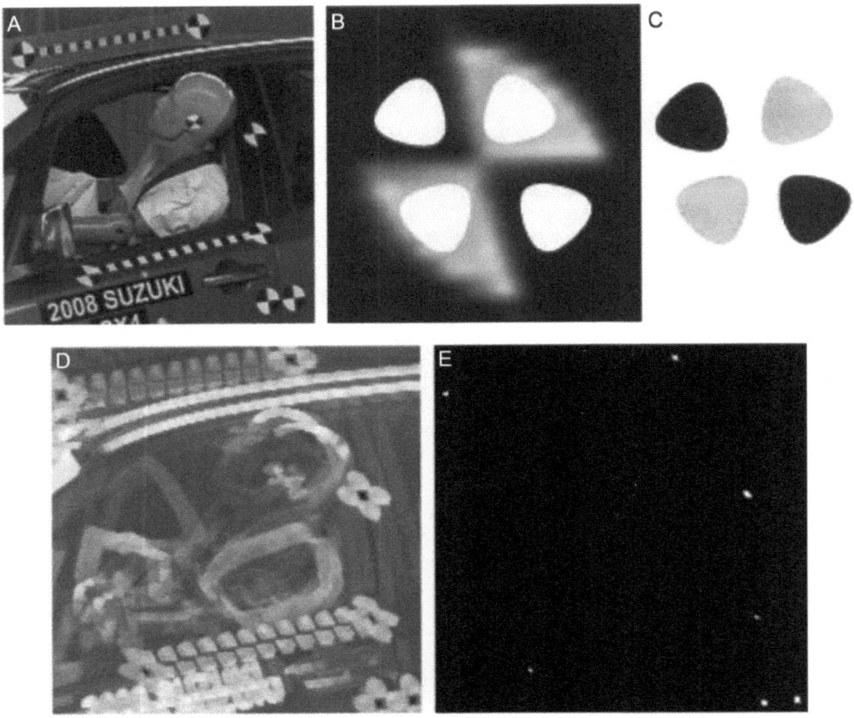

Fig. 14 Target detection based on multiplicative Asplünd's metric. (A) Crash test image. (B) Probing zones inside the target magnified 10 times. (C) Probing function magnified 10 times. (D) Asplünd's map. (E) Thresholded image of (D). (See the color plate.)

If we now come back to the example given in Fig. 7 where the targets were extracted through the logarithmic version $d_1^{\triangle}$ of the metric d_1, the problem can be solved by the multiplicative Asplünd's metric $d_{\mathrm{As}}^{\triangle}$. Fig. 14 presents the chosen target and the corresponding distance map.

Comment on Fig. 14: As remarked for the previous case of a brick wall, it is recommended to select probing zones smaller than the target. The results obtained are very similar to those produced by the metric $d_1^{\triangle}$.

4.2 Neighborhoods Generated by the Multiplicative Asplünd's Metric

The neighborhoods generated by Asplünd's metric have been already interpreted for binary shapes as tolerance tubes.

The same reasoning is possible for gray-level images. Given an image f and a tolerance number $\varepsilon > 0$, let us consider the successive steps:

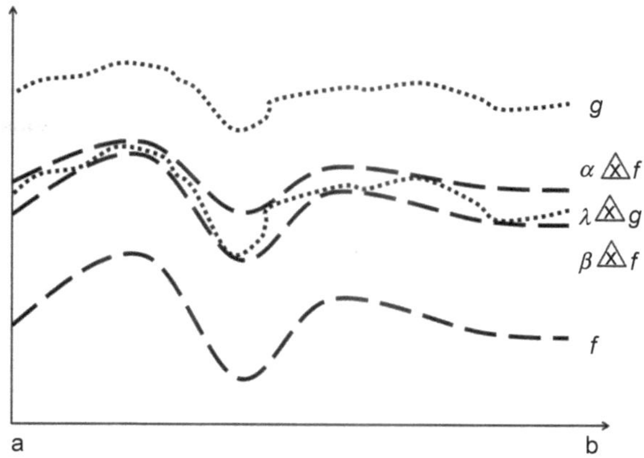

Fig. 15 Example of two images *f* and *g*, which are ε-neighbors in Asplünd's sense.

1. Create a family of tolerance tubes $T_{\alpha,\beta,\varepsilon}(f)$ consisting of regions delimited by two homothetics $\alpha \triangle f$ and $\beta \triangle f$ such that $Ln\left(\frac{\alpha}{\beta}\right) = \varepsilon$
2. Define the neighborhood $N^{\triangle}_{As,\,\varepsilon}(f)$:

$$g \in N^{\triangle}_{As,\,\varepsilon}(f) \Leftrightarrow \exists(\alpha,\beta), \exists\lambda,\ Ln\left(\frac{\alpha}{\beta}\right) = \varepsilon \text{ and } \lambda \triangle g \in T_{\alpha,\beta,\varepsilon}(f)$$

3. Visualize a monodimensional representation (Fig. 15)

Remark 6. In a first approach, it may be surprising to associate an infinite number of tolerance tubes $T_{\alpha,\beta,\varepsilon}(f)$ to a pair (f, ε) constituted of an image and a tolerance $\varepsilon > 0$. This point is clarified based on the Mathematicians Section. In fact, in a rigorous reasoning, an image *g* lies in a ε-neighborhood of the considered image *f* if, and only if, it satisfies the inequality $d^{\triangle}_{As}\left(f^{\triangle}, g^{\triangle}\right) \leq \varepsilon$. This means that an element of the equivalence class $f^{\triangle}$ and another element of the equivalence class $g^{\triangle}$ both lie in the same tolerance tube, as shown in Fig. 15. If we remember the physical meaning of the logarithmic multiplicative law $\lambda \triangle$ (which multiplies the observed object *thickness* by λ), the ε-similarity of *f* and *g* according to $d^{\triangle}_{As}$ means that there exist two thicknesses λ and μ, such that $\lambda \triangle f$ and $\mu \triangle g$ belong to the same ε-tube.

Moreover, it is possible to exhibit a tolerance tube centered on the studied gray-level function. In fact, if we consider a gray-level image *f* and a tolerance value ε related to the Asplünd's metric, the neighborhood

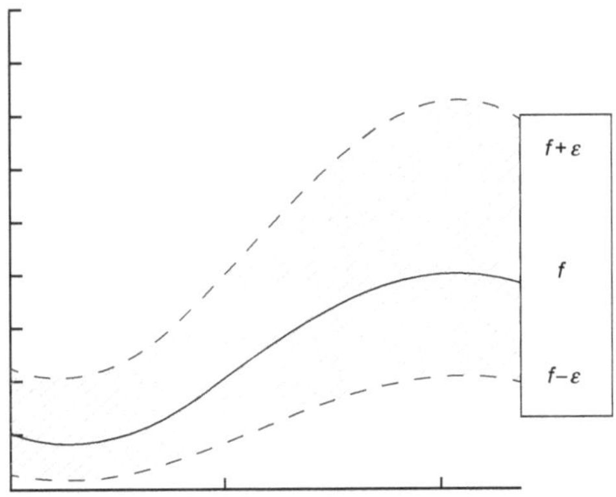

Fig. 16 Visualization of a tolerance tube centered at *f*.

$N^{\triangle}_{As,\varepsilon}(f)$ is the union of all the tubes $T_{\alpha,\beta,\varepsilon}(f)$. Among them, consider the tube $T_{\alpha_c,\beta_c,\varepsilon}(f)$ defined by a pair of real numbers α_c and β_c computed to have symmetric values around the unit integer 1, for example, such that $\alpha_c = 1 + k$ and $\beta_c = 1 - k$. Such a tube is centered at f and the condition $Ln\left(\frac{\alpha_c}{\beta_c}\right) = \varepsilon$ yields:

$$Ln\left(\frac{1+k}{1-k}\right) = \varepsilon \Leftrightarrow \frac{1+k}{1-k} = e^{\varepsilon} \Leftrightarrow k = \frac{e^{\varepsilon}-1}{e^{\varepsilon}+1}$$

which gives the values of α_c and β_c:

$$\alpha_c = \frac{2e^{\varepsilon}}{e^{\varepsilon}+1} \text{ and } \beta_c = \frac{2}{e^{\varepsilon}+1}$$

The resulting tube is displayed in Fig. 16.

We are now focusing on the main drawback of the multiplicative Asplünd's metric: its sensitivity to noise, which is a weakness shared by every atomic metric.

4.3 How to Overcome the Noise Sensitivity of Multiplicative Asplünd's Metric

This point has been studied in Jourlin et al. (2014), thus we will only present a summary of this paper in which some examples of the multiplicative Asplünd's metric have been exposed, such as detection of human skin pores or extraction of dermal papillae from images acquired in confocal microscopy.

In order to highlight the sensitivity of multiplicative Asplünd's metric to noise, let us start with an initial image f whose representative surface is a tilted

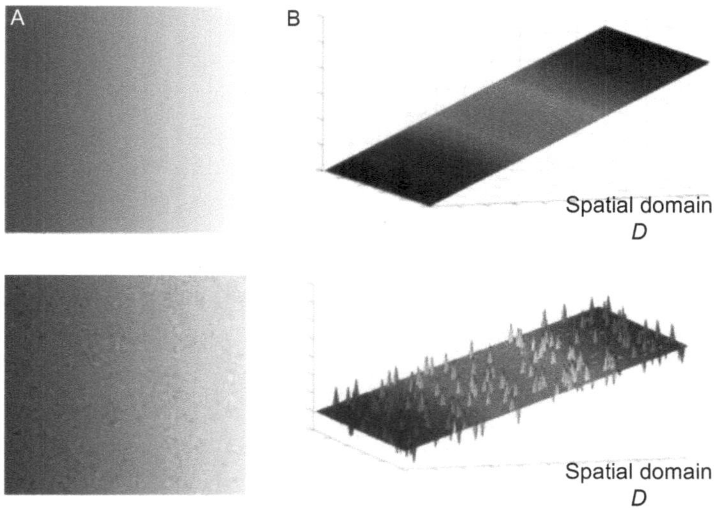

Fig. 17 A synthesis image *f* and its noised version. (A) Initial image *f* with a linear drift. (B) Representative surface of image *f*. (C) Image *g*: image *f* with salt and pepper noise added. (D) Representative surface of image *g*. (See the color plate.)

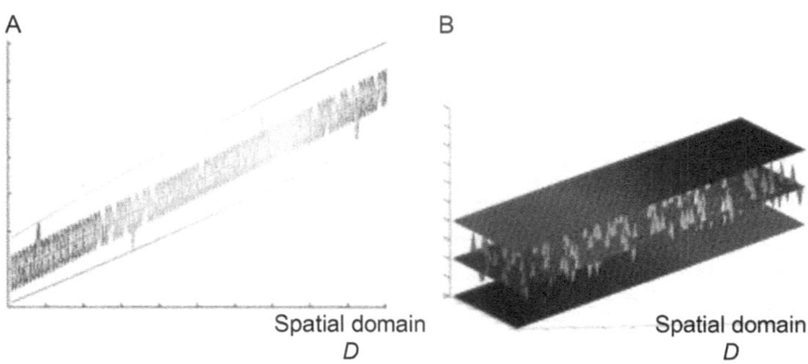

Fig. 18 Sensitivity to noise of $d_{As}^{\triangle}$. (A) Representation of the probing of *g* by *f* (2-D representation). (B) Representation of the probing of *g* by *f* (3-D representation). (See the color plate.)

plane (cf. Fig. 17A and B). A salt and pepper noise is added to f (cf. Fig. 17C and D).

The computation of the multiplicative Asplünd's distance $d_{As}^{\triangle}(f,g)$ between f and g may be done by probing image f with g (cf. Fig. 18). It appears clearly that the local maxima/minima of f determine the probes locations, which show precisely the sensitivity to noise of $d_{As}^{\triangle}(f,g)$.

Let us now expose a solution to overcome this drawback of $d_{As}^{\triangle}$.

We will adapt a metric (later noted M-metric) to our particular situation defined in the context of measure theory. Our goal here is not to deeply enter this theory, so we will limit our discussion to a short recall adapted to the context of gray-level images defined on a subset D of $\mathbb{R}^2$. Given a measure μ on $\mathbb{R}^2$, a gray-level image f, a metric d on the space of gray-level images $I(D, [0, M[)$, and two arbitrary small positive real numbers ε and ε', a neighborhood $N_{\mu, d, \varepsilon, \varepsilon'}(f)$ of f may be defined according to:

$$N_{\mu, d, \varepsilon, \varepsilon'}(f) = \{g, \mu(\{x \in D, d(f(x), g(x)) > \varepsilon\}) < \varepsilon'\}$$

Such a definition implies that the measure of the set of points x where the distance $d(f(x), g(x))$ overcomes the tolerance ε satisfies another tolerance ε'.

Now, let us adapt this approach to the context of Asplünd's metric:

- If a gray-level image is digitized, the number of pixels lying in D is finite, thus the measure of a subset D' of D is directly linked to the cardinal (number of pixels) of this subset; for example, the percentage P of its elements related to D (or a region of interest R included in D). In our case, we search a subset D' of D such that the restrictions $f|_{D'}$ and $g|_{D'}$ of f and g to D' are neighbors for $d^{\triangle}_{As}$ and at the same time, the complementary set $D \setminus D'$ of D' related to D is small when compared with D. This last condition means that the ratio of the cardinals of $D \setminus D'$ and D is smaller than an arbitrary acceptable percentage p:

$$\frac{\mathrm{Card}(D\, D')}{\mathrm{Card}(D)} \leq p$$

In such conditions, the neighborhood $N_{\mu, d, \varepsilon, \varepsilon'}(f)$ becomes $N_{p, a^{\triangle}_{As}, \varepsilon}(f)$:

$$N_{p, d^{\triangle}_{As}, \varepsilon}(f) = \left\{ g, \exists D' \subset D, d^{\triangle}_{As}\left(f|_{D'}, g|_{D'}\right) < \varepsilon \text{ and } \frac{\mathrm{Card}(D\, D')}{\mathrm{Card}(D)} \leq p \right\}$$

- Let us now explain the role played by the M-metric in the case of the oblique plane (cf. Fig. 18). For decreasing the Asplünd's distance between f and g, we need to move the probing functions $\lambda {\triangle} f$ and $\mu {\triangle} f$ closer together with the M-metric. If we refer to the definition of this metric, the set $D \setminus D'$ corresponds to the part of D where the highest noise peaks are located. Before more precisely presenting the method, let us visualize (Fig. 19) the set $D \setminus D'$ emerging through the probing functions for the thresholds values $p = 0.98$ and $p = 0.95$. These values mean that the set $D \setminus D'$ represents 2% of the set D (Fig. 19A) and 5% of the set D (Fig. 19B). It appears that a small-sized restriction of the set D permits strongly decreasing the Asplünd's distance, and thus overcomes the noise effect.

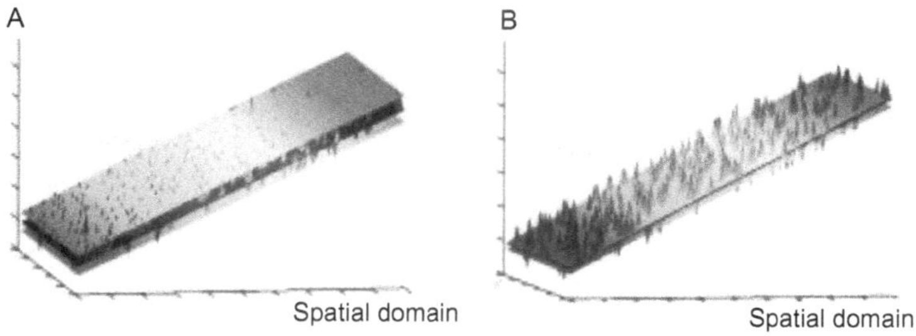

Fig. 19 Getting probing functions closer to the target. (A) Probing of g by f with tolerance $p=0.98$. (B) Probing of g by f with tolerance $p=0.95$. (See the color plate.)

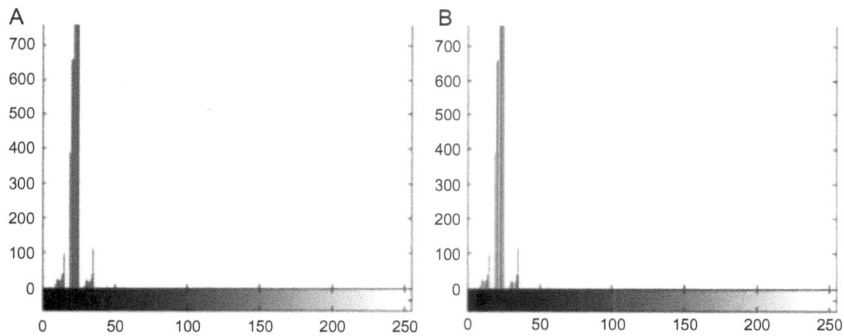

Fig. 20 Differences between g and its probing functions. (A) Histogram of the differences between $\lambda \triangle f$ and g. (B) Histogram of the differences between g and $\mu \triangle f$. (See the color plate.)

Associated with Asplünd's distance, the M-metric permits determining a set D' that satisfies the distance (percentage) condition. The method takes into consideration the differences, pixel per pixel, between two images, and more precisely the histogram of these differences. This subtraction does not imply any problem because the superior probing function $\lambda \triangle f$ is always superior to the image g, which is superior to the inferior probing function $\mu \triangle f$. In such conditions, the resulting differences always remain in the gray-scale. The histograms corresponding to the differences between $\lambda \triangle f$ and g (resp. g and $\mu \triangle f$) are presented in Fig. 20A (resp. 20B).

Let us note that the first bin of each histogram (value 0) represents the number of contact pixels between the two corresponding images. In order to decrease Asplünd's distance, the probing functions $\lambda \triangle f$ and $\mu \triangle f$ must be as close as possible to g. Thus, we decided to disregard the pixels corresponding to the first bin of the actual histogram of differences, until the most relevant bins have been reached. Neglecting the first bin means that the following bin is considered as the first one. This step modifies the selection of the pixels making contacts: they will be closer to the image g.

Now let us define the real numbers k_λ (resp. k_μ) corresponding to the bins making contact between the images g and $\lambda \triangle\!\!\!\!\times f$ (resp. g and $\mu \triangle\!\!\!\!\times f$). Initially, $k_\lambda = k_\mu = 0$ and they are referring to the real first bin of the histograms. In this situation, λ and μ determine the initial Asplünd's distance $d_{\text{As}}^{\triangle\!\!\!\!\times}$. If we increase k_λ for instance, a new $\lambda_1 < \lambda$ will be computed such that the $(k_{\lambda_1})^{\text{th}}$ bin becomes the bin making a contact. In other words, if x_λ is a pixel belonging to the first bin of the histogram of differences between g and $\lambda \triangle\!\!\!\!\times f$, we can write

$$\lambda \triangle\!\!\!\!\times f(x_\lambda) = g(x_\lambda) \text{ and } \lambda_1 \triangle\!\!\!\!\times f(x_\lambda) = g(x_\lambda) - k_{\lambda_1}$$

For k_μ, a similar approach is performed: if x_μ belongs to the first bin of the histogram of differences between g and $\mu \triangle\!\!\!\!\times f$, a number $\mu_1 > \mu$ is computed such that:

$$\mu \triangle\!\!\!\!\times f(x_\mu) = g(x_\mu) \text{ and } \mu_1 \triangle\!\!\!\!\times f(x_\mu) = g(x_\mu) + k_{\mu_1}$$

The scalar k_{λ_1} (resp. k_{μ_1}) is the gray-level number we must subtract (resp. add) from the superior probe $\lambda \triangle\!\!\!\!\times f$ (resp. to the inferior probe $\mu \triangle\!\!\!\!\times f$) in order to get the contact points closer to g. The scalars k_λ and k_μ are increased until the percentage p has not been reached. The set of pixels belonging to the bins of the histograms which are neglected corresponds to $D \setminus D'$. A new Asplünd's distance (smaller than the previous one) is computed for each new pair (λ_1, μ_1).

For evaluating this technique on a real example, let return to the image bricks wall presented in Fig. 12A. We apply salt and pepper noise to the image and we show (cf. Fig. 21) the results obtained based on the M-metric for various values of the percentage p.

Comments on Fig. 21: All the displayed images use the same standard gray scale: a dark area corresponds to a small value of Asplünd's distance, and the lighter the area, the larger the distance. Images (D) and (E) show the detection of the chosen target (image (C)) when applying the standard multiplicative Asplünd's metric: no information can be extracted (image (E)) from the noisy image.

Images (F), (G), and (H) correspond to the target extraction from the noisy image (B). They have been obtained for various values of the percentage p of neglected pixels ($p = 0.98$, $p = 0.95$, $p = 0.90$, for (F), (G), (H), respectively. They prove that the M-metric allows us to improve the result produced on (B) by the standard multiplicative Asplünd's metric and to finally get a quality comparable to that obtained without any noise (image (G) corresponding to the percentage 0.95). However, if the percentage value decreases too much, we can observe the emergence of horizontal black

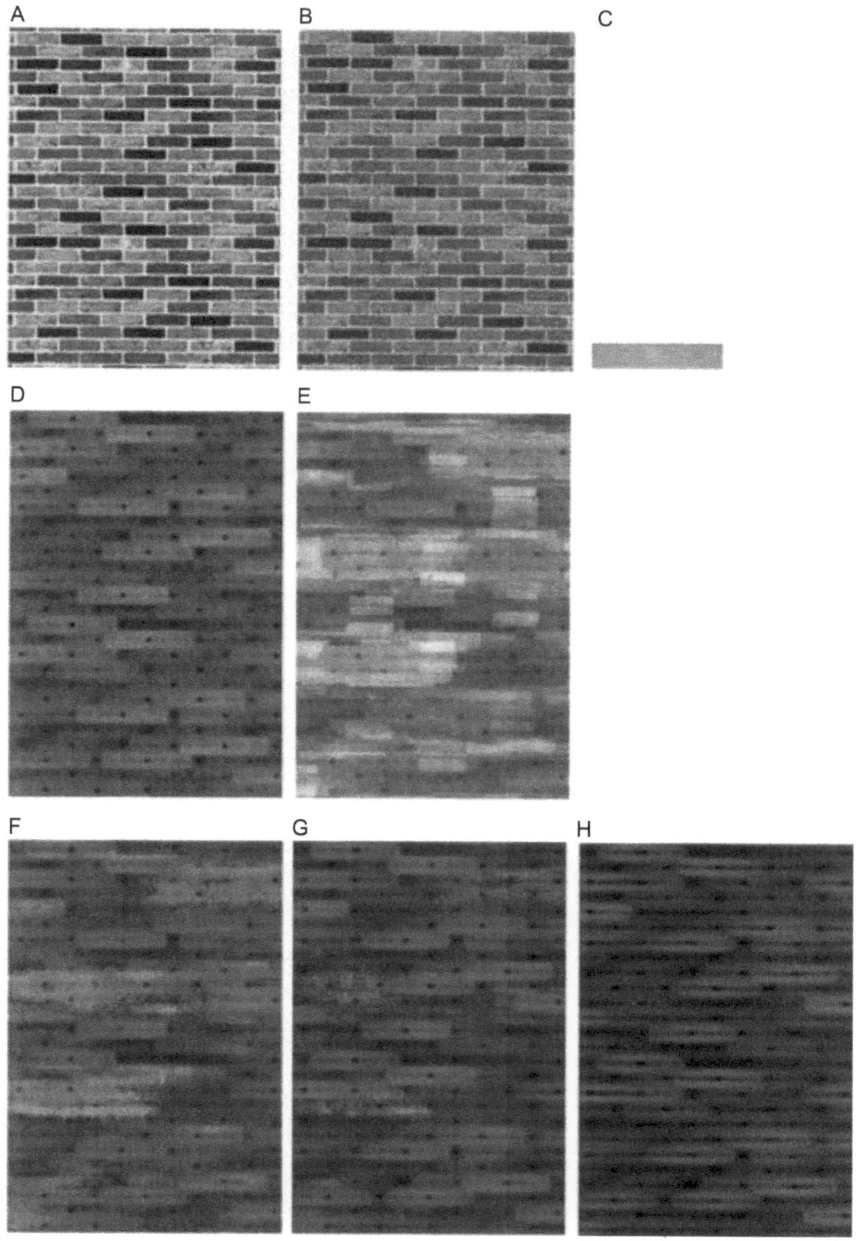

Fig. 21 Detection of a target (here a brick wall) on a noisy image. (A) Initial image f: Brick wall. (B) Image f with Salt and Pepper noise added. (C) Target (magnified). (D) Extraction of the target from image (A) based on $d_{As}^{\triangle}$. (E) Extraction of the target from image (B) based on $d_{As}^{\triangle}$. (F) Extraction of the target from image (B) based on the M-Metric when the percentage p of neglected points equals 0.98. (G) Extraction of the target from image (B) based on the M-Metric when the percentage p of neglected points equals 0.95. (H) Extraction of the target from image (B) based on the M-Metric when the percentage p of neglected points equals 0.90.

lines (image (H)). Such lines are due to the fact that the vertical borders between two successive bricks are small enough to be neglected at the considered percentage. In conclusion, the *M*-metric appears capable of overcoming the problem of noise sensitivity for the multiplicative Asplünd's metric. Nevertheless, research remains to be done in order to find an automated method for determining the optimal percentage adapted to a given noisy image. This would probably necessitate a hypothesis on the noise nature. Another way to approach this problem would consist of studying the curve representing Asplünd's metric decreasing according to the percentage of neglected points.

5. THE ADDITIVE ASPLÜND'S METRIC $d^{\triangle}_{AS}$

Plenty of papers dedicated to the LIP model use it through a minor property: its ability to transform a linear processing into a logarithmic one, ie, to act as a logarithmic look-up table. This remark does not mean that the concerned papers are of limited interest, but they refer to existing methods, which damages their novelty.

I want to warmly thank Dennis Deng, one of the first PhD students to dedicate their research work to the LIP model. In fact, Dennis said to me two years ago: "what limits the notoriety of the LIP model is it fails in proposing one or two *killer applications*." I must admit that this opinion led me think about what could be such a killer application.

I hope that the results exposed in this section contribute partially to answer the question: the introduction of an additive Asplünd's metric appears as a first step for introducing image processing tools able to overcome the difficult situation of uncontrolled lighting. This point appears to be a crucial one and, in our knowledge, general solutions do not exist.

After we had extended the binary Asplünd's metric to gray-level images based on the logarithmic scalar multiplication, I must admit that it was not obvious to imagine an additive variant of this metric founded on the logarithmic additive law. What oriented me in this direction was considering the ability of the LIP addition to produce brightness variations on an image, and consequently to simulate and compensate lighting variations.

The proposed additive Asplünd's metric answers such questions well: in particular, it will be shown that it is a tool theoretically insensitive to exposure time variations (cf. section 2 of Chapter "Ability of the LIP Model to Simulate Variable Acquisition Conditions"). The main remaining weakness is the sensitivity to noise, but a first solution has been mentioned in Section 4.3 for the multiplicative Asplünd's metric, which would be adaptable to the additive one.

5.1 Definition of This Novel Metric

The novel metric we study in this paper is not a proper Asplünd-like metric because it is not defined by means of homothetic functions, but based on the LIP additive law. Nevertheless, it remains in the class of double-sided probing metrics.

Given two gray-level images f and g, we define two real numbers C_1 and C_2 according to:

$$C_1 = \mathrm{Inf}_{C \in R}\{C, g \triangle\!\!\!\!+ C \geq f\}$$
$$C_2 = \mathrm{Sup}_{C \in R}\{C, g \triangle\!\!\!\!+ C \leq f\}$$

where C lies in the interval $]-\infty, M[$ and R designs a region of the spatial support D, possibly D itself (cf. Fig. 22). In such conditions, C_1 is always greater than C_2, C_1 lies in the interval $[0, M[$ and C_2 in the interval $]-M, +M[$, which means that, in the definition of C_2, the probing function $g \triangle\!\!\!\!+ C$ is not always an image.

For evaluating the distance between the probing functions $g \triangle\!\!\!\!+ C_1$ and $g \triangle\!\!\!\!+ C_2$, it seems natural to take into account the difference $C_1 \triangle\!\!\!\!- C_2$. After observing that $C_1 \triangle\!\!\!\!- C_2$ takes its values in $[0, 2M[$, we will compute $\frac{C_1 \triangle\!\!\!\!- C_2}{2}$, which will be interpretable as a gray level.

We are now able to evaluate the additive Asplünd's distance $d^{\triangle\!\!\!\!+}_{\mathrm{As}, R}(f, g)$ between f and g according to the equation:

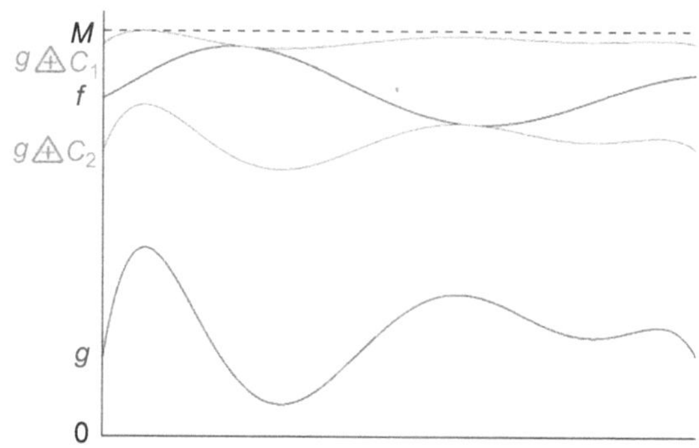

Fig. 22 Additive Asplünd's distance between *f* and *g*, where *g* is the probing function. (See the color plate.)

$$d^{\triangle}_{\mathrm{As},R}(f,g)=\frac{C_1 \triangle C_2}{2} \tag{2}$$

Remark 7. Such a value remains unchanged when f (or g) is replaced by a translated function $f \triangle K$. In fact, the constant C_1 becomes $C_1 \triangle K$ and C_2 becomes $C_2 \triangle K$.

The previous remark implies that:

$$\forall f, \forall K, d^{\triangle}_{\mathrm{As},R}(f, f \triangle K)=0$$

Remark 8. Instead of defining $d^{\triangle}_{\mathrm{As},R}$ on the space $I(D, [0, M[)$ of gray-level images, it is more appropriate to associate each image f with its equivalence class, noted $f^{\triangle}$, constituted of images g such that $g = f \triangle K$ for some constant K lying in $[0, M[$.

Thus, we consider the space

$$I^{\triangle}(D, [0,M[) = \{f^{\triangle}, f \in I(D, [0,M[)\}$$

and the following result holds: *The application $d^{\triangle}_{\mathrm{As},R}$ associating to a pair of equivalence classes $(f^{\triangle}, g^{\triangle})$ the distance $d^{\triangle}_{\mathrm{As},R}(f^{\triangle}, g^{\triangle}) = d^{\triangle}_{\mathrm{As},R}(f,g) = \frac{C_1 \triangle C_2}{2}$ is a metric on the space $I^{\triangle}$ $(D, [0,M[)$.*

Proof. In fact, $d^{\triangle}_{\mathrm{As},R}$ satisfies the properties of a metric:

(1) $d^{\triangle}_{\mathrm{As},R}$ takes its values in $\mathbb{R}^+$:

The constants C_1 and C_2 correspond, respectively, to the superior and inferior probes, thus $C_1 \geq C_2$ and $\frac{C_1 \triangle C_2}{2} \geq 0$

(2) $d^{\triangle}_{\mathrm{As},R}(f,g) = d^{\triangle}_{\mathrm{As},R}(g,f)$ (Symmetry condition)

The rigorous expression of this condition is:

$$d^{\triangle}_{\mathrm{As},R}(f^{\triangle}, g^{\triangle}) = d^{\triangle}_{\mathrm{As},R}(g^{\triangle}, f^{\triangle})$$

Nevertheless, it has been seen that such values remain unchanged when the equivalence classes $f^{\triangle}$ and $g^{\triangle}$ are replaced by any of their elements, say f and g, for example.

In a first step, g is considered as the probing function, and the constants C_1 and C_2 are defined in order to satisfy:

$$\forall x \in R, g \triangle C_2(x) \leq f(x) \leq g \triangle C_1(x)$$

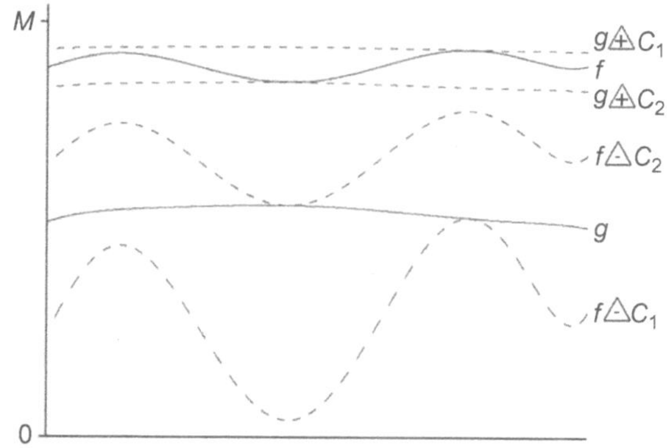

Fig. 23 Illustration of the symmetry property. (See the color plate.)

If f now represents the probing function, let us consider the functions $g\blacktriangle C_1$ and f. We apply to them the subtraction $\triangle C_1$ and get the resulting functions:

$(g\blacktriangle C_1)\triangle C_1 = g$ and $f\triangle C_1$ (cf. Fig. 23)

Then $f\blacktriangle(\triangle C_1) = f\triangle C_1 \leq g$

Moreover, $\triangle C_1$ satisfies:

$$\triangle C_1 = \mathrm{Sup}\{C, f\blacktriangle C \leq g\}$$

This means that $\triangle C_1$ is the constant C'_2 performing the inferior probing of g by f.

In the same way,

$$f\blacktriangle(\triangle C_2) = f\triangle C_2 \geq (g\blacktriangle C_2)\triangle C_2 = g$$

and $\triangle C_2$ appears as the constant C'_1 performing the superior probing of g by f.

Finally,

$$d^{\blacktriangle}_{As,R}(g,f) = \frac{C'_1 \triangle C'_2}{2} = \frac{\triangle C_2 \blacktriangle (\triangle C_1)}{2} = \frac{C_1 \triangle C_2}{2} = d^{\blacktriangle}_{As,R}(f,g)$$

(3) $d^{\blacktriangle}_{As,R}$ satisfies the separation property:

We have to prove the following equivalence:

$$d^{\blacktriangle}_{As,R}(f,g) = 0 \Leftrightarrow f^{\blacktriangle} = g^{\blacktriangle}$$

– Suppose $f^{\triangle\!\!\!\!\triangle} = g^{\triangle\!\!\!\!\triangle}$. It implies that $f = g \triangle\!\!\!\!\triangle K$ for some constant K and we have seen (cf. Remark 7) that $d^{\triangle\!\!\!\!\triangle}_{\mathrm{As},R}(f, f \triangle\!\!\!\!\triangle K) = 0$

– To establish the other implication, $d^{\triangle\!\!\!\!\triangle}_{\mathrm{As},R}(f, g) = 0$ yields:

$$\frac{C_1 \triangle\!\!\!\!\triangle C_2}{2} = 0 \Rightarrow C_1 = C_2 \Rightarrow g \triangle\!\!\!\!\triangle C_1 \leq f \leq g \triangle\!\!\!\!\triangle C_1 \Rightarrow f = g \triangle\!\!\!\!\triangle C_1 \Rightarrow f^{\triangle\!\!\!\!\triangle} = g^{\triangle\!\!\!\!\triangle}$$

(4) $d^{\triangle\!\!\!\!\triangle}_{\mathrm{As},R}$ satisfies the triangular inequality:

Let us consider three images $f, g,$ and h. We must establish the following inequality:

$$d^{\triangle\!\!\!\!\triangle}_{\mathrm{As},R}(f, h) \leq d^{\triangle\!\!\!\!\triangle}_{\mathrm{As},R}(f, g) + d^{\triangle\!\!\!\!\triangle}_{\mathrm{As},R}(g, h)$$

To the distance $d^{\triangle\!\!\!\!\triangle}_{\mathrm{As},R}(f, g)$, we associate the probing constants C_1 and C_2 of f by g (cf. Fig. 24). Such constants satisfy the double inequality:

$$g \triangle\!\!\!\!\triangle C_2 \leq f \leq g \triangle\!\!\!\!\triangle C_1$$

To the distance $d^{\triangle\!\!\!\!\triangle}_{\mathrm{As},R}(g, h)$, we associate the probing constants C_3 and C_4 of g by h (cf. Fig. 24), which satisfy:

$$h \triangle\!\!\!\!\triangle C_4 \leq g \leq h \triangle\!\!\!\!\triangle C_3$$

Now we perform the probing of image f by h, resulting in two constants C_s and C_i corresponding to the superior and inferior probes.

We can write:

$$f \leq g \triangle\!\!\!\!\triangle C_1 \leq (h \triangle\!\!\!\!\triangle C_3) \triangle\!\!\!\!\triangle C_1 = h \triangle\!\!\!\!\triangle (C_3 \triangle\!\!\!\!\triangle C_1)$$
$$\Rightarrow C_s \leq C_3 \triangle\!\!\!\!\triangle C_1 \qquad (\alpha)$$

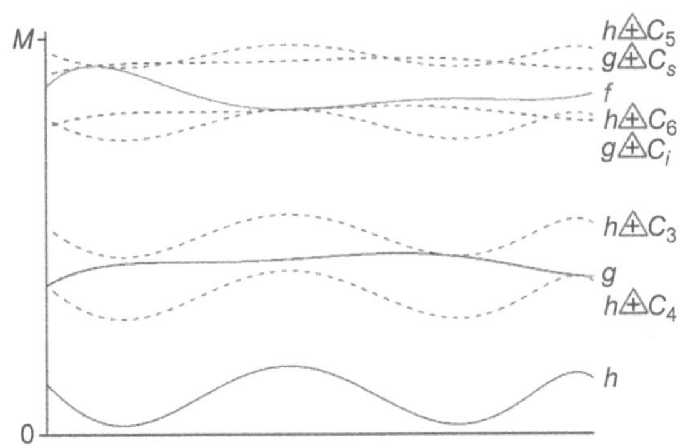

Fig. 24 Triangular inequality property. (See the color plate.)

On another hand:

$$h \triangle\!\!\!+ (C_4 \triangle\!\!\!+ C_2) = (h \triangle\!\!\!+ C_4) \triangle\!\!\!+ C_2 \leq g \triangle\!\!\!+ C_2 \leq f \\ \Rightarrow C_4 \triangle\!\!\!+ C_2 \leq C_i \qquad (\beta)$$

From (α) and (β), we deduct:

$$d^{\triangle\!\!\!+}_{\text{As},R}(f,h) = \frac{C_s \triangle\!\!\!- C_i}{2} \leq \frac{(C_3 \triangle\!\!\!+ C_1) \triangle\!\!\!- (C_4 \triangle\!\!\!+ C_2)}{2} \text{ and}$$

$$d^{\triangle\!\!\!+}_{\text{As},R}(f,h) \leq \frac{C_1 \triangle\!\!\!- C_2}{2} \triangle\!\!\!+ \frac{C_3 \triangle\!\!\!- C_4}{2} \leq \frac{(C_1 \triangle\!\!\!- C_2) + (C_3 \triangle\!\!\!- C_4)}{2}$$

This result is obtained by the definition of $A \triangle\!\!\!+ B$:

$$A \triangle\!\!\!+ B = A + B - \frac{1}{M}(A \cdot B) \leq A + B$$

Finally:

$$d^{\triangle\!\!\!+}_{\text{As},R}(f,h) \leq d^{\triangle\!\!\!+}_{\text{As},R}(f,g) + d^{\triangle\!\!\!+}_{\text{As},R}(g,h)$$

□

5.2 Main Property: Insensitivity of the Additive Asplünd's Metric to Exposure Time Variations

5.2.1 Theoretical Case: Simulated Exposure Times

In a previous paper (Carré & Jourlin, 2014), we established that the LIP addition of a constant to a gray-level image f permits the precise estimatation of images of the same scene acquired under other exposure times. Such a property has been refined and extended to color images (Deshayes, Guilbert, & Jourlin, 2015). This subject will be detailed in Chapter "Ability of the LIP Model to Simulate Variable Acquisition Conditions." If we refer to Remark 7 of Section 5.1, the following result holds:

$$\forall f, \forall K, d^{\triangle\!\!\!+}_{\text{As},R}(f, f \triangle\!\!\!+ K) = 0$$

Remark 9. In such conditions, considering an image f and images $f \triangle\!\!\!+ K$ of the same scene simulating the acquisition of f under variable exposure times (Fig. 25A), it becomes possible to perform the recognition of a target T extracted from f (Fig. 25B) in every image $f \triangle\!\!\!+ K$. In fact, if T is defined on a region $R \subset D$, for each image $f \triangle\!\!\!+ K$, we move the target inside D and for each location of T at a pixel x, we compute the additive Asplünd's distance between T and the restriction of $f \triangle\!\!\!+ K$ to R, noted

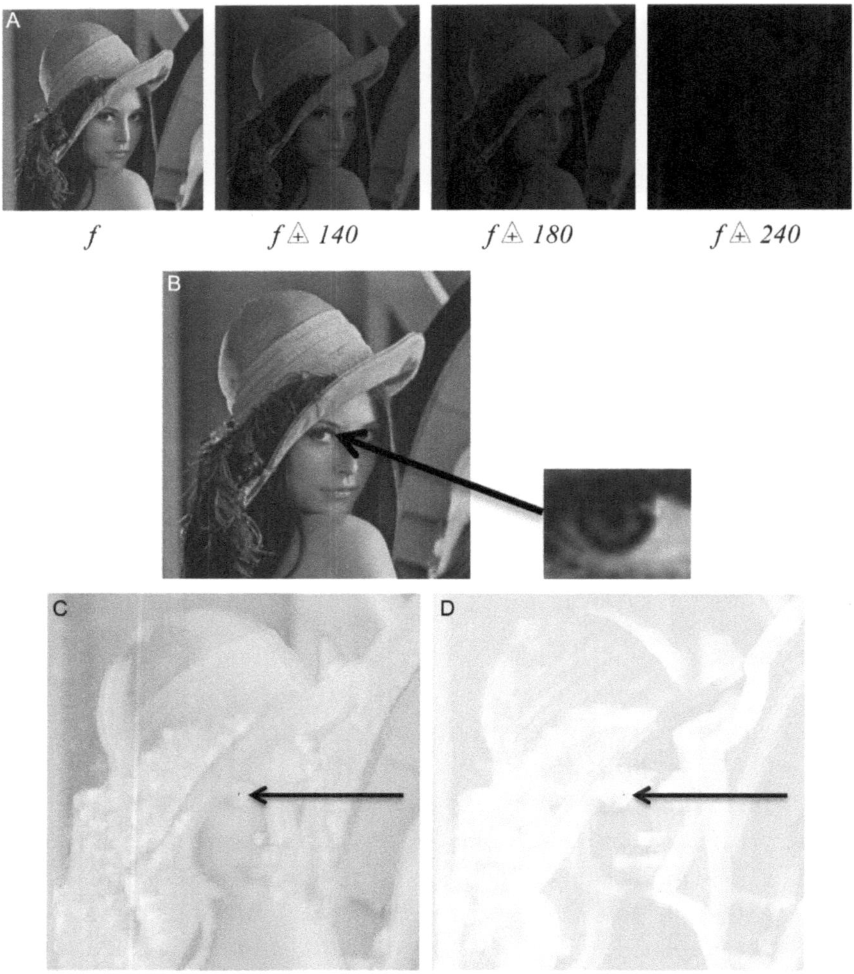

Fig. 25 Location of a target independently of simulated exposure times. (A) Initial image f. Images $f \mathbin{\triangle\!\!\!\!\times} K$, $K=140$, $K=180$, $K=240$. (B) Target T extracted from f: the eye of Lena magnified four times. (C) and (D) Asplünd's maps for f and $f \mathbin{\triangle\!\!\!\!\times} 240$: the location of the target corresponds to a black pixel.

$(f \mathbin{\triangle\!\!\!\!\times} K)|_R$. Such a distance $d^{\triangle}_{\mathrm{As},R}(f \mathbin{\triangle\!\!\!\!\times} K|_R, T)$ takes its values in the standard gray scale $[0, \ldots, 255]$ and is then a gray level assigned to x. In such a way, we get a map of additive Asplünd's distances between the target and the corresponding region of $(f \mathbin{\triangle\!\!\!\!\times} K)$ (Fig. 25C and D). The location of the target corresponds to the values 0 observed inside the map (Fig. 25C and D).

5.2.2 Real Case

We know that probing metrics are very sensitive to noise. It is the reason why we proposed a solution to overcome this drawback in Jourlin et al. (2014). Section 5.2.1 has shown the major interest of the additive Asplünd's metric at a theoretical level. We will now focus our study on real acquisitions of the same scene under variable exposure times.

Let $f(t_1), f(t_2), f(t_3), f(t_4)$ represent images of a same scene acquired, respectively, at exposure times $t_1 = 50\,\text{ms}$, $t_2 = 90\,\text{ms}$, $t_3 = 130\,\text{ms}$, $t_4 = 170\,\text{ms}$ (Fig. 26A), and T a target selected inside $f(t_4)$, for example (Fig. 26B).

We proceed as for the theoretical case: for each location of T, we compute the additive Asplünd's distance between T and the restriction of $[f(t_i)]_{i=1,2,3,4}$ to R, noted $f(t_i)|_R$. We get a map of additive Asplünd's distances between the target and the corresponding region of $f(t_i)$ (Fig. 26C). We must then extract the minimum of each map to locate the target (Fig. 26D for $f(t_i)$).

First comments on this metric: a novel "Asplünd-like" metric has been proposed, called additive Asplünd's metric. Precisely, this metric is defined on the space of equivalence classes

$$I^{\triangle\!\!\!+}(D, [0, M[) = \{f^{\triangle\!\!\!+}, f \in I(D, [0, M[)\}$$

which means, in a mathematical formulation, on the quotient space of $I(D, [0, M[)$ by the equivalence relation R:

$$f R g \Leftrightarrow \exists K \in [0, \ldots, 255], f = g \triangle\!\!\!+ K \text{ or } g = f \triangle\!\!\!+ K$$

If we remember that the logarithmic addition (resp. subtraction) of a constant to (resp. from) an image permits simulating images of the same scene acquired under arbitrary exposure times, we can conclude that the additive Asplünd's metric is absolutely insensitive to exposure time variations. This point is mathematically established at a theoretical level (case where exposure time variations are simulated). When considering real cases, where images are acquired at various exposure times, we could fear that this insensitivity property fails due to quantification effects and decreasing of SNR on dark images. Nevertheless, the detection of a target has been successfully performed (cf. Fig. 26).

5.3 Other Examples of Application

In this section, we selected some examples illustrating the interest of the additive Asplünd's metric insensitivity to exposure time variations.

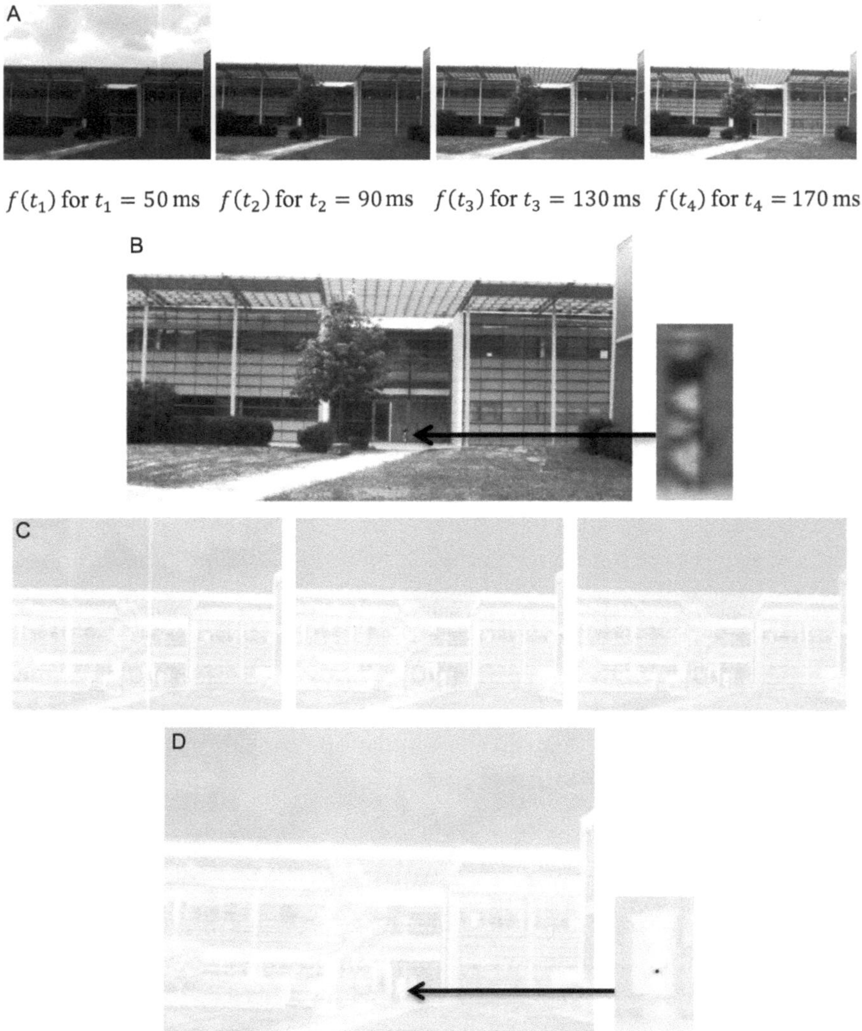

Fig. 26 Location of a target independent of exposure times. (A) Images $f(t_1)$, $f(t_2)$, $f(t_3)$, $f(t_4)$. (B) Initial target T (magnified six times) selected in $f(t_4)$ at the extremity of the black arrow. (C) Maps of additive Asplünd's distances between the target and a corresponding region of $f(t_1)$, $f(t_2)$, $f(t_3)$. (D) Asplünd's map for $f(t_1)$. Magnification of the target location inside the image $f(t_1)$): it corresponds to the minimum of the additive Asplünd's distance (black pixel).

5.3.1 Region Growing Algorithm

Let us recall the basic region growing algorithm (RGA), called single linkage (cf. Haralick & Shapiro, 1985, for example). Given a gray-level image f, we start with a *seed* x selected in D and we consider the N_8-neighborhood of x. The first step consists of aggregating to x the elements of $N_8(x)$ similar to x.

Similarity is classically satisfied when the gray-level difference between x and a candidate y is less than a fixed threshold ε. This step results in a region R_1. For the following step, we consider a pixel y lying in the external N_8-boundary of R_1 and the condition to aggregate y to R_1 is that there exists a pixel z similar to y in the intersection $N_8(y) \cap R_1$. The process is iterated until the convergence step, which means when no more new pixels may be aggregated to the current region.

Remark 10. It is possible to evaluate similarity by means of a distance: two pixels are similar if their distance is less than ε. This is what we are presenting now, using the additive Asplünd's metric, but we must take into consideration that by definition the additive Asplünd's distance between two gray levels always equals zero. For that reason we propose considering the N_8-neighborhoods of the studied pixels x and y. If the nine elements of $N_8(x)$ are noted $\{v_i(x)\}_{i=1,\ldots,9}$ we define:

$$d^{\triangle\!\!\!\!+}_{\mathrm{As},N_8}(f(x),f(y)) = \frac{C_1 \triangle\!\!\!\!- C_2}{2}$$

where

$$C_1 = \mathrm{Inf}\{C, \forall i = 1,\ldots,9, f(v_i(x) \triangle\!\!\!\!+ C \geq f(v_i(y))\}$$
$$C_2 = \mathrm{Sup}\{C, \forall i = 1,\ldots,9, f(v_i(x) \triangle\!\!\!\!+ C \leq f(v_i(y))\}$$

5.3.1.1 Theoretical Case: Simulated Exposure Times

As established for $d^{\triangle\!\!\!\!+}_{\mathrm{As},R}(f,g)$ in Section 5.2.1, the local metric $d^{\triangle\!\!\!\!+}_{\mathrm{As},N_8}$ is theoretically insensible to exposure time variations. To establish this property, let us consider the initial image f (Lena, Fig. 27A) and we simulate exposure time variations of $f: f \triangle\!\!\!\!+ 140, f \triangle\!\!\!\!+ 180, f \triangle\!\!\!\!+ 240$ (Fig. 27B).

Then we choose a seed (white cross, Fig. 27A), and we perform a RGA by means of the additive Asplünd's distance, with a similarity threshold $\varepsilon = 12$. The region obtained (represented in white, Fig. 27C) is independent of the considered exposure time.

Let us note that other metrics fail in performing RGA on images acquired under variable exposure times. This point is illustrated (cf. Fig. 28) for RGA applied to the same sequence of images used in Fig. 27. In order to drive the aggregation process, we used the Euclidean distance between the gray levels of two pixels (ie, the absolute value of their difference) and we chose a value of the similarity parameter ε $(\varepsilon = 4)$ to get, starting from the same seed, a region comparable to that obtained in Fig. 26C for the initial image of Lena. Then we applied RGA to the darkened images $f \triangle\!\!\!\!+ 140, f \triangle\!\!\!\!+ 180, f \triangle\!\!\!\!+ 240$ with the same value of ε.

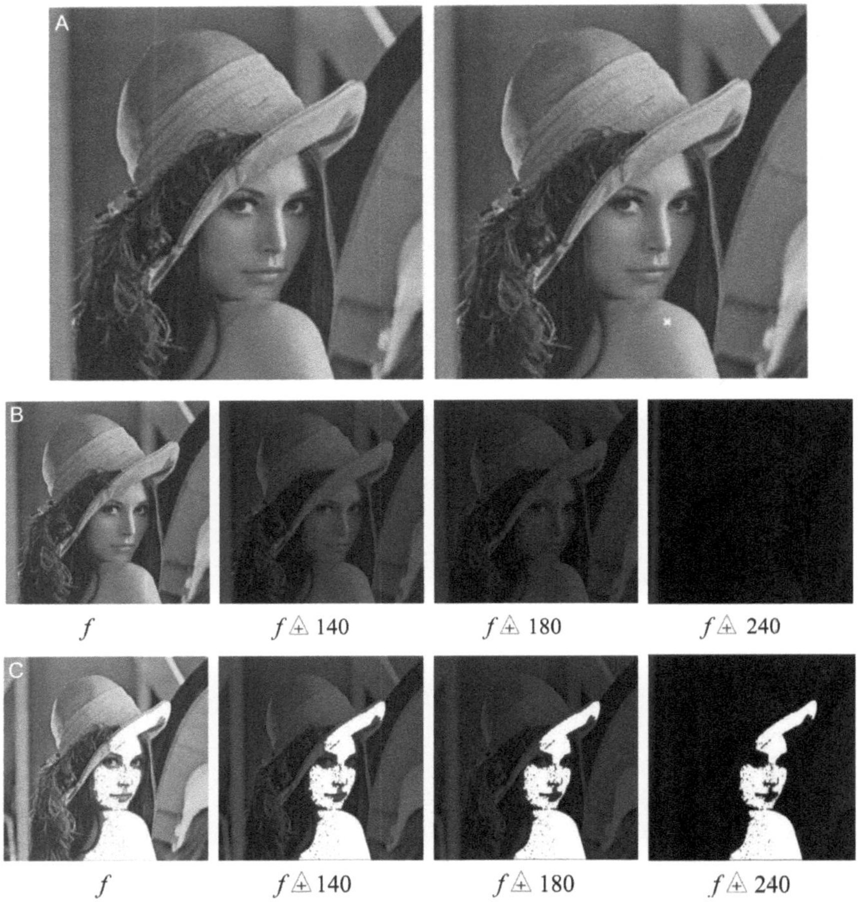

Fig. 27 Insensitivity of region growing to exposure time variations based on the additive Asplünd's metric. (A) Initial image (Lena) and the initial seed (white cross). (B) Simulations of exposure time variations: f, $f \triangle\!\!\!\!+ 140$, $f \triangle\!\!\!\!+ 180$, $f \triangle\!\!\!\!+ 240$. (C) Regions obtained with the initial seed, $\varepsilon = 12$ and the additive Asplünd's metric.

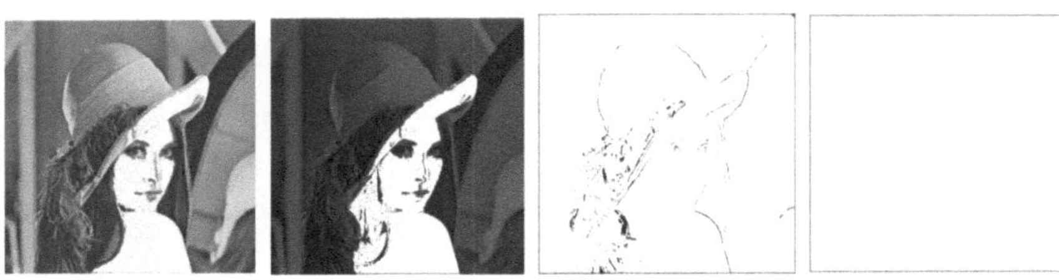

Fig. 28 Region growing applied to Lena (image f) with the same initial seed and with aggregation parameter $\varepsilon = 4$. In *white*, the obtained regions for $f, f \triangle\!\!\!\!+ 140, f \triangle\!\!\!\!+ 180, f \triangle\!\!\!\!+ 240$.

Remark 11. The previous result concerning the insensitivity of additive Asplünd's metric to lighting variations is not surprising because the logarithmic addition (resp. subtraction) of a constant to (resp. from) an image permits together to simulate exposure time variations and to define the additive Asplünd's metric itself. Nevertheless, it provides a new way to process images under variable lighting, which consists of defining and applying invariant tools instead of trying to enhance/stabilize such variable images with more or less efficient algorithms.

5.3.1.2 Real Case

We proceed as for the theoretical case, but the considered images of a gray-level chart are acquired at various exposure times, here 7.5, 15, 30, and 50 ms (Fig. 29A). Then we choose an initial seed (white cross, Fig. 29B) and we perform the region growing by means of Asplünd's metric with $\varepsilon = 12$ (Fig. 29C) and by means of Euclidean's metric with $\varepsilon = 4$ (Fig. 29D). The values of ε are the same as those in the theoretical case.

Comment on Fig. 29: The proposed example shows clearly that the Asplünd's driven RGA is very stable compared with the Euclidean case. Nevertheless, as previously said, probing metrics are very sensitive to noise, thus it is possible that this problem occurs in other situations (darker images, bad quality of the sensor, etc.). We proposed a solution to attenuate this drawback in the case of the multiplicative Asplünd's metric (cf. Section 4.3 or Jourlin et al., 2014) by neglecting a percentage of pixels: the most penalizing ones to perform the probing and compute the Asplünd's distance. Such an approach is obviously adaptable to the case of the additive Asplünd's metric.

Another possible solution would be to apply a noise filtering algorithm to the studied images as a preprocessing. Note that this technique is more efficient if we have determined an appropriate model of the noise nature.

5.3.2 Contour Detection

To put in evidence the contours present inside an image *f*, it is possible to use the local metric $d^{\triangle\!\!\!+}_{\mathrm{As},N_8}$ previously defined in Section 5.3.1 (Remark 10).

Such a metric permits computing the distance $d^{\triangle\!\!\!+}_{\mathrm{As},N_8}(f(x), f(v_i(x))$ between a pixel x and each of its eight neighbors $\{v_i(x)\}_{i=1,\ldots,8}$.

From these eight values lying in the gray scale $[0, 255[$, we can derive a gray level $\mathrm{GL}(x)$ according to:

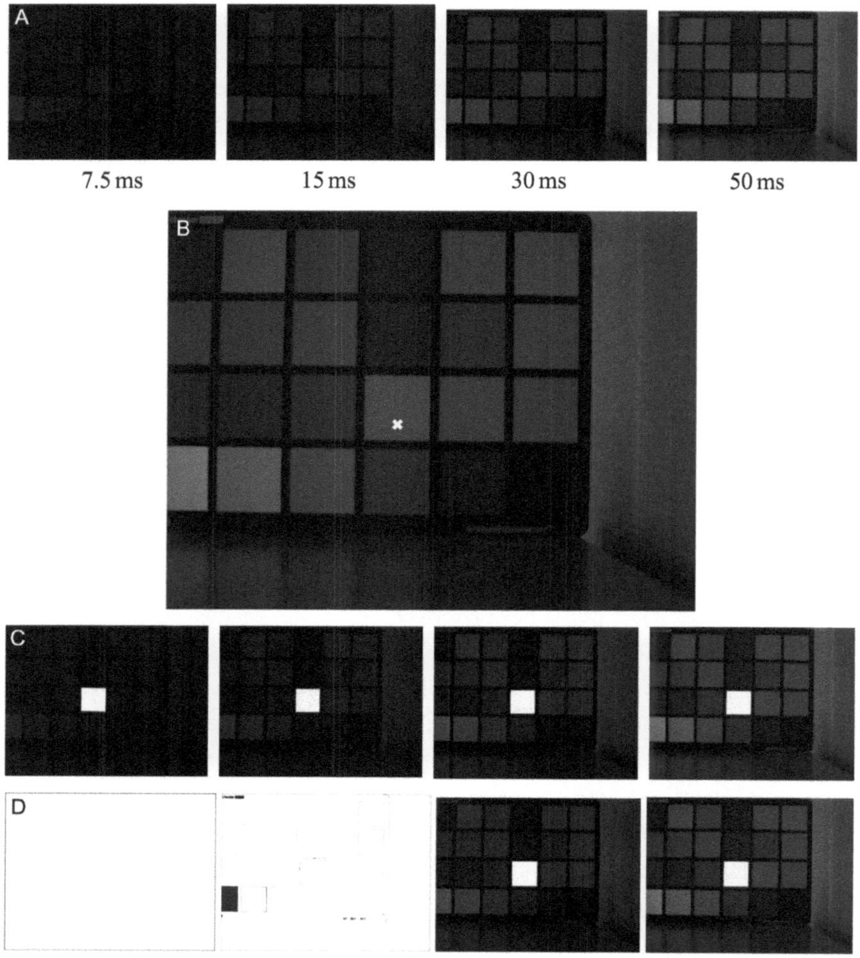

Fig. 29 Compared results of region growing algorithm driven by Asplünd's and Euclidean's metrics. (A) Acquisition of the gray level chart under various exposure times. (B) Initial seed (white cross). (C) Results of region growing by means of Asplünd's metric (exposure times: 7.5, 15, 30, and 50 ms). (D) Results of region growing by means of Euclidean's metric (exposure times: 7.5, 15, 30, and 50 ms).

$$\mathrm{GL}(x) = \mathrm{Max}_{i=1,\ldots,8}\left\{ d^{\triangle}_{\mathrm{As},N_8}(f(x), f(v_i(x)))\right\}$$

In this way, we get a gradient additive Asplünd map (GAAM) representing, for each pixel x, the gray-level variations of any image g of the equivalence class $f^{\triangle}$ (cf. Remark 8), in a neighborhood of x. This means that the GAMM is the same for all images $f \triangle K$: in other words, the proposed contour detection method will be theoretically independent of the exposure time value.

Remark 12. A variant of the previous gradient consists of replacing the term "Max" by an average value "Mean" in the definition, computed, for example, in a logarithmic manner:

$$\frac{1}{8} \triangle \sum_{i=1,\ldots,8}^{\triangle} d_{\mathrm{As},N_8}^{\triangle}(f(x),f(v_i(x)))$$

where $\sum_{i=1,\ldots,8}^{\triangle}$ represents the LIP-sum.

Remark 13. We will not develop this point. Similar approaches are proposed in Jourlin et al. (2012), section 2.3.2, where the Asplünd's distance $d_{\mathrm{As},N_8}^{\triangle}$ was simply replaced by the additive contrast. For all these approaches, the interested reader will take into consideration their sensitivity to noise.

This section aspires essentially to propose a novel tool and prove its efficiency in several situations. Thus, we will limit our discussion to the examples previously cited, even if many other applications of the additive Asplünd's metric could be presented, such as characterization of pseudo-periodic textures by means of autocorrelation, covariograms, etc. (cf. Fig. 8 dedicated to the approach of this problem by means of the multiplicative Asplünd's metric).

5.4 Conclusion of Section 5 and Perspectives

The notion of "metric" is one of the most useful in image processing. In fact, a metric is able to efficiently replace correlation tools, contrast, or gradient concepts (interpreted as the distance between two gray levels).

It permits, for example:

- Performing pattern recognition
- Comparing remote sensing images in order to detect what has changed between two acquisitions
- Locating a given target inside an image or following it in a video stream
- Performing autocorrelation of an image in order to detect, for example, pseudoperiodic structures

Nevertheless, most existing metrics do not possess strong physical justifications. The additive Asplünd's metric defined in this section takes place in the LIP framework and is thus built on the transmittance law, which makes it optimal to process images acquired in transmission. Moreover, the consistency of the LIP model with human vision permits us to interpret images in reflection, as the human eye would do. Finally, the additive Asplünd's metric is actually insensitive to lighting changing and particularly to exposure time variations. To show such properties we have chosen, among plenty of possible subjects, to focus on the location of a target and on the RGA.

Concerning future works, we plan:

- To establish another property of the additive Asplünd's metric: its ability to simulate variations of diaphragm aperture. This point is partially done and some results are exposed in Chapter "Ability of the LIP Model to Simulate Variable Acquisition Conditions."
- To extend the results to color and multivariate images (partially done, cf. Section 6).

6. EXAMPLES OF METRICS FOR COLOR IMAGES

We explored three ways to develop this point. The first one consists of considering the LIP color (LIP-C) model, the second to build metrics associated to contrasts and the third to create Asplünd-like color metrics. This section will mainly focus on the third approach for two reasons: on one hand, it is the most original, and on the other hand, the Asplünd-like metrics will possess very strong properties based on the LIP laws.

6.1 LIP-C Approach

We presented an extended paper on the LIP-C Model (Jourlin, Breugnot, Itthirad, Bouabdellah, & Closs, 2011) where the interested reader will probably find some points that require further research. This model has been founded on visual experimentations (tables of Stiles and Burch: http://cvrl.ucl.ac.uk) in order to be consistent with human vision. LIP-C permits generalizing Asplünd's metrics to color images. Nevertheless this model produces a rather complex environment. For this reason we will propose simplified approaches.

6.2 Extension of LAC to Color Images and Associated Metrics

This point has been initialized in section 3.2 of Chapter "Various Contrast Concepts," where we have shown how various color metrics can be deducted from the LACs defined in each channel R, G, B. Such metrics are probably interesting to study in depth, but this is not our objective here.

6.3 Asplünd-Like Metrics for Color Images

Throughout this section, the various Asplünd-like metrics dedicated to color images will be noted with a subscript As, C: $d_{\mathrm{As},C}$ and a superscript $\triangle\!\!\!\times$ or $\triangle\!\!\!+$ according to the nature multiplicative or additive of the metric: $d^{\triangle\!\!\!\times}_{\mathrm{As},C}$ or $d^{\triangle\!\!\!+}_{\mathrm{As},C}$.

Note that the present study of Asplünd-like metrics is an extension of section 3 of Chapter "Various Contrast Concepts," where we shortly introduced the multiplicative Asplünd's metric between two colors and the logarithmic additive contrast (LAC) for color images.

6.3.1 Multiplicative Asplünd-Like Metric

Between two colors

Before extending the multiplicative Asplünd's metric to color images, let us define it between two colors $C_1 = (R_1, G_1, B_1)$ and $C_2 = (R_2, G_2, B_2)$. We will note it $d^{\triangle\!\!\!\times}_{\mathrm{As},C}(C_1, C_2)$ and we propose the following definition:

$$\left.\begin{array}{ll} & d^{\triangle\!\!\!\times}_{\mathrm{As},C}(C_1, C_2) = Ln\left(\dfrac{\alpha}{\beta}\right) \\ \text{where} & \alpha = \inf_{k\in\mathbb{R}}\{k, k\triangle\!\!\!\times R_1 \geq R_2, k\triangle\!\!\!\times G_1 \geq G_2, k\triangle\!\!\!\times B_1 \geq B_2\} \\ & \beta = \sup_{k\in\mathbb{R}}\{k, k\triangle\!\!\!\times R_1 \leq R_2, k\triangle\!\!\!\times G_1 \leq G_2, k\triangle\!\!\!\times B_1 \leq B_2\} \end{array}\right\} \quad (3)$$

Remark 14. The previous approach (3) seems an interesting compromise between the use of LIP-C and a trivial solution consisting of computing a distance in each channel, as we have done for gray levels, and gathering them in different manners.

Remark 15. Let us highlight a strong advantage of the color multiplicative Asplünd's metric: it remains unchanged when one color (C_1, for example) is replaced by any homothetic $k\triangle\!\!\!\times C_1$ for $k > 0$. Remember that $k\triangle\!\!\!\times C_1$ is brighter than C_1 when $k \in]0, 1]$ and darker than C_1 when $k \in [1, +\infty[$. This property shows the insensitivity of $d^{\triangle\!\!\!\times}_{\mathrm{As},C}$ to strong illumination variations modeled by the logarithmic multiplicative law.

From the definition of $d^{\triangle\!\!\!\times}_{\mathrm{As},C}$ and the Remark 15, it would be more rigorous to present the color multiplicative Asplünd's metric as in Section 3.2, by means of an equivalence relation R on the space of colors.

Given a pair of colors $C_1 = (R_1, G_1, B_1)$ and $C_2 = (R_2, G_2, B_2)$, we say they are "in relation" if they satisfy:

$$(C_1 R C_2) \Leftrightarrow \exists k > 0, C_1 = \mathrm{k}\triangle\!\!\!\times C_2$$

ie, $R_1 = k\triangle\!\!\!\times R_2$, $G_1 = k\triangle\!\!\!\times G_2$ and $B_1 = k\triangle\!\!\!\times B_2$

Then we associate each C with its equivalence class $C^{\triangle}$ grouping the colors that are in relation with C. From a rigorous point of view, $d^{\triangle}_{\mathrm{As},C}$ is defined on the set of color equivalence classes (see the "Mathematician" of Section 3.2).

Between two color jmages

Knowing the distance between two colors, we can deduce various distances between two color images $f=(f_R,f_G,f_B)$ and $g=(g_R,g_G,g_B)$ on a subset $S\subset D$. We adopt here the notation S instead of R for the region of interest in order to avoid confusion with the red channel R.

First solution: distance $d^{\triangle}_{\mathrm{As},C,S}(f,g)$ (cf. Noyel & Jourlin, 2015)

$$d^{\triangle}_{\mathrm{As},C,S}(f,g)=Ln\left(\frac{\alpha}{\beta}\right)$$

with:

$$\alpha=\mathrm{Inf}_{k\in\mathbb{R}}\{k,\forall x\in S, k\triangle f_R(x)\geq g_R(x), k\triangle f_G(x)\geq g_G(x), k\triangle f_B(x)\geq g_B(x)\}$$
$$\beta=\mathrm{Sup}_{k\in\mathbb{R}}\{k,\forall x\in S, k\triangle f_R(x)\leq g_R(x), k\triangle f_G(x)\leq g_G(x), k\triangle f_B(x)\leq g_B(x)\}$$

Remark 16. If the whole spatial support D is taken in consideration instead of a subset S, the notation for the previous distance will be simply:

$$d^{\triangle}_{\mathrm{As},C}(f,g)$$

Remark 17. The previous definition is easily extended to multivariate images by applying the Inf and Sup onto all the corresponding channels.

Remark 18. As for gray-level images, this probing metric is sensitive to noise and the solution we exposed here (cf. Section 4.3) has been demonstrated to be applicable for color and multivariate images.

Now let us propose an application example similar to that exposed in Section 4.1 dedicated to target tracking. We consider an initial color image f representing a "Bricks Wall" (cf. Fig. 30A) and we select a target t constituted of one brick (cf. Fig. 30B). It is thus possible to create the Asplünd's map associated to the multiplicative Asplünd's metric $d^{\triangle}_{\mathrm{As},C}\left(f|_{D_t},t\right)$ where $f|_{D_t}$ designs precisely the restriction of f to the subset D_t of D corresponding to the target location (cf. Fig. 30C). From this Asplünd's map, it is quite easy to extract each brick of the wall (dark points in Fig. 30C).

Another example is given in Noyel and Jourlin (2015) concerning the detection of a target inside a low-light image. This point is illustrated in Fig. 31 where a ball has been selected on the initial image "Christmas Tree"

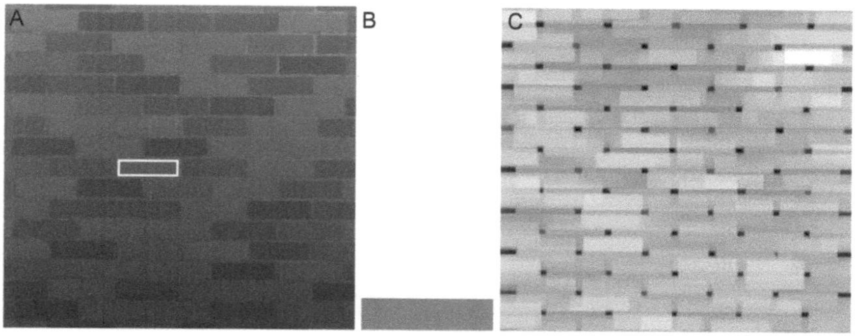

Fig. 30 Example of color target tracking based on multiplicative Asplünd's metric. (A) Initial color image f (brick wall) with selected target. (B) Color target t (magnified two times). (C) Asplünd's map of (B) inside (A).

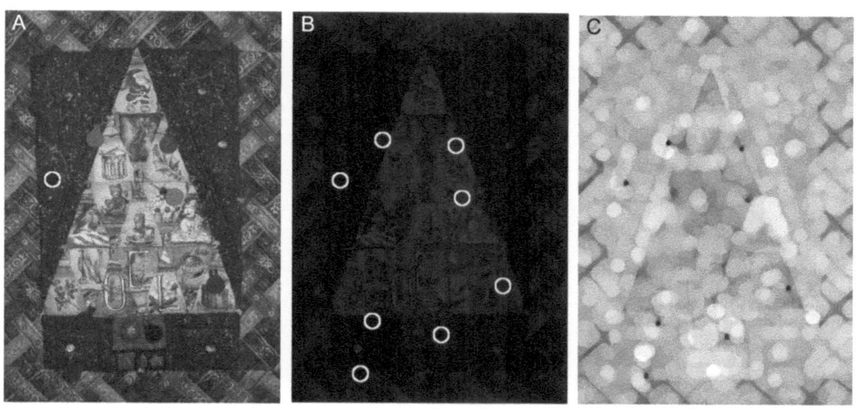

Fig. 31 Color target tracking on a dark image. (A) Initial color image f (Christmas Tree) with selected target t. (B) Detected Christmas balls on a low-light version of (A). (C) Asplünd's map of t inside the low-light version of (A). (See the color plate.)

(31A). The goal is to detect all the Christmas balls on a low-light version of image (A). The final result is shown in Fig. 31B and the balls' locations have been extracted without difficulty from the Asplünd's map $d_{\text{As},C}^{\triangle\!\!\!\triangle}$ (Fig. 31C) of t inside the dark version of (A).

Remark 19. In Noyel and Jourlin (2015), Noyel makes a comparison in the situation of Fig. 31, between the multiplicative Asplünd's approach and color metrics classically used in the lab context, such as:

$$d_{\text{Lab}}(f(x), t(x)) = \sqrt{\frac{1}{3}\left[(f_L(x) - t_L(x))^2 + (f_a(x) - t_a(x))^2 + (f_b(x) - t_b(x))^2\right]}$$

and the associated ones:

$$\mathrm{Sup}_{x\in D_t}\left\{d_{\mathrm{Lab}}\left(f|_{D_t}(x),t(x)\right)\right\}\text{or}\sqrt{\frac{1}{N}d_{\mathrm{Lab}}\left(f|_{D_t}(x),t(x)\right)^2}$$

It appears that the superiority of Asplünd's metric is unquestionable.

Second solution: Knowing the distance between two colors in one point of the region of interest S, it becomes possible to take into account all the points lying in S and to cumulate the information in the manner of d_1 (resp. d_∞) according to:

$$d^{\triangle\!\!\!\triangle}_{1,\mathrm{As},C,S}(f,g)=\frac{1}{\mathrm{Card}(S)}\sum\nolimits_{x\in S}d^{\triangle\!\!\!\triangle}_{\mathrm{As},C}(f(x),g(x))$$

or

$$d^{\triangle\!\!\!\triangle}_{\infty,\mathrm{As},C,S}(f,g)=\mathrm{Sup}_{x\in S}\left\{d^{\triangle\!\!\!\triangle}_{\mathrm{As},C}(f(x),g(x))\right\}$$

6.3.2 Additive Asplünd-Like Metric

As did for the multiplicative Asplünd's metric, let us first define the additive Asplünd's metric between two colors $C_1=(R_1,G_1,B_1)$ and $C_2=(R_2,G_2,B_2)$.

We will note it $d^{\triangle\!\!\!\triangle}_{\mathrm{As},C}(C_1,C_2)$ and we propose the following definition:

$$\left.\begin{aligned}&d^{\triangle\!\!\!\triangle}_{\mathrm{As},C}(C_1,C_2)=\frac{K_1\triangle\!\!\!\triangle K_2}{2}\\ \text{where}\quad &K_1=\inf_{K\in\mathbb{R}}\{K,R_1\triangle\!\!\!\triangle K\ge R_2,G_1\triangle\!\!\!\triangle K\ge G_2,B_1\triangle\!\!\!\triangle K\ge B_2\}\\ &K_2=\sup_{K\in\mathbb{R}}\{K,R_1\triangle\!\!\!\triangle K\le R_2,G_1\triangle\!\!\!\triangle K\le G_2,B_1\triangle\!\!\!\triangle K\le B_2\}\end{aligned}\right\}\tag{4}$$

Remark 20. This color additive Asplünd's metric $d^{\triangle\!\!\!\triangle}_{\mathrm{As},C}(C_1,C_2)$ is not a color, but a real number. More precisely $d^{\triangle\!\!\!\triangle}_{\mathrm{As},C}(C_1,C_2)$ lies in the interval $[0,M[$ and is thus interpretable as a gray level without any normalization. In fact, as for gray-level images, we have always:

$$K_2\le K_1,K_1\in[0,+M[\text{ and }K_2\in]-M,+M[$$

This implies that $K_1\triangle\!\!\!\triangle K_2\in[0,2M[$ and justifies the definition of $d^{\triangle\!\!\!\triangle}_{\mathrm{As},C}(C_1,C_2)$ where $K_1\triangle\!\!\!\triangle K_2$ is divided by 2.

Example of application: As we have already done for gray-level images, we can apply this metric in various situations like region growing, contour detection, etc., and we know that the results will be faintly sensitive to exposure time variations, to lighting variations and lighting drift. We limit

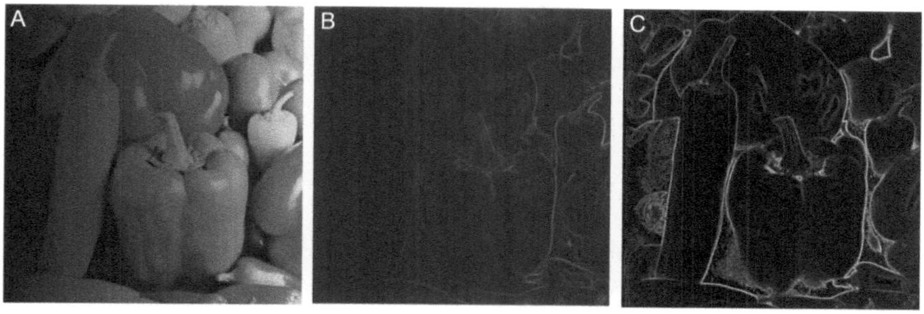

Fig. 32 Sobel applied to the luminance image compared to color additive Asplünd gradient. (A) Initial image "Peppers" with lighting drift. (B) Sobel filter applied to (A). (C) Result of contour detector defined in Formula (5). (See the color plate.)

ourselves to an example of contour detection on the image f "Peppers" in presence of lighting drift, compared with the classical Sobel filter applied to the luminance image (Fig. 32). To do that, we need only to define, for example, the average distance (gray level) $G(x)$ between a pixel x and its eight neighbors $n_i(x)$ according to:

$$G(x) = \frac{1}{8} \triangle\!\!\!\!\times \sum_{i=1,\ldots,8}^{\triangle\!\!\!+} d^{\triangle\!\!\!+}_{\mathrm{As},C}(f(x), f(n_i(x))) \tag{5}$$

Now, starting from the additive Asplünd's distance between two colors, if we aim to define the color additive Asplünd's metric between two color images f and g, we can adopt the same reasoning as for the multiplicative Asplünd metric (Section 6.3.1) where three solutions were proposed. We leave the interested reader to make these developments.

7. OTHER NOTIONS AROUND METRICS

7.1 Recalls

It is well known that topology is the "neighborhoods science" and is available in many configurations depending on the structure of the space on which it is defined:

- *General topology* does not require any hypothesis and is then usable on a simple set S without structure. It consists of a family $\mathcal{T}$ of subsets of S satisfying the following basic properties:
  - $\emptyset$ and $S \in \mathcal{T}$
  - Any union of elements of $\mathcal{T}$ is an element of $\mathcal{T}$
  - A finite intersection of elements of $\mathcal{T}$ is an element of $\mathcal{T}$

A pair $(S, \mathcal{T})$ constituted of a set S and a family $\mathcal{T}$ satisfying the previous hypotheses is called a *topological space*, the elements of $\mathcal{T}$ are the open sets of $(S, \mathcal{T})$ and a subset N of S verifying $x \in \Omega \subset N$ where Ω is an open set is a neighborhood of x

A very complete reference on the subject has been written by Dugundji (1966).

A particular example of *Topological Spaces* is constituted of *Metric Spaces*:

- A metric space is a set S equipped with a metric (distance) which means: d is defined on SxS with values in $\mathbb{R}\,(d : SxS \longrightarrow \mathbb{R})$ satisfying the properties recalled in Section 2.1. In such case, a neighbor of $x \in S$ is a set $N \subset S$ containing an open ball including x.

Fundamental Remark: Some particular concepts stronger or weaker than metrics require a vector space structure to be defined. It is the case, for example, of scalar products and norms (stronger concepts) and gauges (weaker notion). The space $[F(D,]-\infty, M[), \triangle\!\!\!\!+, \triangle\!\!\!\!\times]$ is a *Real Vector Space* (cf. section 2.2 of Chapter "Gray-Level LIP Model. Notations, Recalls and First Applications") and we are thus justified to study these topological notions in the context of images. Nevertheless, our objective here is not to develop these points in detail, even if we are convinced of their interest, especially concerning gauges. We will simply mention them in the following two sections.

7.2 Notions Stronger than Metrics

Pinoli (1992) proposed a definition of scalar product between two gray-level images f and g according to:

$$(f|g)^{\Delta} = M^2 \iint\limits_{(x,y)\in D} \ln\left(1 - \frac{f(x,y)}{M}\right) \ln\left(1 - \frac{g(x,y)}{M}\right) dxdy$$

The discrete version is obtained from the previous equation by replacing the double integral by a double sum and the points (x, y) with pixels (i,j).

It is well known that we can derive a norm $||\cdot||^{\Delta}$ and a correlation coefficient $\mathrm{Cor}^{\Delta}(f,g)$ from a scalar product:

$$||f||^{\Delta} = \sqrt{(f|f)^{\Delta}}$$

$$\mathrm{Cor}^{\Delta}(f,g) = \frac{(f|g)^{\Delta}}{||f||^{\Delta}||g||^{\Delta}}$$

Remark 21. Such a correlation coefficient offers an applicative interest: it permits, for example, novel approaches of target tracking or covariograms (autocorrelation). In this instance, we focus on the difference between a correlation approach, which tries to optimize the overlapping of functions (in fact of their subgraphs) and a probing approach (like Asplünd's metrics), which uses one image to surround the other by means of ordinary or logarithmic scalar multiplication or addition of a constant.

7.3 Notions Weaker than Metrics

7.3.1 Recalls on Gauges Theory in Vector Spaces and Topological Vector Spaces

This part mainly concerns mathematicians or readers curious about novel notions in image processing. In fact, the notion of gauge is classical in the field of vector spaces and it seemed interesting to propose a short summary on this subject, in direct link with Asplünd's metrics.

Before introducing the gauge concept, we need some recalls:

- The notion of vector space has been explored in Chapter "Gray-Level LIP Model. Notations, Recalls and First Applications," when we studied the set $[F(D,]-\infty, M[), \triangle\!\!\!+, \triangle\!\!\!\times]$
- If E is a real vector space and A is a subset of E:
  - A is said to be balanced if $\forall \lambda \in \mathbb{R}, |\lambda| \leq 1 \Rightarrow \lambda A \subset A$
  - A is said to be absorbing if $\forall x \in E, \exists \alpha_x > 0, \forall \lambda : |\lambda| \geq \alpha_x \Rightarrow x \in \lambda A$

It is easy to verify that a balanced set A contains necessarily the origin 0_E of the vector space E and that A is symmetric with respect to 0_E.

In the same way, 0_E is always an element of an absorbing set A.

Now let us define a gauge:

Definition. A being a subset of E, we call gauge of A the following application (function) j_A defined on E with values in $[0, +\infty]$:

$$x \to j_A(x) = \begin{cases} \text{Inf}\{\alpha > 0 / x \in \alpha A\} \text{ if it exists some } \alpha \text{ such that } x \in \alpha A \\ +\infty \text{ if for every } \alpha > 0, x \notin \alpha A \end{cases}$$

Remark 22. Here are some basic properties of gauges:

1. If A is absorbing, $\forall x \in E, j_A(x) < +\infty$
2. If $0_E \in A, j_A(0_E) = 0$ and if $0_E \notin A, j_A(0_E) = +\infty$
3. If $A \subset B, j_B \leq j_A$
4. If A is a subset of E, with $A \neq \emptyset$:
   a. $\forall \lambda > 0, \forall x \in E, j_A(\lambda x) = \lambda j_A(x)$
   b. If A is convex: $\forall (x, y) \in E^2, j_A(x + y) \leq j_A(x) + j_A(y)$
   c. If A is balanced: $\forall \lambda \in \mathbb{R}, \forall x \in E, j_A(\lambda x) = |\lambda| j_A(x)$
   d. If A is convex, balanced, and absorbing (ie, a *disk*), $x \longrightarrow j_A(x)$ is a seminorm on E

Remark 23. The concept of gauge exists in particular in topological vector spaces (cf. Schaefer, 1999), ie, vector spaces E equipped with a topology compatible with the vectorial structure, which means that the two laws (addition and scalar multiplication) of this vectorial structure are continuous:

$$(x, y) \longrightarrow x + y \quad \text{from } ExE \longrightarrow E \text{ is continuous}$$
$$(\lambda, x) \longrightarrow \lambda x \quad \text{from } \mathbb{R}xE \longrightarrow E \text{ is continuous}$$

In this situation, ExE and $\mathbb{R}xE$ are equipped with the corresponding product topologies.

For mathematicians familiar with topological vector spaces, it is well-known that in such a space E, each neighbor $N(0_E)$ is absorbing, thus it generates a gauge $j_{N(0_E)}$.

7.3.2 Definition of Gauges in Image Processing

In the case of the subspace $[I(D, [0, M[), \triangle\!\!\!+, \triangle\!\!\!\times]$ of the vector space $[F(D,]-\infty, M[), \triangle\!\!\!+, \triangle\!\!\!\times]$, it is obvious that the subgraph of any image $g > 0$ is a neighborhood of 0_F where 0_F is the origin of $[F(D,]-\infty, M[)$ or $[I(D, [0, M[)$, ie, the image null in every point of D (cf. Fig. 33).

Comment on Fig. 33: If we consider an image $f \in I(D, [0, M[)$, the homothetic $\lambda_0 \triangle\!\!\!\times f$ which absorbs g for the first time is defined by:

$$\lambda_0 = \text{Inf}\{\lambda, \forall x \in D, \lambda \triangle\!\!\!\times f(x) \geq g(x)\}$$

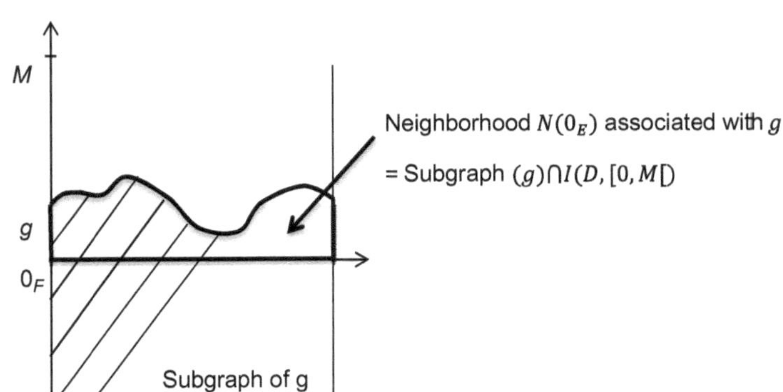

Fig. 33 Neighborhood of 0_E generated by an image $g > 0$ in the space of images $I(D,[0,M[)$.

Note that it corresponds precisely to the superior probe of image f by g permitting to define the multiplicative Asplünd's distance between f and g.

This remark gives a novel interpretation of the probing approach and in particular of the Asplünd-like metrics. Moreover, we are justified to ask if it would be interesting (and possible) to introduce inferior probing in topological vector spaces. If so, we can plan to define Asplünd's metrics in such spaces.

8. CONCLUSION AND PERSPECTIVES

In this chapter, we introduced novel metrics in link with LIP operations $\triangle\!\!\!\!+$ and $\triangle\!\!\!\!\times$ and we proposed a special focus on Asplünd-like metrics, due to their high interest in the processing of images acquired under low lighting, or variable lighting.

In Section 1, I explained the reasons that encouraged me to introduce the LIP addition. The motivation was mainly to develop a gray-level mathematical morphology where structuring elements could be gray-level functions themselves. However, the logarithmic operators provided many of novel research processes with efficient applications, so that the initial objective was temporarily forgotten.

The Asplünd's metrics I introduced refer back to the founding ideas. In fact, we have seen how the probing and moreover the double-sided probing techniques consist of moving a gray-level image (the probe) in upper and lower contact with the representative surface of the studied image. This point of view appears as the next episode of the abundant literature devoted to 3D structuring elements and particularly gray-level ones (see, for example, Serra, 1983).

Considering we dispose now of various metrics, I want to return to the application "automated thresholding" presented in sections 1.6.1 and 2.2.1 of Chapter "Various Contrast Concepts," where we used only LIP additive contrast and LIP multiplicative contrast to propose novel versions of Köhler's method. My goal here is not explore the numerous existing automated thresholding techniques I studied 30 years ago. At this time, I introduced a new algorithm based on functional metrics.

The principle of this algorithm uses the concept of metrics to find the binary image, which most closely resembles the initial image. Let us recall some notations:

f is the initial image and t a threshold within the interval $[0, 255]$. To t we associate a binary image constituted of two classes $C_0^t(f)$ and $C_1^t(f)$ defined as:

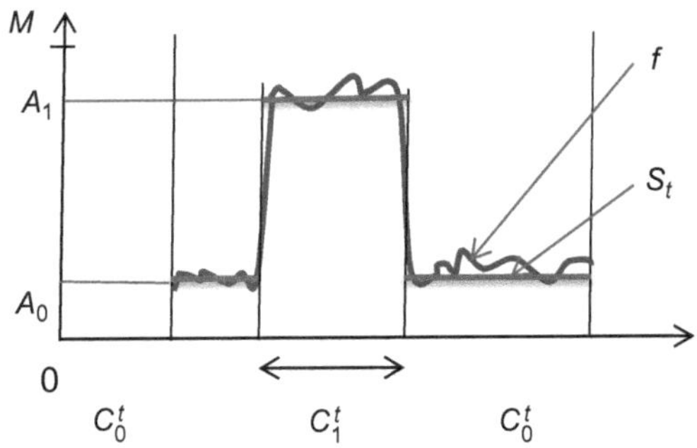

Fig. 34 Initial image *f* (in *blue*) and associated step function (in *red*). (See the color plate.)

$$C_0^t(f) = \{x \in D, f(x) \leq t\} \text{ and } C_1^t(f) = \{x \in D, f(x) > t\}$$

Such classes permit building a step function S_t according to:

$$\forall x \in C_0^t(f), S_t(x) = A_0 \text{ and } \forall x \in C_1^t(f), S_t(x) = A_1$$

where A_0 and A_1 represent the gray-level average values of $C_0^t(f)$ and $C_1^t(f)$, respectively (cf. Fig. 34).

Now let us consider a functional metric d, in order to evaluate the similarity between f and S_t. With such notations, we propose the following automated thresholding method:

1. For each threshold $t \in [0, 255]$, compute $d(f, S_t)$
2. Select the threshold t_0 satisfying:

$$d(f, S_{t_0}) = \text{Min}_{t \in [0,255]} d(f, S_t)$$

3. Apply t_0 to f to get a binary image

Comment on the method: The major interest of this approach is that it generates different results according to the choice of the metric:

- If d is a global distance like d_1, the corresponding method will be close to that of Otsu (interclass variance maximization), with the same drawback: it will be insensitive to very small (in terms of area) defects.
- If d is an atomic distance like d_∞, the method becomes sensitive to very small defects, such as the entropy maximization.
- If we apply an intermediate metric (cf. Section 2.2.2.1) we are able to create intermediate methods between the two extreme previous solutions.

REFERENCES

Asplund, E. (1960). Comparison between plane symmetric convex bodies and parallelograms. *Mathematica Scandinavica*, *8*(1), 71–180.

Barat, C., Ducottet, C., & Jourlin, M. (2003). Pattern matching using morphological probing. *IEEE International Conference on Image Processing (ICIP): 1* (pp. 369–372). .

Barat, C., Ducottet, C., & Jourlin, M. (2010). Virtual double-sided image probing: A unifying framework for non-linear grayscale pattern matching. *Pattern Recognition*, *43*(10), 3433–3447.

Beucher, S. (1991). The watershed transformation applied to image segmentation. *Scanning Microscopy International*, *1991*, 299–314.

Beucher, S., & Lantuejoul, C. (1979). Use of watersheds in contour detection. In *International workshop on image processing: Real-time edge and motion detection/estimation, Rennes, France.*

Carré, M., & Jourlin, M. (2014). LIP operators: Simulating exposure variations to perform algorithms independent of lighting conditions. In *International conference on multimedia computing and systems (ICMCS), Marrakech.*

Deshayes, V., Guilbert, P., & Jourlin, M. (2015). How simulating exposure time variations in the LIP Model. Application: Moving objects acquisition. In *14th International congress for stereology and image analysis, Liège.*

Dugundji, J. (1966). *Topology.* Boston: Allyn and Bacon. 447 p.

Grünbaum, B. (1963). Measures of symmetry for convex sets. *Proceedings of Symposium in Pure Mathematics*, 7, 233–270.

Haralick, R. M., & Shapiro, L. G. (1985). Image segmentation techniques (Survey). *Computer Vision, Graphics and Image Processing*, *29*, 100–132.

Jourlin, M., Breugnot, J., Itthirad, F., Bouabdellah, M., & Closs, B. (2011). Logarithmic image processing for color images. *Advances in Imaging and Electron Physics*, *168*(2), 65–107.

Jourlin, M., Carre, M., Breugnot, J., & Bouabdellah, M. (2012). Logarithmic image processing: Additive contrast, multiplicative contrast, and associated metrics. In P. W. Hawkes (Ed.), *Advances in imaging and electron physics: 171* (pp. 357–406). NewYork: Elsevier.

Jourlin, M., Couka, E., Abdallah, B., Corvo, J., & Breugnot, J. (2014). Asplünd's metric defined in the Logarithmic Image Processing (LIP) framework: A new way to perform double-sided image probing for non-linear grayscale pattern matching" *Pattern Recognition*, *47*(9), 2908–2924.

Jourlin, M., & Montard, N. (1998). A logarithmic version of the top-hat transform in connection with the Asplünd distance. *Acta Stereologica*, *16*(3), 201–208.

Köhler, R. (1981). A segmentation system based on thresholding. *Computer Graphics and Image Processing*, *15*(4), 319–338.

Noyel, G., & Jourlin, M. (2015). Asplünd's metric defined in the logarithmic image processing framework for color and multivariate images. In *IEEE international conference on image processing, Québec (Canada).*

Otsu, N. (1979). A threshold selection method from grey-level histograms. *IEEE Transactions on Systems, Man, and Cybernetics*, *9*(1), 62–66.

Pinoli, J.-C. (1992). Metrics, scalar product and correlation adapted to logarithmic images. *Acta Stereologica*, *11*(2), 157–168.

Schaefer, H. H. (1999). *Topological vector spaces* (2nd ed.). New York, NY: Springer-Verlag. 362 p.

Serra, J. (1983). *Image analysis and mathematical morphology.* London: Academic Press.

CHAPTER FOUR

Dynamic Range Expansion, Night Vision. Stabilization, Centering. Industrial and Biomedical Applications

Contents

Advances in Imaging and Electron Physics, Volume 195
ISSN 1076-5670
http://dx.doi.org/10.1016/bs.aiep.2016.04.004

This chapter, along with Chapter "Ability of the LIP Model to Simulate Variable Acquisition Conditions," are among the most innovative.

In fact, from the beginning of the logarithmic image processing (LIP) model, it was clear that one of its major interests would be in with its ability to modify, and potentially control, the brightness of an image through one of the logarithmic operators $\triangle\!\!\!+$ or $\triangle\!\!\!\times$, while taking into account the human visual system. Due to the nonlinearity of the model, it was easy to predict that the efficiency of LIP enhancement would be greatest when applied to brighten low-light images.

Before entering the subject, I begin with a physical interpretation of negative gray levels.

Note concerning the organization of this chapter

Section 1 provides an original interpretation of the generalized gray scale $]-\infty, M[$, representing all possible values of the source intensity. The selection of a particular intensity in this interval is equivalent to determining the origin of the LIP gray scale. The negative part $]-\infty, 0[$ is the set of gray levels' opposites, interpreted as light intensifiers.

Such results prepare the use of negative gray levels to attain full dynamic range-enhanced images associated with a low-light image.

Section 2 (resp. Section 3) is dedicated to image enhancement based on LIP scalar multiplication (resp. LIP addition). These two sections are structured in the same way:

- In Sections 2.1 and 3.1, we will present enhancement methods remaining in the gray scale, ie, not using negative gray levels.
- In Sections 2.2 and 3.2, we expose new methods to get full dynamic range-enhanced images.
- Finally, we suggest a simple extension to color images (Sections 2.3 and 3.3).

Section 4 is a short complement to the enhancement methods previously presented. In fact, most of enhancement algorithms produce images rather well centered in the gray scale: if it was not the case, ie, if the enhanced image is globally dark or globally bright, it would mean that the applied enhancement can be easily improved. Section 4 mainly suggests very simple approaches to stabilize images of a scene with variable lighting, for example, a transformation of such images so that they keep a fixed average gray-level value. At the same time, the simplicity of that solution guarantees its execution at the camera speed.

In Section 5, I present some significant applications to underline the efficiency of the LIP tools. However, I regret that plenty of other biomedical or industrial examples were covered by confidentiality agreements.

Since many of situations require local corrections, Section 6 is dedicated to this problem. The proposed solutions are mainly drawn from Maxime Carré. I shortened the algorithms' presentation because the interested reader can refer to Maxime's results.

Finally, in Section 7, comparisons with existing techniques are made. Nevertheless, I decided to present only the most basic ones for two reasons: first, considering the number of existing papers on the subject, a detailed comparison would be very fastidious. Second, we are convinced of the better scientific justification of LIP tools (mathematical and optical aspects), which give them a real superiority at the confidence and reliability levels, even if some existing algorithms present punctually excellent results.

1. A "KILLER" PROPERTY OF THE LIP MODEL: NEGATIVE GRAY LEVELS AND NEGATIVE THICKNESS OF AN OBJECT MAY BE INTERPRETED AS LIGHT INTENSIFIERS

Considering the number of situations where it is useful to perform image enhancement, it is normal that many papers propose algorithms dedicated to this problem. The difficulty is that most of these papers concern very simple and heuristic methods, such as histogram equalization, dynamic range expansion, logarithmic or powers law transformations used as look-up tables, and many others. Unfortunately, the common weakness of most of them is they are not founded on sound arguments in combination with physics, optics, mathematics, or human vision models. For a simple survey of these basic techniques, see for example, Maini and Aggarwall (2010).

The interest of this section is obvious at the applicative level, which we will see through the numerous presented examples.

Moreover, the results we will present take a specific interest due to their mathematical and physical meaning: on one hand, we will take advantage of the vector space structure generated on the space of functions $F(D,]-\infty,M[)$ by logarithmic laws $\triangle\!\!\!\!+$ and $\triangle\!\!\!\!\times$ and on the other hand, this mathematical structure will provide an optical interpretation.

More precisely, we have already seen that the interval $[0,M[$ is the set of gray levels and the interval $]-\infty,M[$ is the set of gray levels opposites. When we defined the opposite $\triangle\!\!\!\!-f(x)=\dfrac{-f(x)}{1-\dfrac{f(x)}{M}}$ of a gray level $f(x)$, our objective was that the equality $(\triangle\!\!\!\!-f(x))\triangle\!\!\!\!+f(x)=0$ would hold. Such

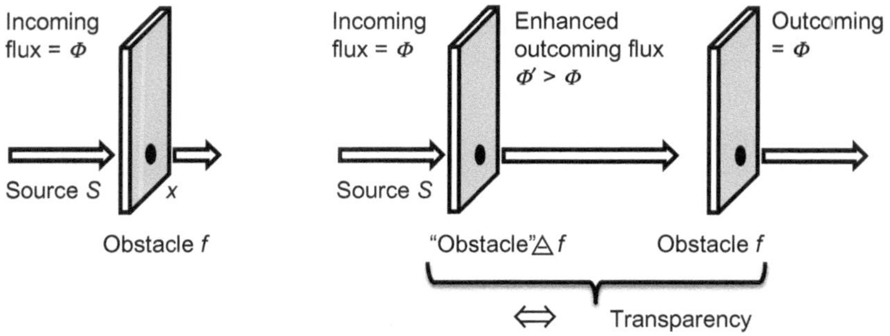

Fig. 1 Interpretation of the opposite $\triangle f$ as a "photons generator."

equality allows, in the set of images acquired in transmission, interpreting $(\triangle f(x))$ as an "obstacle" placed between the source and the obstacle $f(x)$ so that the sensor observes the gray level 0 at x, ie, the initial source intensity (Fig. 1). This means that the union of the two obstacles $\triangle f(x)$ and $f(x)$ becomes transparent. In other words: $\triangle f(x)$ is interpretable as a light generator, which emits the necessary photons (electron, X-rays, etc., according to the situation) to produce an outgoing flux at x corresponding to the flux of the initial source.

The same reasoning holds for the scalar multiplication $\lambda \triangle f$ for negative values of λ: the interpretation previously done of this law in terms of thickness variation of the virtual obstacle proportional to λ adapts to negative values if we consider "negative" thicknesses.

Thus, we formulate the following fundamental properties (summarized in Fig. 2A–C):

- The generalized gray scale $]-\infty, M[$ represents the set of all possible values of the source intensity: $-\infty$ corresponds to an infinite intensity, while M corresponds to a null one.
- When we choose an initial source intensity I_S in the interval $]-\infty, M[$, it consists of selecting an origin 0_S inside this interval.
- At this moment, $[0_S, M[$ represents the reference gray scale and $]-\infty, 0[$ the negative gray levels. The space of actual images has been noted $I(D, [0_S, M[)$.
- The negative gray levels correspond to source intensities greater than the reference one, while the gray scale $[0_S, M[$ is the set of source intensities lower than the reference I_S.
- Negative gray levels (and consequently negative thicknesses) are light intensifiers.

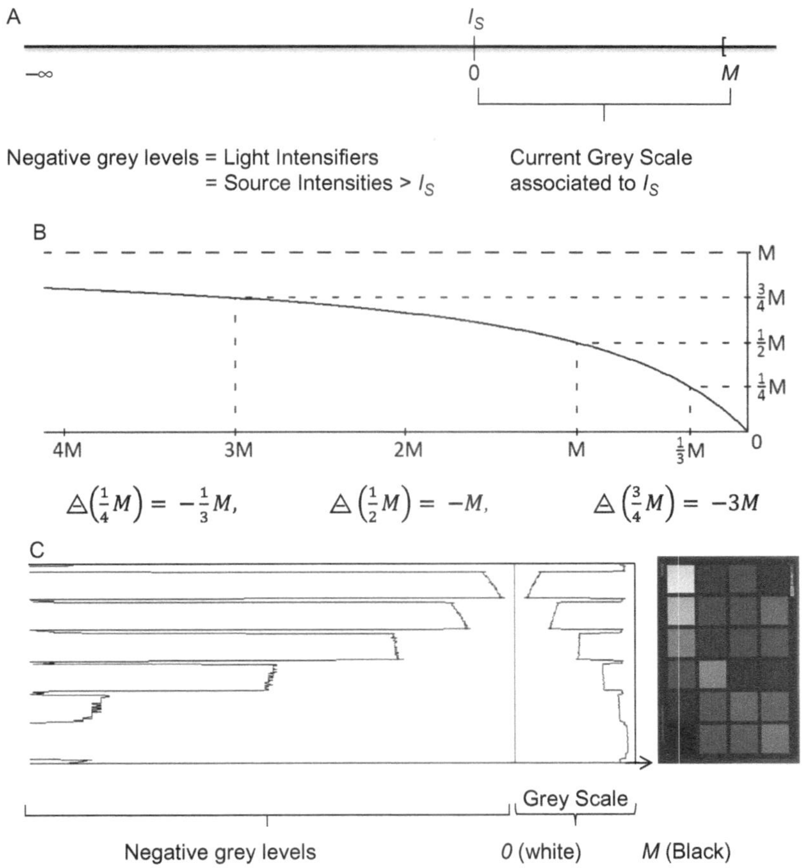

Fig. 2 (A) Gray-scale and negative gray levels. (B) Representative curve of the gray-levels opposites. (C) Gray-levels profile of the white straight line on the chart and their opposites.

- The over-space $F(D,]-\infty, M[)$ represents the set of all possible images when the reference source moves in $]-\infty, M[$. In other words:

$$F(D,]-\infty, M[) = \cup_{S\in]-\infty, M[} I(D, [0_S, M[)$$

This representation can be clarified by drawing the curve of the gray-levels opposites (Fig. 2B): $f(x) \to \triangle f(x) = \dfrac{-f(x)}{1 - \dfrac{f(x)}{M}}$.

Finally, we propose (cf. Fig. 2C) to visualize the gray-level profile of a straight line represented in white on the chart, and the profile of its opposite in the negative part of the gray scale.

2. LOW-LIGHT IMAGES ENHANCEMENT THROUGH LIP SCALAR MULTIPLICATION

2.1 Remaining in the Gray Scale

2.1.1 Dynamic Range Maximization

We have already emphasized the ability of LIP scalar multiplication of a low-light image f by a real number $\lambda \in]0,1]$ to brighten f, knowing that, in such a case, the virtual thickness of the obstacle generating f decreases. In order to automate such a lightening, the simplest parameter is the dynamic range of f, noted DR(f). Suppose a (respectively, b) designs one point of the spatial support D, corresponding to the darkest gray-level value of image f (respectively, the brightest one). It means that:

$$f(a) = \mathrm{Sup}\{f(x), x \in D\}, f(b) = \mathrm{Inf}\{f(x), x \in D\} \text{ and}$$
$$\mathrm{DR}(f) = f(a) - f(b)$$

In the following, we will suppose $f(a) \neq f(b)$, which means that the image to be enhanced presents a strictly positive dynamic range.

Considering the set of homothetic images $\{\lambda \triangle f\}$ for values of λ varying from 1 to 0, we observed that the dynamic range $\mathrm{DR}(\lambda \triangle f)$ first increases before converging to 0. From this observation, it is natural to focus on a possible maximal dynamic range.

Jourlin and Pinoli (1995) established that there exists a unique value λ_0 satisfying:

$$\mathrm{DR}(\lambda_0 \triangle f) = \mathrm{Sup}\{\mathrm{DR}(\lambda \triangle f), \lambda \in]0,1]\}$$

Such a value λ_0 is expressed according to:

$$\lambda_0 = \frac{Ln\left[\dfrac{Ln\left(1 - \dfrac{f(a)}{M}\right)}{Ln\left(1 - \dfrac{f(b)}{M}\right)}\right]}{Ln\left[\dfrac{\left(1 - \dfrac{f(b)}{M}\right)}{\left(1 - \dfrac{f(a)}{M}\right)}\right]} \tag{1}$$

Comment about Eq. (1)

The optimal scalar λ_0 exclusively depends on the upper and lower bounds $f(a)$ and $f(b)$ of image f and it corresponds to the theoretical optimal thickness of the real (in transmission) or virtual (in reflection) observed object to yield a maximal dynamic range.

Proof of Eq. (1). With our usual notations, the dynamic range of the image f is defined according to:

$$\mathrm{DR}(f) = f(a) - f(b)$$

and the dynamic range of a homothetic $\lambda \triangle f$ is done with the equation:

$$\begin{aligned}\mathrm{DR}(\lambda \triangle f) &= \lambda \triangle f(a) - \lambda \triangle f(b) \\ &= M - M\left(1 - \frac{f(a)}{M}\right)^{\lambda} - \left[M - M\left(1 - \frac{f(b)}{M}\right)^{\lambda}\right] \\ &= M\left[\left(1 - \frac{f(b)}{M}\right)^{\lambda} - \left(1 - \frac{f(a)}{M}\right)^{\lambda}\right]\end{aligned}$$

Each of the previous terms represents an exponential law, whose derivative is easy to compute:

$$\begin{aligned}&\mathrm{DR}(\lambda \triangle f)'_{\lambda} \\ &= M\left[\left(1 - \frac{f(b)}{M}\right)^{\lambda} \cdot Ln\left(1 - \frac{f(b)}{M}\right) - \left(1 - \frac{f(a)}{M}\right)^{\lambda} \cdot Ln\left(1 - \frac{f(a)}{M}\right)\right]\end{aligned}$$

Our goal is to determine the real number λ_0 for which this expression is null. It satisfies: $Ln\left[\dfrac{\left(1 - \frac{f(b)}{M}\right)}{\left(1 - \frac{f(a)}{M}\right)}\right]^{\lambda_0} = \dfrac{Ln\left(1 - \frac{f(a)}{M}\right)}{Ln\left(1 - \frac{f(b)}{M}\right)}$. This implies taking the logarithm of each member: $\lambda_0 Ln\left[\dfrac{\left(1 - \frac{f(b)}{M}\right)}{\left(1 - \frac{f(a)}{M}\right)}\right] = Ln\left[\dfrac{Ln\left(1 - \frac{f(a)}{M}\right)}{Ln\left(1 - \frac{f(b)}{M}\right)}\right]$ which achieves the proof. □

An example of application is given in Fig. 3 for two images "castle" and "cells preparation."

Fig. 3 Examples of dynamic range maximization (parameter λ_0). (A) Initial image f "Castle." (B) Enhanced image $\lambda_0 \triangle\!\!\!\times f$. (C) Initial image g (cells preparation). (D) Enhanced image $\lambda_0 \triangle\!\!\!\times g$.

Comments on Fig. 3: The result presented is satisfactory from the visual point of view. Nevertheless, we would underline two limitations, which will justify further improvements. The first one is linked to the sensitivity of standard sensors in very dark-gray levels. Most of them produce noisy images and the proposed method enhances the signal and the noise simultaneously. Another problem is linked to digitized images: a weak dynamic range is represented by a small number n of gray levels and the enhanced image also possesses n gray levels. This is the reason why, in a certain number of situations, one can observe a quantification effect on the resulting image. The enhancement performed by $\lambda_0 \triangle\!\!\!\times f$ is due both to a centering of f in the gray scale and to the increasing of the distance (difference) between two successive levels. Applying denoising algorithms to the initial image as a preprocessing step would certainly produce enhanced images of better quality, particularly for very low-light images.

Real-time aspects: When this technique was introduced (Jourlin & Pinoli, 1995), personal computers did not have sufficient power to perform it in real-time, in the sense of the standard camera speed (25 images per second). As I headed the Electrical Engineering Department at CPE-Lyon Engineer School, I required electricians to speed up the algorithm and they developed a prototype of a circuit board, which allowed me to prove the feasibility of real time. More recently, I proposed that a Masters student create a program based on look-up tables: in fact, for 8-bits digitized images, the number of pairs $(f(a), f(b))$ used in Eq. (1) is finite and it is then possible to precompute a double-entry table in which we store with the desired precision the value λ_0 associated with each pair $(f(a), f(b))_{a\in[0,\,255],\,b\in[0,\,255]}$. Another table is precomputed for each value of λ_0 in order to store:

- In input, the possible 255 gray-level values of the initial image
- In output the corresponding gray levels of the enhanced image. In this way, images are easily processed at camera speed. Knowing that we have reached real time for high-resolution color images, we will not focus on this problem now.

Drawback of the method: We observed that the method fails when bright points are present inside the initial image. In such a case, the dynamic range presents an already large size and its expansion is negligible, even becoming a reduction in some cases. An example of this last case is given in Fig. 4 representing a low-light image with a street lamp.

Nevertheless, a possible solution to improve the result of Fig. 4 consists of suppressing the bright region. In Fig. 5, we propose to threshold the image f at the level $t = 50$ (in a standard gray scale where $0 = \text{black}$), ie, to affect the value 0 to pixels x verifying $f(x) > 50$ (result presented in Fig. 5A): image $f|_{[0,50]}$, restriction of f to the gray-level interval [0,50]). Then, the corresponding optimal number λ_0 is computed and the produced image $\lambda_0 \triangle\!\!\!\times f|_{[0,50]}$ is displayed (Fig. 5B). The interest of this approach is limited by the thresholding step: it is uncertain if an automated thresholding performs exactly the expected suppression of bright points, and one can observe this phenomenon here with the presence of artifacts (black zones under the street lamp).

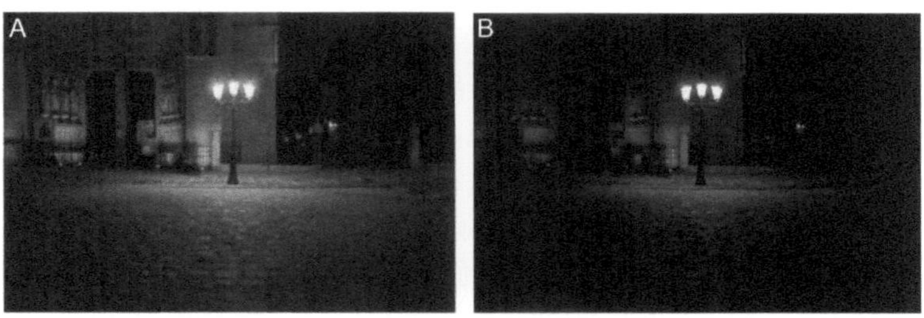

Fig. 4 Example where the dynamic range expansion fails. (A) Initial image f (Night street scene) with bright points with permission of Richard Vantielcke. (B) "Enhanced" image $\lambda_0 \triangle\!\!\!\times f$.

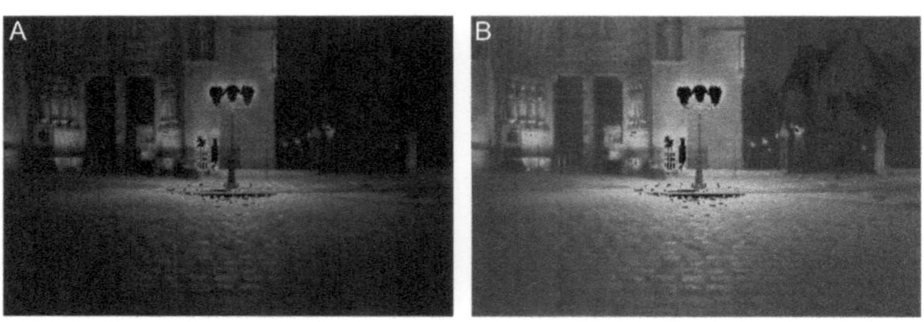

Fig. 5 Improvement of dynamic range expansion in presence of bright points. (A) Image "Night Street Scene" thresholded at $t = 50$: $f|_{[0,50]}$. (B) Image $\lambda_0 \triangle\!\!\!\times f|_{[0,50]}$.

The situation illustrated in Fig. 4 incited us to imagine another approach, which we will discuss in Section 2.1.2.

2.1.2 Maximization of the Standard Deviation of the Histogram of $\lambda \triangle f$

We consider a situation where the initial low-lighted image contains a few bright points. The idea consists of expanding the dark region of the image, considering that the information brought by the bright points is not essential. Thus, we propose the following steps to process the studied image f:

1. Compute the logarithmic homothetic $\lambda \triangle f$
2. Create the histogram $h(\lambda \triangle f)$
3. Compute the standard deviation $\sigma[h(\lambda \triangle f)]$
4. Compute λ_1 such that

$$\sigma[h(\lambda_1 \triangle f)] = \text{Sup}_{\lambda \in]0,1]}\{\sigma[h(\lambda \triangle f)]\} \tag{2}$$

5. Display $\lambda_1 \triangle f$

In Fig. 6, we show the result of this method, to be compared with the previous "expansion" $\lambda_0 \triangle f$ of Fig. 4.

Comment on Fig. 6: The proposed method appears to be a good compromise if we compare it with a classical histogram equalization whose enhancement is often too strong and has no physical meaning. Fig. 7 permits comparison between the two methods and between the resulting histograms.

Remark 1. The main interest of the two methods discussed in this section lies in the processing of low-light images. They are absolutely ineffective when applied, for example, to images of weak dynamic range, but well centered in the gray scale. They apply sometimes in the situation of over-lighted images (cf. Fig. 8), but this is not their natural use: since, as we know, the nonlinearity of LIP scalar multiplication favors the dark side of the gray scale.

Now let us introduce the most original part of this chapter:

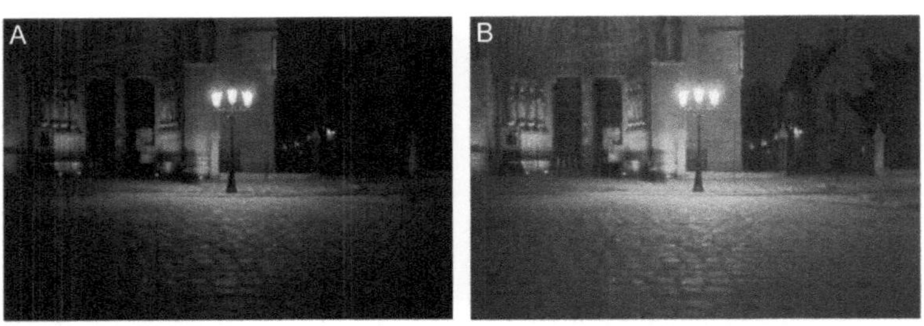

Fig. 6 Low-light image enhancement in presence of bright points (parameter λ_1). (A) Initial image f "Night Street Scene." (B) Enhanced image $\lambda_1 \triangle f$.

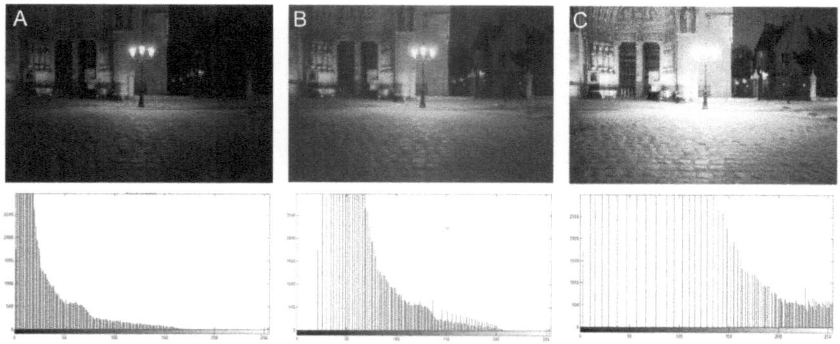

Fig. 7 Comparison of λ_1-enhancement and histogram equalization. (A) Initial image f "Night Street Scene" and its histogram. (B) Enhanced image $\lambda_1 \triangle\!\!\!\!\times f$ and its histogram. (C) Enhancement of (A) by histogram equalization and equalized histogram.

Fig. 8 Enhancement of an over-lighted image. (A) Initial over-lighted image. (b) Enhanced image $\lambda_1 \triangle\!\!\!\!\times f$.

2.2 Full Dynamic Range Expansion Based on Negative Gray Levels

When applying the scalar multiplication $\lambda \triangle\!\!\!\!\times f$ to a low-light image f, we first increase its dynamic range as long as λ varies in the interval $[1,\lambda_0]$, then $\mathrm{DR}(\lambda \triangle\!\!\!\!\times f)$ decreases until 0 when λ varies from λ_0 to 0. This situation occurs when we force λ to remain positive.

Now, if we decide to work in the vector space of functions $F(D,]-\infty, M[)$ it becomes possible to consider negative values of λ, in such a way that the resulting function $\lambda \triangle f$ lies in the interval $]-\infty, 0]$. Moreover, when λ varies from 0 to $-\infty$, $DR(\lambda \triangle f)$ varies from 0 to $+\infty$. In fact, using the notations from Section 2.1.1 yields:

$$\begin{aligned}
&DR(f) = f(a) - f(b) \quad \text{and for } \lambda < 0: \\
&DR(\lambda \triangle f) = \lambda \triangle f(b) - \lambda \triangle f(a) \\
&\qquad = M - M\left(1 - \frac{f(b)}{M}\right)^{\lambda} - \left[M - M\left(1 - \frac{f(a)}{M}\right)^{\lambda}\right] \\
&\qquad = M\left[\left(1 - \frac{f(a)}{M}\right)^{\lambda} - \left(1 - \frac{f(b)}{M}\right)^{\lambda}\right]
\end{aligned}$$

In such conditions, there exists a unique value $\lambda_2 < 0$ satisfying

$$DR(\lambda_2 \triangle f) = M - 1 = 255 \,(\text{for } 8 - \text{bits images}) \tag{3}$$

which means

$$\begin{aligned}
&(\lambda_2 \triangle f)(b) - (\lambda_2 \triangle f)(a) = M - 1 \\
&\Leftrightarrow M - M\left(1 - \frac{f(b)}{M}\right)^{\lambda_2} - \left[M - M\left(1 - \frac{f(a)}{M}\right)^{\lambda_2}\right] = M - 1 \\
&\Leftrightarrow M\left[\left(1 - \frac{f(a)}{M}\right)^{\lambda_2} - \left(1 - \frac{f(b)}{M}\right)^{\lambda_2}\right] = M - 1
\end{aligned}$$

$$\Leftrightarrow \left(1 - \frac{f(a)}{M}\right)^{\lambda_2} - \left(1 - \frac{f(b)}{M}\right)^{\lambda_2} = 1 - \frac{1}{M} = \frac{255}{256} \tag{4}$$

To achieve the full dynamic range expansion of *f*, two steps remain:

Step 1: Computation of λ_2

From Eq. (4), it is not possible to compute explicitly λ_2. Nevertheless, the result of Annex 1 proves the unicity of λ_2: in fact, $DR(\lambda_2 \triangle f)$ being expressed as a difference of two exponential functions is continuous in λ_2 and then reaches each value lying in $[0, +\infty[$ only once, in particular the value $1 - \frac{1}{M}$.

In order to get an approximated solution, we propose to perform a dichotomy to reduce an interval surrounding λ_2. First, we choose the initial size of the interval, for example, 1, and we determine the first value $\lambda = -1, -2, -3, \ldots$ for which the following inequality holds:

$$\left(1-\frac{f(a)}{M}\right)^{\lambda} - \left(1-\frac{f(b)}{M}\right)^{\lambda} > 1 - \frac{1}{M}$$

It means that the optimal value λ_2 belongs to the interval $]\lambda, \lambda+1]$. The principle of dichotomy consists then of dividing this interval by two and deciding if λ_2 lies in $]\lambda, \lambda+\frac{1}{2}]$ or in $]\lambda+\frac{1}{2}, \lambda+1]$. The operation is iterated until the required precision for λ_2 is reached.

Step 2: Displaying full dynamic range expansion $\lambda_2 \triangle f$ of image f
When λ_2 is computed, it remains to display the function $\lambda_2 \triangle f$ as an image. This step is performed through the translation of $\lambda_2 \triangle f$ by the vector $-(\lambda_2 \triangle (f(a)))$, which results in a full dynamic range expansion of f.

Comment on the method: The proposed full dynamic range expansion (FDRE) clearly associates a highly contrasted result to a low-light image whose visual quality sometimes appears degraded for various reasons: first, the enhanced image does not present more gray levels than the initial one, which implies a "quantization" effect. Second, the noise is enhanced with the signal, which concerns plenty of low-light images enhancements. This drawback may be widely solved by associating a noise-filtering algorithm with the FDRE. However, we will not develop this point in depth here. Third, the quality of low-lighted images strongly depends on the sensitivity of the sensor in the dark part of the gray scale. Nevertheless, the proposed algorithm permits a significantly better interpretation (automated or visual) of very low-lighted images. Using the image "Castle" presented in Fig. 3A, Fig. 9 illustrates this point.

In order to evaluate the efficiency of the method, Fig. 10 presents three other applications: one concerns a back-light acquisition (A) and (B) where λ_2 is computed on a window of the initial image, the second concerns an image with a significant lighting drift (C) and (D), and the third concerns an enhancement inside a window located in the dark part of the image (E) and (F). Note that for enhanced images (B) and (F), the full available dynamic is used inside the considered window.

Real-time aspects: The computation of λ_2 with a high precision (1/1000th, for example) is very fast, even applied to high-resolution images (1280×800 pixels). We had no issues performing it at camera speed

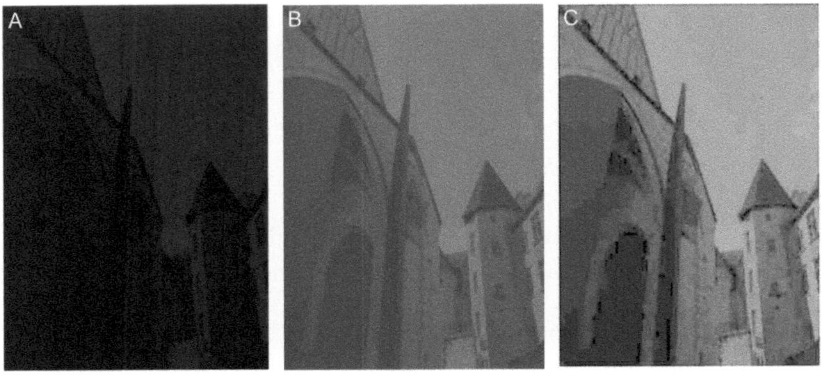

Fig. 9 Full dynamic range expansion compared with λ_0-enhancement. (A) Image f "Castle." (B) For memory, image $\lambda_0 \triangle\!\!\!\times f$. (C) Image $\lambda_2 \triangle\!\!\!\times f$.

Fig. 10 Efficiency of the full dynamic range expansion. (A) Initial image f (back light + low light) with authorization of Claude Sadik. (B) Result of $\lambda_2 \triangle\!\!\!\times f$ computed on a window of f. (C) Initial image g (back light + lighting drift). (D) Result of $\lambda_2 \triangle\!\!\!\times g$. (E) Initial image h "Night Street Scene." (F) Result of $\lambda_2 \triangle\!\!\!\times h$ computed on a window of h.

(25 images per second) with a standard personal computer. Note also that we can realize a precomputation of estimated values of λ_2 associated with the pairs of gray levels (integers of the interval [0,255]) and store them in a table. In fact, Eq. (4) shows that λ_2 depends only of the maximal and minimal gray levels of the image to process.

Annex 1. The aim of this Annex is to establish that $DR(\lambda \triangle f)$ varies from 0 to $+\infty$ when λ varies from 0 to $-\infty$. Let us recall the expression of $DR(\lambda \triangle f)$:

$$DR(\lambda \triangle f) = \lambda \triangle f(b) - \lambda \triangle f(a) = M\left[\left(1 - \frac{f(a)}{M}\right)^{\lambda} - \left(1 - \frac{f(b)}{M}\right)^{\lambda}\right]$$
$$= M\left[(A)^{\lambda} - (B)^{\lambda}\right]$$

This last term appears as the difference of two exponential functions of the variable λ, knowing that M, A, and B do not depend on λ.

Moreover, the inequalities $0 \le f(b) < f(a) < M$ imply $0 < A < B < 1$ and $\frac{A}{B} < 1$. In such conditions:

$$(A)^{\lambda} \to +\infty \text{ when } \lambda \to -\infty \text{ and } \left(\frac{B}{A}\right)^{\lambda} \to 0 \text{ when } \lambda \to -\infty$$

Thus, $DR(\lambda \triangle f) = M\left[(A)^{\lambda}\left(1 - \left(\frac{B}{A}\right)^{\lambda}\right]\right. \to +\infty$ when $\lambda \to -\infty$

2.3 What About Color Images?

Since this book is mainly focused on the logarithmic processing of gray-level images, we have already said that the reader can refer to Jourlin, Breugnot, Itthirad, Bouabdellah, and Closs (2011) for LIP color (LIPC) images. These "true" LIPC operators, defined from human visual experiences (Stiles & Birch, 1959), take into account different eye's sensitivities in the three channels R, G, B. The consequence is that the LIPC operators present a certain difficulty level.

Thus, it seemed interesting to propose a simplified extension to color images of our enhancement techniques.

Given a color image f defined on the spatial support D, we can consider its components f_R, f_G, f_B in the three channels red, green, blue. Each component is similar to a gray-level function and could be processed separately by means of the enhancement algorithms we have defined in Sections 2.1

and 2.2. Such a technique sometimes presents an applicative interest, for example, if one component of f carries the information we look for. Nevertheless, it is not recommended to group the three enhanced channels to create a color image. In fact, it is well known that the result would generate false colors and thus destroy the global information.

The simple version we are going discuss proceeds differently: it will consist of grouping the information from the three channels before applying each of the "optimal" enhancements we have defined. In order to do that, we need some definitions:

- Global Dynamic Range GDR(f) of image f:

$$\begin{aligned}\mathrm{GDR}(f) &= \mathrm{Max}\{\mathrm{DR}(f_R), \mathrm{DR}(f_G), \mathrm{DR}(f_B)\}\\ &= \mathrm{Max}_{x\in D}\{f_R(x), f_G(x), f_B(x)\} - \mathrm{Min}_{x\in D}\{f_R(x), f_G(x), f_B(x)\}\end{aligned} \tag{5}$$

- *Global Histogram* GH_f of image f:

  The histogram H_g of a gray-level image g is defined on $[0, M[$ according to:

$$\forall i \in [0, M[,\ H_g(i) = \mathrm{Card}\{x \in D, g(x) = i\}$$

The three histograms $H_{f_R}, H_{f_G}, H_{f_B}$ of f_R, f_G, f_B are obviously defined in the same way, which permits building the global histogram GH_f of f, verifying:

$$\forall i \in [0, M[,\ \mathrm{GH}_f(i) = \sum\nolimits_{i=0,\ldots,255} H_{f_R}(i) + H_{f_G}(i) + H_{f_B}(i) \tag{6}$$

An example of global histogram grouping $H_{f_R}, H_{f_G}, H_{f_B}$ is given in Fig. 11:

We are now able to transfer the enhancement algorithms previously established for gray-level images to color images. In fact, definitions (5) and (6) permit extending the computations of λ_0, λ_1, and λ_2.

Case of λ_0: To the color image f, we associate its GDR (Eq. 5), which allows computing of (Eq. 1) the scalar λ_0 presented in Section 2.1.1. Then the scalar multiplication is performed on each channel, and we obtained an enhanced color image f_0 of f, whose components are $(\lambda_0 \triangle f_R, \lambda_0 \triangle f_G, \lambda_0 \triangle f_B)$. An example is given in Fig. 12 where the initial image is a color chart acquired in low-light conditions.

Comment on Fig. 12: When using the logarithmic scalar multiplication by λ_0 to enhance an image, we know that it does not exploit the full dynamic

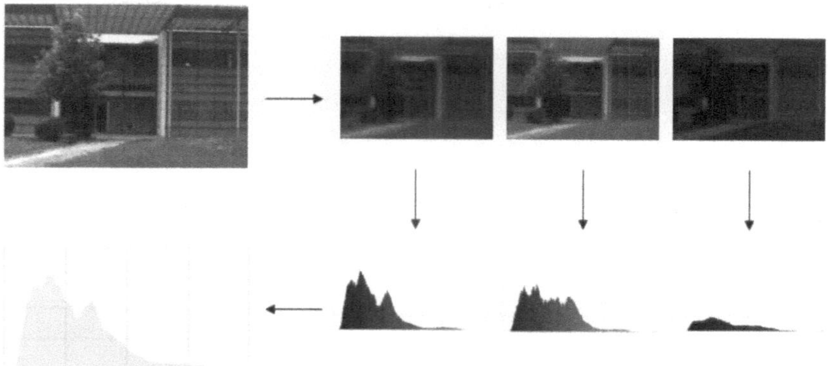

Fig. 11 Global histogram of a color image. Initial image f (upper *left*) and according to successive arrows: Images f_R, f_G, f_B; Histograms H_{f_R}, H_{f_G}, H_{f_B}; Finally: global gistogram GH_f. (See the color plate.)

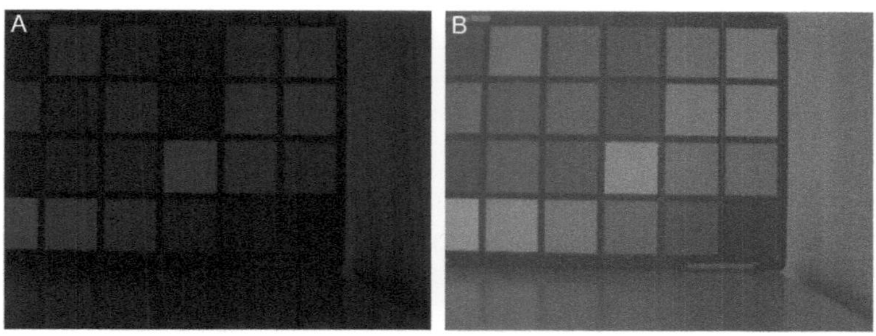

Fig. 12 λ_0-Enhancement of a color image. (A) Initial low-light image f "color chart." (B) Enhanced image $\lambda_0 \triangle f$. (See the color plate.)

range available in the gray scale. Thus, it results in a rather "smooth" enhancement. More precisely, such a technique preferably refers to "stabilization" and "centering" ones presented at the end of this chapter (Section 4).

Case of λ_1: To the low-light color image f, one associates its global histogram GH_f (Eq. 6 and Fig. 11). The "optimal" scalar λ_1 is defined by the following equation:

$$\sigma(\lambda_1 \triangle GH_f) = \mathrm{Sup}_{\lambda \in]0,1]}\{\sigma(\lambda \triangle GH_f)\}$$

Finally, the scalar multiplication by λ_1 is applied to each of the three components f_R, f_G, f_B, resulting in an enhanced image of f (cf. examples in Fig. 13).

Comment on Fig. 13: The results shown in this figure are not of the same visual quality. In fact, in case (A) we start from an image uniformly dark

Fig. 13 λ_1-Enhancement of a color image. (A) and (C) Initial color images. (B) and (D) Enhanced images of (A) and (B) (λ_1 parameter applied on the global histogram). (See the color plate.)

(small dynamic range concentrated toward the black side). In such conditions, the global enhancement is totally supported by the dark part, so the result is disturbed by noise. In case (C), the initial image is globally dark but possesses bright parts (sky, lights) and the enhancement applies to both dark and bright regions, so that the dark part is less bright than in case (A) and the noise is less visible.

Case of λ_2: For applying a full dynamic range expansion to a low-light color image *f*, we define its global dynamic range GDR(*f*) (cf. Eq. 5). Then one proceeds as for gray-level images: computing the unique negative scalar λ_2 such that:

$$\lambda_2 \triangle [\mathrm{Min}_{x\in D}\{f_R(x), f_G(x), f_B(x)\}] - \lambda_2 \triangle [\mathrm{Max}_{x\in D}\{f_R(x), f_G(x), f_B(x)\}] = M - 1$$

Then the three components f_R, f_G, f_B are multiplied by λ_2, giving three negative functions $\lambda_2 \triangle f_R$, $\lambda_2 \triangle f_G$, $\lambda_2 \triangle f_B$. In order to bring back these functions in the gray scale $[0, M[$, it remains to apply them a translation by the vector

$$-\lambda_2 \triangle [\mathrm{Max}_{x\in D}\{f_R(x), f_G(x), f_B(x)\}]$$

Finally one displays the enhanced color image (cf. Fig. 14).

Fig. 14 Examples of λ_2-full dynamic range expansion. (A) Low-light color image f "Castle." (B) Enhanced image $\lambda_2 \triangle f$. (C) Color image g "Campus" (back lighted). (D) $\lambda_2 \triangle g$ inside a window. (E) Very low-light color image h "Young Girl." (F) Enhanced image $\lambda_2 \triangle h$. (See the color plate.)

Comment on Fig. 14: The first image (14A) is a low-light color version of the image "Castle" realized by a short exposure time. The second image (14C) represents our "Campus" (Saint-Etienne University) acquired in back light. The full dynamic range expansion has been performed on a window located in the dark part of the image. For images presenting both bright and dark regions, it would be possible to apply them, for example, an automated thresholding to limit the enhancement to the dark parts. Contrary to the "soft" technique based on λ_0, we dispose now of a strong enhancement permitting the interpretation of very dark zones in good conditions. The third example (image "Young Girl" 14E) is a very dark one on which the FDRE shows its efficiency. Nevertheless, we observe that the enhanced image (14F) is noisy due to the initial very low-light level.

2.4 Conclusion of Section 2

We have presented very fast and efficient techniques of low-light images enhancement based on the scalar multiplicative law of the LIP model. This model possesses strong advantages due to its physical and mathematical properties: based on the transmittance law, the LIP operators are first dedicated to images in transmission and second, to images in reflection because of the consistency of the model with human vision. Moreover, the addition and scalar multiplication laws permit obtaining a functional space structured as a vector space in which the space of images is immersed. The negative gray-level values of the vector space appear as intensities brighter than that of the initial source and thus may be used as "light intensifiers." The expansion of an image through the negative part of the gray scale takes advantage of this property.

Section 3 is developed like Section 2, but uses the LIP additive law (more precisely the LIP subtraction) instead of the LIP scalar multiplication to perform low-light images enhancement.

3. LOW-LIGHT IMAGES ENHANCEMENT BASED ON LIP SUBTRACTION

3.1 Remaining in the Gray Scale

3.1.1 Dynamic Range Maximization

Now let us consider an acquisition system (source + sensor) whose intensity S and diaphragm aperture are fixed. The intensity associated to a gray level lying in the gray scale $[0, M[$ is less than the reference one $I(S)$. More precisely, if $f \in I([D, [0, M[)$ is a low-light image with superior bound $f(a) = \text{Max}_{x \in D} f(x)$ and inferior bound $f(b) = \text{Min}_{x \in D} f(x)$, for each constant gray level $C \in [0, f(b)]$, the value $f(b) \triangle C$ remains positive, ie, lies in $[0, M[$. Thus, $f \triangle C$ is an image, and this image is brighter than f (cf. Fig. 15). In the case of a constant greater than $f(b)$, negative values can occur for $f \triangle C$. In such a situation, each negative value is constrained to 0, which

Fig. 15 Enhancement due to logarithmic subtractions of a constant constrained to remain in the gray scale. From *left* to *right*: initial image *f* and various logarithmic subtractions of a constant.

means that some saturated regions appear (cf. background in images $f \triangle 224$ and $f \triangle 240$ in Fig. 15).

Furthermore, it is easy to show that the dynamic range $\mathrm{DR}(f \triangle C)$ increases with C for $C \in [0, f(b)]$. In particular $\mathrm{DR}(f \triangle C)$ is greater than the dynamic range DR(f) of f. In fact:

$$\mathrm{DR}(f \triangle C) = (f(a) \triangle C) - (f(b) \triangle C)$$

Using the equation that makes explicit the LIP subtraction yields:

$$\mathrm{DR}(f \triangle C) = \frac{f(a) - C}{1 - \frac{C}{M}} - \frac{f(b) - C}{1 - \frac{C}{M}}$$

$$= \frac{f(a) - f(b)}{1 - \frac{C}{M}}$$

$\geq f(a) - f(b)$ when $C \in [0, f(b)]$, this inequality being strict if $C \neq 0$.

Remark 2. The previous property implies that for the logarithmic subtraction of a constant, we do not properly have a method equivalent to what we proposed for scalar multiplication with the optimization "$\lambda_0 \triangle\!\!\!\times f$." The maximal dynamic is always obtained for $C_0 = f(b)$.

In the limit case where $C_0 = f(b)$, the enhanced image $f \triangle C_0$ reaches the 0 value at $b \in D$. This situation corresponds clearly to the maximal dynamic range we can produce under the constraint to remain in the gray scale.

The obtained effect is generally a strong lightening of image f, except if the initial image contains rather bright points. The method is illustrated Fig. 16A and B. In this figure, the initial image "Laboratory" has been captured in the evening, with low-light condition and with a small diaphragm aperture. Thus, the initial image is not easy to interpret and (B) represents a significant enhancement. The histogram evolution is displayed in (C) and (D).

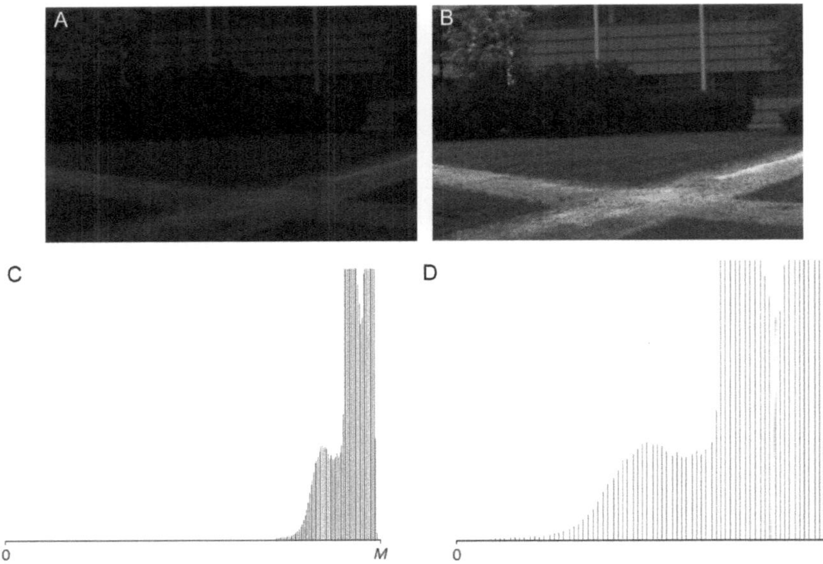

Fig. 16 Dynamic range maximization remaining in the gray scale (method $f \triangle C_0$). (A) Initial low-light image f "Laboratory" (exposure time 12 ms). (B) Enhanced image of f according to $f \triangle C_0$ with $C_0 = f(b) = 17$. (C) and (D) Histograms of (A) and (B).

3.1.2 Maximization of the Standard Deviation of the Histogram of $f \triangle C$

As for the scalar multiplication (Section 2.1.2), the enhancement method suggested here will work even in presence of some bright points in the initial low-lighted image. The dark region of f is expanded by subtraction of a constant $C \in [0, M[$, which can generate negative values of $f \triangle C$. Considering that the information brought by the bright points is not essential, the negative values of $f \triangle C$ are constrained to zero. The following steps are proposed:

1. Compute the logarithmic subtraction $f \triangle C$ (constrained to remain in the gray scale)
2. Create the histogram $h(f \triangle C)$
3. Compute the standard deviation $\sigma[h(f \triangle C)]$
4. Compute C_1 such that

$$\sigma[h(f \triangle C_1)] = \mathrm{Sup}_{C \in [0, M[}\{\sigma[h(f \triangle C)]\} \tag{7}$$

5. Display $f \triangle C_1$

In Fig. 17, we show the result of this method on the same image "Laboratory" of Fig. 16 to make the comparison easier.

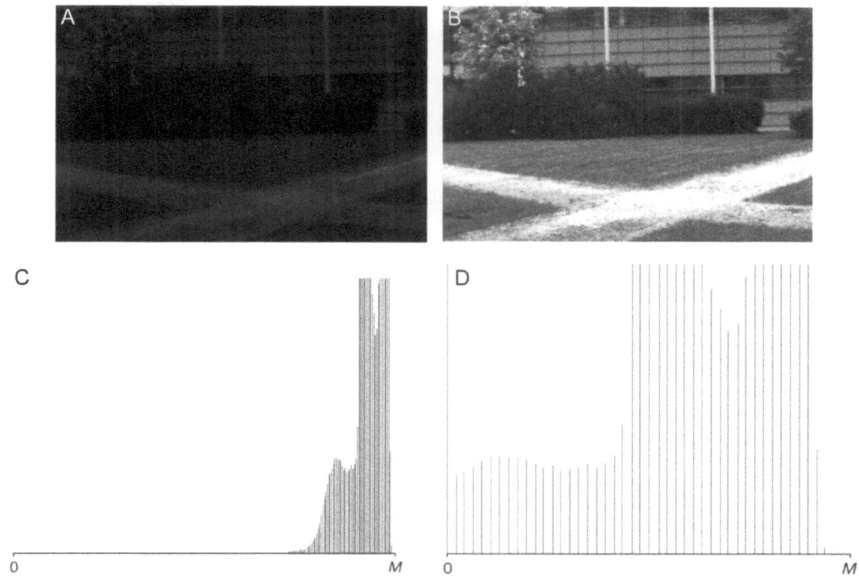

Fig. 17 Maximization of the standard deviation of the histogram of $f\Delta C$ (method $f\Delta C_1$). (A) Initial low-light image f "Laboratory" (exposure time 12 ms). (B) Enhanced image of f according to $f\Delta C_1$, with $C_1 = 213$. (C) and (D) Histograms of (A) and (B).

Remark 2 incites us to consider values of C greater than the low bound $f(b)$, resulting in a function $f\Delta C$ of the space $F(D, -\infty, M)$. In such a way, it becomes possible to apply a full dynamic range expansion to f.

3.2 Full Dynamic Range Expansion Based on Negative Gray Levels

In the previous section, we have seen how the subtraction of a constant C from an image f increases the dynamic range of f. Such a property remains true when the values of C are not limited to the interval $[0, f(b)]$, but are extended to the interval $[0, M[$. In such conditions, the function $f\Delta C$ may present negative gray levels. Its dynamic range is expressed according to

$$\mathrm{DR}(f\Delta C) = \frac{f(a) - f(b)}{1 - \dfrac{C}{M}}$$

If we suppose f is not uniform (constant) on the spatial support D, ie, $f(a) \neq f(b)$, the dynamic range $\mathrm{DR}(f\Delta C)$ increases with C and tends to $+\infty$ if C approaches M because the denominator $1-\frac{C}{M}$ tends to 0. It is then possible to choose C_2 such that:

$$\mathrm{DR}(f \triangle C_2) = M - 1$$

which represents the dynamic range of the whole gray scale. It will remain to apply a translation of $f \triangle C_2$ to superimpose it with the gray scale and then give a full dynamic range to f. The calculus is very simple:

$$\begin{aligned} &\mathrm{DR}(f \triangle C_2) = M - 1 \\ &\Leftrightarrow \frac{f(a) - f(b)}{1 - \frac{C_2}{M}} = M - 1 \\ &\Leftrightarrow C_2 = M\left[1 - \frac{f(a) - f(b)}{M - 1}\right] \end{aligned}$$

The minimum (negative) value of $f \triangle C_2$ being $f(b) \triangle C_2$, the full dynamic range image associated to f is expressed as the translated function of $f \triangle C_2$ by $-f(b) \triangle C_2)$. An example is presented in Fig. 18.

Remark 3. The LIP subtraction associates to a given gray level $f(x)$ the value:

$$f(x) \triangle C = \frac{f(x) - C}{1 - \frac{C}{M}}$$

which can be expressed as an affine function $a \cdot f(x) + b$.

In such conditions, it is surprising that the proposed full expansion corresponds to a classical linear expansion, which consists of applying a bijection of the observed dynamic range $[f(b), f(a)]$ onto the whole gray scale $[0, M-1]$.

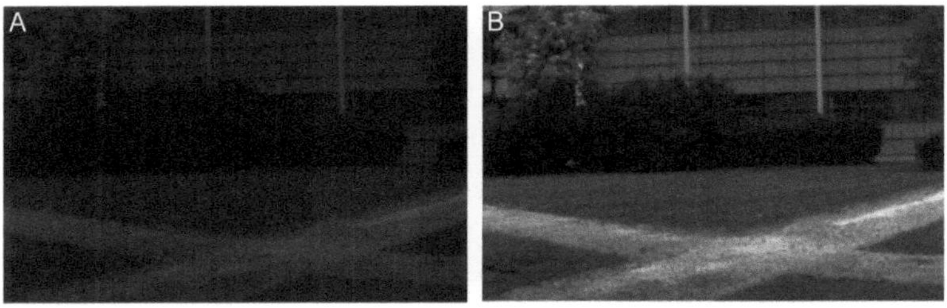

Fig. 18 Full dynamic range expansion of an image based on LIP-subtraction of a constant (case $M = 256$). (A) Initial low-light image f "Laboratory" (exposure time 12 ms). (B) Enhanced image of f according to $(f \triangle C_2) - f(b) \triangle C_2$, with $C_2 = 178.7$.

3.3 What About Color Images?

For this section, one adopts the same reasoning as in Section 2.3 for the LIP scalar multiplication. In particular, for a given color image f, one needs the definitions of the global dynamic range of f, noted GDR(f) (cf. Eq. 5) and the global histogram of f, noted GH_f (cf. Eq. 6).

Case of C_0: We have seen the weak interest for a gray-level image g of the constant C_0 ($C_0 = g(b)$), which produces the image $g \triangle C_0$ of maximal dynamic range remaining in the gray scale. The same approach is possible for the color image f, by replacing $g(b)$, ie, the minimal value of g, with $C_0 = \mathrm{Min}_{x \in D}\{f_R(x), f_G(x), f_B(x)\}$. Then we subtract C_0 from each channel R, G, B, resulting in an enhanced color image $f \triangle C_0$.

An example is given in Fig. 19 with the color initial image "Laboratory." The resulting enhancement is lower than the same method applied to gray-level images, which is not surprising because the definition of C_0 is the minimum value on the three channels.

Case of C_1: To the color image f, we apply the following steps:

1. Compute its global histogram GH_f
2. Compute the logarithmic subtraction $GH_f \triangle C$ (constrained to remain in the gray scale)
3. Compute the standard deviation $\sigma[GH_f \triangle C]$
4. Compute C_1 such that

$$\sigma[GH_f \triangle C_1] = \mathrm{Sup}_{C \in [0, M[}\{\sigma[GH_f \triangle C]\} \tag{8}$$

5. Display the resulting enhanced image $f \triangle C_1$ whose components in the three channels are $f_R \triangle C_1, f_G \triangle C_1, f_B \triangle C_1$

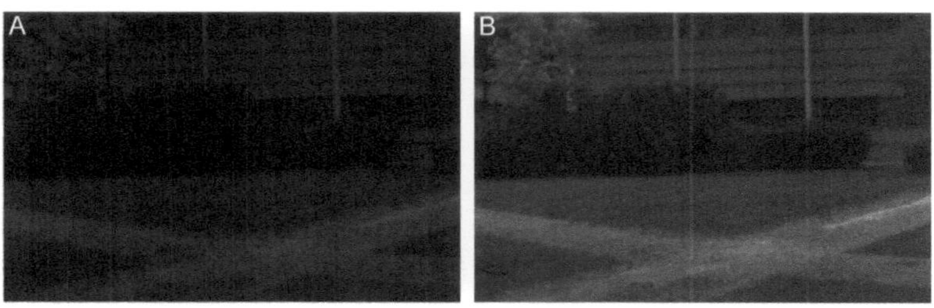

Fig. 19 Enhancement of a color image (method $f \triangle C_0$). (A) Initial low-light image f "Laboratory" (exposure time 12 ms). (B) Enhanced image of f according to $f \triangle C_0$, with $C_0 = 144$. (See the color plate.)

Fig. 20 Enhancement of a color image by maximization of the standard deviation of $GH_f \triangle C$ (method $f \triangle C_1$). (A) Initial color chart f (acquisition time 20 ms). (B) Enhanced image $f \triangle C_1$. (See the color plate.)

An example of this method is proposed in Fig. 20.

Case of C_2: We use the same reasoning as for a gray-level image. The dynamic range of the color image f is now defined by taking into consideration the three channels R, G, B, ie, the global histogram, sum of the histograms associated to each channel:

$$\begin{aligned} DR(f) &= f(a) - f(b) = \text{Max}_{i \in [0, \ldots, 255]} \left[GH_f(i) \right] - \text{Min}_{i \in [0, \ldots, 255]} \left[GH_f(i) \right] \\ &= \text{Max}_{x \in D} \{ f_R(x), f_G(x), f_B(x) \} - \text{Min}_{x \in D} \{ f_R(x), f_G(x), f_B(x) \} \end{aligned}$$

Then the constant C_2 is computed in order to satisfy:

$$\begin{aligned} & DR(f \triangle C_2) = M - 1 \\ \Leftrightarrow & \frac{f(a) - f(b)}{1 - \frac{C_2}{M}} = M - 1 \\ \Leftrightarrow & C_2 = M \left[1 - \frac{f(a) - f(b)}{M - 1} \right] \end{aligned}$$

Finally, the global histogram of f is translated by the value $-f(b) \triangle C_2$, which means that the three channels components of the enhanced image $f \triangle C_2$ are $(f_R \triangle C_2, f_G \triangle C_2, f_B \triangle C_2)$. An example of the method is given in Fig. 21 for the same image "Laboratory" as in Fig. 19 in order to make the comparison easier. The resulting enhancement is not as strong as we could expect, and this remark is confirmed by the comparison of the values of the corresponding constants to subtract: $C_0 = 144$ and $C_2 = 145,6$.

The values are in the same magnitude order, but the techniques are different: in the case of C_0, the initial global histogram is simply translated by $\text{Min}_{x \in D} \{ f_R(x), f_G(x), f_B(x) \}$. If this value is not null, all the gray values of

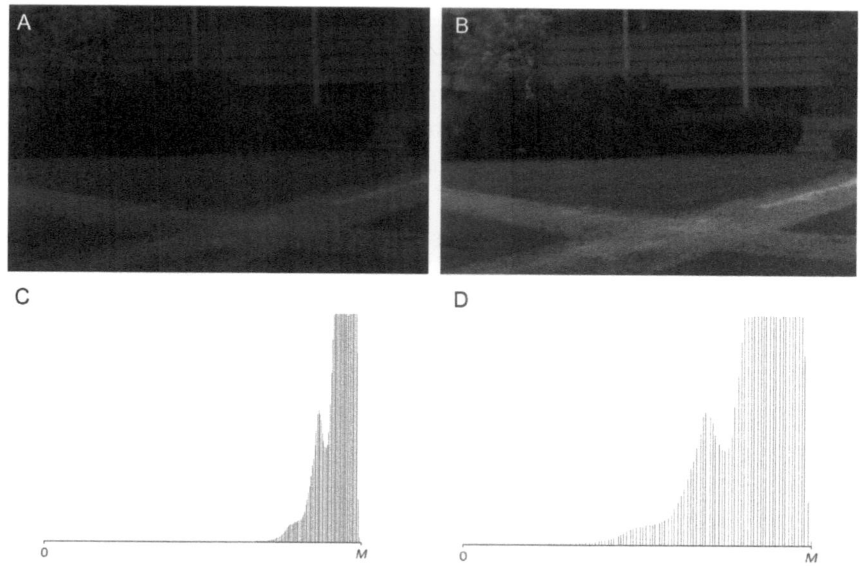

Fig. 21 Enhancement of a color image (method $f \triangle C_2$+translation). (A) Initial low-light image f "Laboratory" (exposure time 12 ms). (B) Enhanced image of f according to the method based on C_2 with $C_2 = 145.6$. (C) and (D) Histograms of (A) and (B). (See the color plate.)

the image are brightened, which is the desired effect. Moreover, the enhanced global histogram does not include any value greater than $\text{Max}_{x \in D}\{f_R(x), f_G(x), f_B(x)\} - \text{Min}_{x \in D}\{f_R(x), f_G(x), f_B(x)\}$ and the initial image is enhanced as a whole.

On the opposite side, the method based on C_2 gives the image a maximal dynamic range, which means that the enhanced image covers the dark part of the scale.

Nevertheless, the use of C_2 can be limited, if necessary, to a dark window of the considered image, as we have done with λ_2 in Fig. 14D.

Fundamental Remark 4. As we will see in Chapter "Ability of the LIP Model to Simulate Variable Acquisition Conditions," the subtraction of a constant from an image permits precise simulation of an increasing exposure time. Thus, applying the method based on C_2 to an image f (or to a subimage $f|_R$ of f restricted to a region R of D), consists of virtually adjusting the exposure time in order that f (or $f|_R$) covers the whole scale (dynamic range equals $M-1$).

Comments and partial conclusion: In the three first sections of this chapter, we have exposed various enhancement techniques, which proved their efficiency when applied to low-light (even very low light) images.

Major advantages of the proposed algorithms

- Since they are developed inside the LIP framework, the algorithms are strongly justified from the optical point of view (based on the transmittance law). Moreover, the LIP model permits processing and analyzing images in compliance with the human eye, due to the consistency of the model with human vision.
- It has been established that the logarithmic addition $\triangle\!\!\!\!+$ and the logarithmic scalar multiplication $\triangle\!\!\!\!\times$ structure the space of images $I(D, [0, M[)$ as the positive cone of the vector space $F(D,]-\infty, M[)$.
- One of the major reasons for the efficiency of our LIP-enhancements resides in the interpretation of negative gray levels as light intensifiers.
- We presented first (Section 2) the algorithms based on the law $\triangle\!\!\!\!\times$, considering historical arguments: the only algorithm that existed previously concerned the dynamic range maximization method $(\lambda_0 \triangle\!\!\!\!\times f)$. Nevertheless, the enhancement algorithms based on the subtraction of a constant possess stronger physical meaning: they perfectly simulate variations of the source intensity.
- In Chapter "Ability of the LIP Model to Simulate Variable Acquisition Conditions" we will produce new arguments in favor of the addition/subtraction of a constant compared with the scalar multiplication: its efficiency in simulating variable exposure times and even variable diaphragm aperture.

4. STABILIZATION AND CENTERING

Let us begin with a simple remark: except in the case of a low dynamic image already located in the middle part of the gray scale, most of existing enhancement algorithms result in a centering effect, which means that the distribution inside the gray scale of the enhanced image histogram is better balanced than that of the initial image. The problem is: what do we mean concretely by "balanced?" In fact, various parameters carry some information about this question: if we consider the three parameters $\lambda_0, \lambda_1, \lambda_2$, the first produces a centering effect as we remarked that the dynamic range of $\lambda \triangle\!\!\!\!\times f$ tends to 0 when λ tends to 0: in such conditions, the optimal value λ_0 locates $\lambda_0 \triangle\!\!\!\!\times f$ somewhere between the initial low-light image and 0. The same argument holds for λ_1, because the standard deviation of the histogram of $\lambda \triangle\!\!\!\!\times f$ also tends to 0 if λ tends to 0. Finally the centering effect of λ_2 is obvious by definition, because the dynamic range of $\lambda_2 \triangle\!\!\!\!\times f$ fills the whole gray scale.

Analogous arguments hold for parameters C_0, C_1, C_2.

That being said, we can imagine other ways to achieve the centering of an image: one of the simplest is proposed later.

4.1 Assigning a Given Value to the Image Mean Gray Level

This technique concerns both logarithmic scalar multiplication and logarithmic addition/subtraction of a constant.

Consider an image f submitted to varying acquisition conditions, such as variable intensity of the source, exposure time, and/or diaphragm aperture. We have already said that such variations will be easily counterbalanced by a LIP addition/subtraction of a constant.

In Fig. 22, an example of stabilization to the gray level 128 is given for an 8-bits gray-level image f acquired under various values of diaphragm aperture. For each of them, the gray-level mean value is computed, say $\mu(f)$. The method consists of determining the constant C such that:

If $\mu(f) > 128$

$$\mu(f) \triangle C = 128 \Leftrightarrow \frac{\mu(f) - C}{1 - \frac{C}{256}} = 128 \Leftrightarrow C = 2 \cdot (\mu(f) - 128)$$

If $\mu(f) < 128$

$$\mu(f) \mathbin{\blacktriangle} C = 128 \Leftrightarrow \mu(f) + C - \frac{\mu(f) \cdot C}{256} = 128 \Leftrightarrow C = 2 \cdot (128 - \mu(f))$$

Note that there is no difficulty in processing images at the camera speed of 25 images per second (see "Comments on Fig. 23" for the case of color images).

In Fig. 23, an example is given for 3×8 bits color images f_i acquired at variable exposure times. To get both a stabilization and centering effect on these images, we propose to add (resp. subtract) in the LIP sense a constant C to (resp. from) their global histogram GH_{f_i} according to:

If $\mu(GH_{f_i}) > 128$

$$\mu(GH_{f_i}) \triangle C = 128 \Leftrightarrow C = 2 \cdot (\mu(GH_{f_i}) - 128)$$

Then C is subtracted to each channel of f_i.

If $\mu(GH_{f_i}) < 128$

$$\mu(GH_{f_i}) \mathbin{\blacktriangle} C = 128 \Leftrightarrow C = 2 \cdot (128 - \mu(GH_{f_i}))$$

Then C is added to each channel of f_i.

Fig. 22 Examples of real-time stabilization. On the *left*, images acquired with decreasing diaphragm aperture. On the *right*, corrected images with a mean gray-level value = 128.

Comments on Fig. 23

1. The proposed stabilization is very efficient for dark images and slightly worse for bright ones (image on the left).
2. We already announced that the LIP subtraction of a constant will be an excellent tool to simulate exposure time variations, even for color

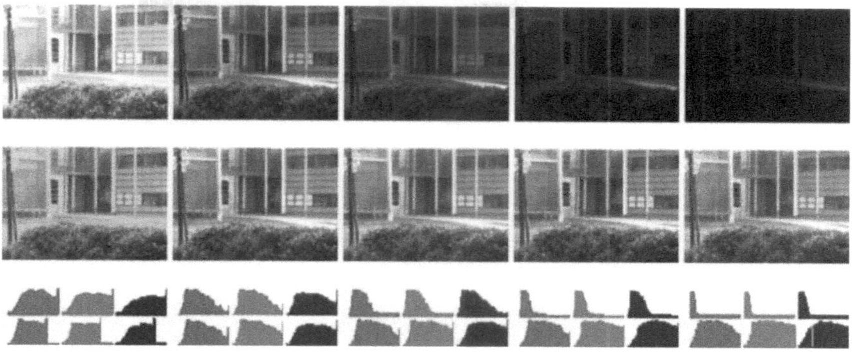

Fig. 23 First line: images acquired under variable aperture. Second line: corrected images. Third line: corresponding histograms. (See the color plate.)

Fig. 24 Brightness enhancement of a low-light image. (See the color plate.)

images. This point refers to Chapter "Ability of the LIP Model to Simulate Variable Acquisition Conditions," where one will study the quality of resulting images, for example, in terms of "distance" between a simulated image and the true one.

3. There is no issues in performing such stabilization at the camera speed (25 images per second) for high-resolution (1280×800 pixels) color images with a simple personal computer.

Remark 5. Assigning a given value to the image's mean gray level is also possible with logarithmic scalar multiplication. In fact, it consists of computing the scalar λ such that $\lambda \triangle \text{mean}(f) = 128$.

This simple method is also applicable to color images and especially to the luminance of such images. An example is given in Fig. 24, where the luminance mean of a low-light color image is centered at 128.

5. APPLICATIONS

This section is dedicated to some applications not protected by confidentiality agreements. The given examples have been selected among various industrial or biomedical studies on which we successfully tested the LIP tools. We hope that their diversity and their tangible interest will convince the reader of their effectiveness.

5.1 Confocal Microscopy

In collaboration with L'Oreal Company, a modified tandem confocal scanning microscope (Fig. 25) was developed and used in the early 1990s to obtain in vivo images of the human skin in real time (Corcuff, Bertrand, & Lévêque, 1993; Corcuff & Lévêque, 1993), especially during the movement of blood cells in the capillaries. This apparatus, working with a Nipkow disk, allowed the observation of living specimens, preserving the natural state of the tissue. The focus was performed by a vertically sliding gauge piloted by a stepping motor, giving optical slices parallel to the surface of the skin with an axial and lateral resolution of approximately one micron. This TSM enabled the whole thickness of the epidermis (approximately 200 μm) to be acquired in less than 5 s.

An automated method for three-dimensional (3D) image reconstruction and visualization has been proposed (Bertrand, 1994; Bertrand & Corcuff, 1994). It includes a Fourier-filtering based algorithm for removing stripped

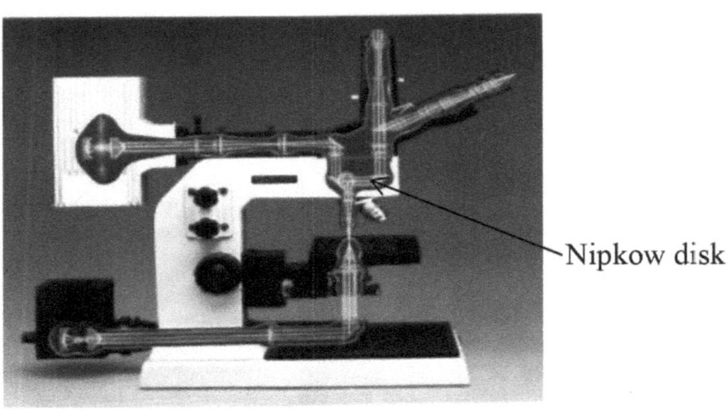

Fig. 25 A confocal TSM showing light trajectory. (See the color plate.)

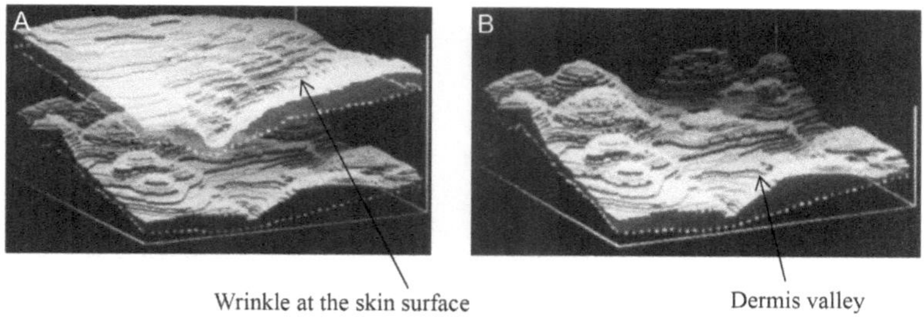

Fig. 26 (A) 3D imaging of the stratum corneum (upper epidermal layer). The two bright surfaces represent the upper and lower limits of the stratum corneum. (B) The dermis upper surface. (See the color plate.)

pattern effect caused by the Nipkow disk and image restoration algorithms: nonlinear enhancement and illumination correction by means of the LIP tools. At this time, the enhancement was made with the logarithmic scalar multiplication by λ_0.

The image acquisition described previously required the use of a light-intensified camera (sensitivity 10^{-5} lux). Moreover, the greater the focalization depth, the greater the needed light enhancement (role of LIP model). Then 3D measurements can be easily performed, such as the thickness distribution of the stratum corneum (Fig. 26).

Comment on Fig. 26: From this study, it appears that the wrinkles at the skin surface correspond to valleys at the dermis surface.

With the same company (L'Oreal), we realized a 3D reconstruction of a human hair (cf. Corcuff, Gremillet, et al., 1993). Internal structures of a human hair (cortex, medulla) were classically studied by means of serial sections. Replacing physical cuts by optical sectioning preserves the integrity of the sample; even though it is limited by the opacity of the specimen, light absorbing structures, surface reflections, and refraction/scattering. In our study, we used a confocal laser scanning microscope (CLSM): the Molecular Dynamics Sarastro instrument. The fluorescent hairs were glued to microscope slides coated with double-sided adhesive tape. Between 40 and 100 optical sections were stored depending on the diameter of the hair shaft, for complete volume reconstruction (cf. Fig. 27).

To conclude this section focused on confocal microscopy, let us present an application developed in collaboration with Morris Goldner (Université Laval) concerning the *Bacteroides fragilis bacteria*, sometimes responsible for peritonitis (de Oliveira Ferreira et al., 2011).

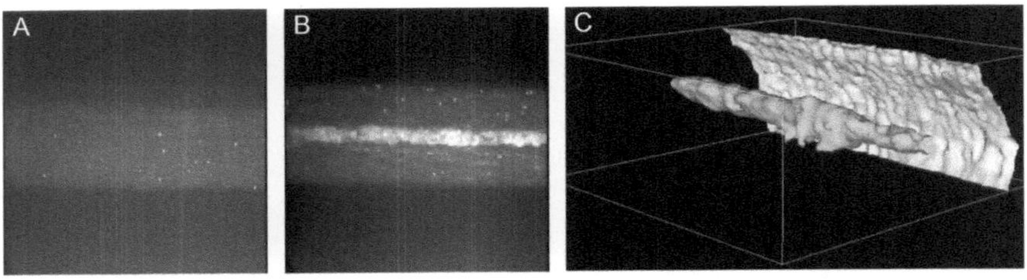

Fig. 27 (A) An optical cut of the hair. (B) A cut crossing the medulla. (C) 3D reconstruction: here the medulla and the hair surface. (See the color plate.)

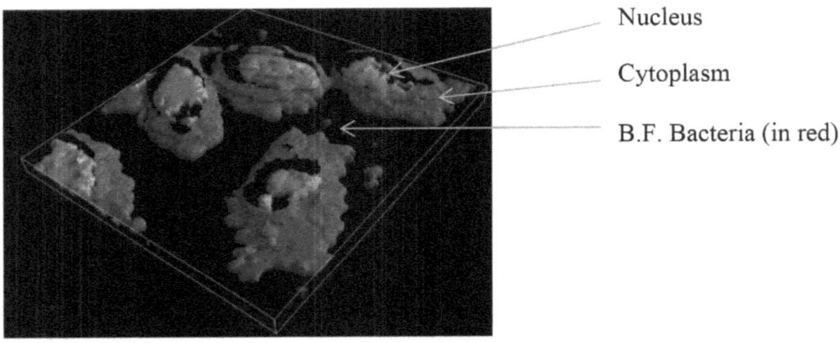

Fig. 28 3D reconstruction of Hela cells and *Bacteroides fragilis* bacteria. (See the color plate.)

The question was: do *B. fragilis* bacteria penetrate Hela cells? The question has been answered positively with CLSM acquisitions (cf. Fig. 28).

Abrupt displacement of *B. fragilis* from an intestinal-reducing to a peritoneal-oxidizing environment can result in peritonitis. This clinical observation suggested that *B. fragilis* would respond to changes in oxidation–reduction (redox) conditions. Previous attempts to clarify this observation based on redox levels, designated by Eh in millivolts (mV) adjusted to pH 7 (Eh_7), have revealed a potential pathogenic characteristic. 3D CLSM indicated *B. fragilis* penetration into Hela cells. This redox effect occurred for the bacteria poised for growth under oxidizing, but not reducing conditions. When the *B. fragilis* bacteria were cultured under oxidizing conditions, examination by scanning electron microscopy had shown that the bacteria were widely dispersed, but under reducing conditions they were densely aggregated. Using appropriate software which applied a bespoke algorithm that separated the Hela cell image into regions of nucleus and cytoplasm, CLSM examination allowed the necessary subtle information to be extracted, thus shaping and limiting the cytoplasmic area for treatment of consecutive laser cuts, which enabled this image to be reassembled in

3D (Stiles & Birch, 1959). Then, it could be demonstrated that *B. fragilis* bacteria grown under oxidizing conditions were located inside the Hela cell.

5.2 Mosaïque Software (NT2I Company)

A drone carries the acquisition system (Fig. 29) in order to inspect civil engineering structures, such as dams, bridges, coolers of nuclear power stations (Fig. 30), canals, etc. In the case of a dam, for example, the drone moves vertically and acquires images partially overlapping, allowing us to correlate them and thus produce a continuous film of the inspected strip. Images are

Fig. 29 Image of the drone in action. (See the color plate.)

Fig. 30 Coolers of a nuclear power station. (See the color plate.)

Fig. 31 The initial recomposition (on the *left*) and recomposition with corrected images (on the *right*). (See the color plate.)

stored during the acquisition and processed later (cf. Patent no 07 58302, Pascal, Jourlin, & Moreaud, 2007).

The goal of such inspection is the detection of various defects, like cracks. The main difficulties of this control are due to lighting variations, the drone instability, and the presence of shadowed areas.

Several brightness corrections are performed on untreated images: first, they are stabilized at a fixed mean (Fig. 31). Second, a correction is applied on the intersection of two consecutive images: the images means are taken into account only in their intersection; then the correction is applied on the wholeness of one of the two images, resulting in a perfect continuity of the brightness (Fig. 32).

Once the corrections are applied, the extraction of cracks, calcite traces, and other disturbances becomes easier (see fig. 45 in section 4 of Chapter "Transfer of Classical Tools to LIP Context" for an example of cracks detection). Finally, in Mosaïque software, the detected defects are reported on the structure plan, which facilitates the restoration, when necessary.

5.3 Braille Alphabet

This project was headed by the Centre Technique du Papier (CTP, Grenoble, France). The aim was to control Braille texts embossed on drugs' cardboard boxes in order to indicate the name of the drug and the dosage to blind persons.

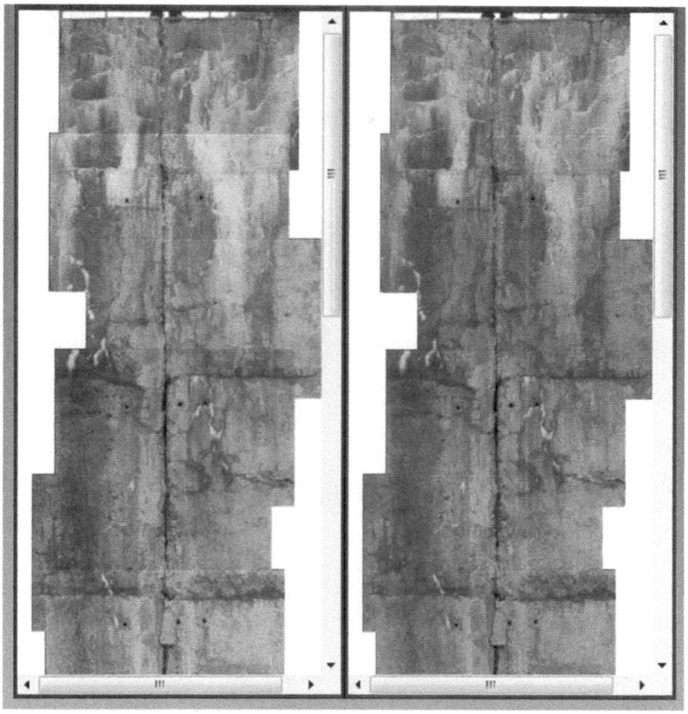

Fig. 32 Correction on successive images (on the *right*, the transitions between successive images are less visible). (See the color plate.)

The system conceived by the CTP was based on laser triangulation: a fixed thin laser line is projected on the flat cardboards (before packaging), which move on a conveyor. A slanted CCD camera acquires this laser line in order to reveal its deformation when it crosses embossed regions of the box. Then the set of deformations allows for reading of the Braille text and finally controlling it.

The main problem is that there are no specifications describing the embossing constraints. Thus, it may be applied on a matte dark background or a bright white one. Such imprecise conditions result in images of the laser line showing extremely variable intensities (we observed variations in the ratio 1/100) and the camera acquires over-lighted lines or imperceptible ones.

Using the LIP tools, we performed an enhancement of the dark segments of the laser line (cf. Fig. 33).

Remark 6. In the LIP Top-Hat version (Jourlin & Montard, 1998) from Fig. 33, the height of the Top-Hat is estimated in the sense of a logarithmic

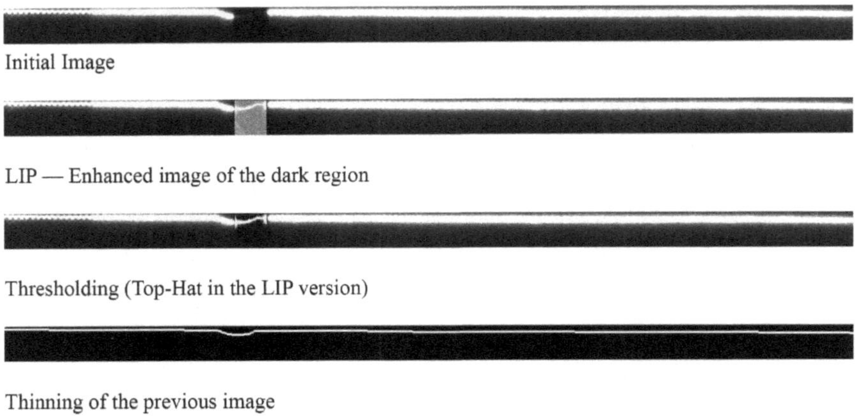

Fig. 33 Detection of an embossed region located in a dark part of the box.

additive contrast, which allows extracting the peaks and crest lines independently of the background intensity.

5.4 Controlled Nanostructures Formation by Ultrafast Laser Pulses for Color Marking

This application does not use LIP enhancement or stabilization. The LIP model was involved in this project because of its consistency with the human visual system.

The initial idea is from E. Audouard who proposed it to the ATT (Advanced Track and Trace) Company as an anticounterfeiting tool (cf. Patent no 2009/090324, Dusser, Audouard, Jourlin, Soder, & Foucou, 2009).

In fact, it is well known that ultrafast laser irradiations generate a precise nanostructuration at the surface of certain materials observable as pseudo-periodic ripples (cf. Sipe, Young, Preston, & van Driel, 1983; Young, Sipe, & van Driel, 1984). Various examples of ripples aspect are shown in Figs. 34 and 35, for beam power from 25 to 100 mW, corresponding to pulse fluence from 0.4 to 1.6 J cm^{-2} and for each beam power, the obtained structures for scan speed from 10 to 130 mm s^{-1}.

What is important to note is that the ripples orientation depends on the laser polarization.

Color effect due to the diffraction of light by periodic structures on metallic surfaces has already been evidenced (Lochbihler, 2009; Vorobyev & Guo, 2008). Beyond these results, it has been demonstrated how to control such properties in order to transfer information on a metallic surface using this specific effect (cf. Dusser et al., 2010).

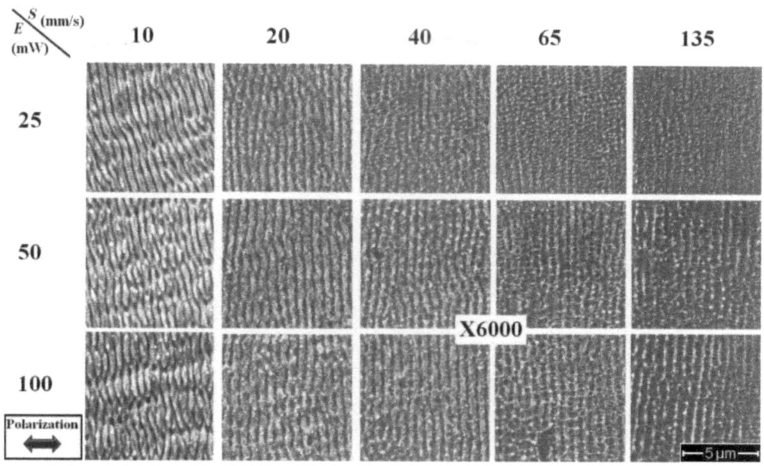

Fig. 34 SEM X6000 images of ripples obtained with the "optical way 2" femtosecond laser chain (150 fs, 800 nm, 5 kHz, achromat doublet lenses: $f=50.8$ mm) on stainless steel 316L. The different ripples aspects are obtained with a beam power from 25 to 100 mW (fluence from 0.4 to 1.6 J cm^{-2}) for rotating mirrors speed from 20 to 130 mm s^{-1}.

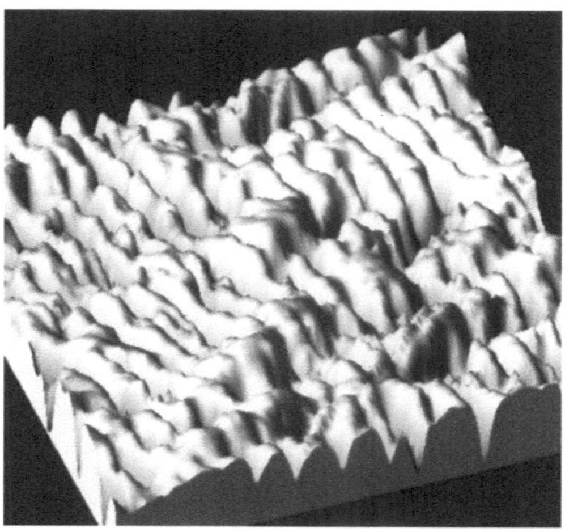

Fig. 35 Ripples 3D view. (See the color plate.)

Fig. 36 shows the colors we are able to obtain by scanning (1200 dpi) a stainless steel sample (316L) marked by nine lines of 19 squares obtained with the "optical way 1" femtosecond laser chain ($P=25$ mW, $F=0.4$ J cm^{-2}). Each square has been marked with varying ripples orientations: 5 by 5 degree from 0 to 90 degree.

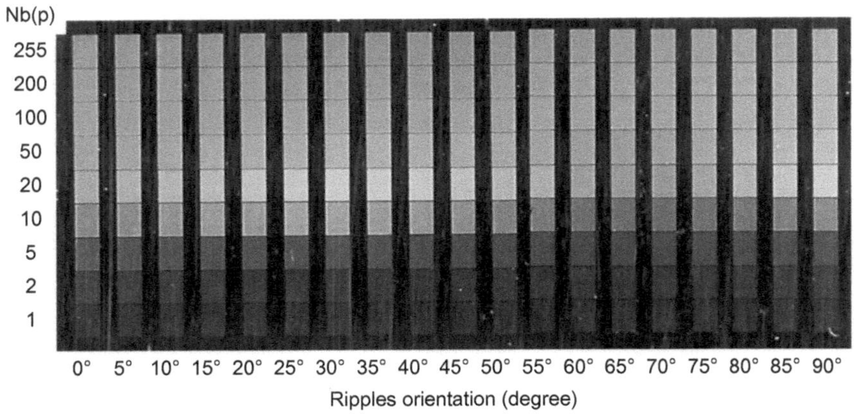

Fig. 36 Colors resulting of light diffraction due to ripples of various orientations. (See the color plate.)

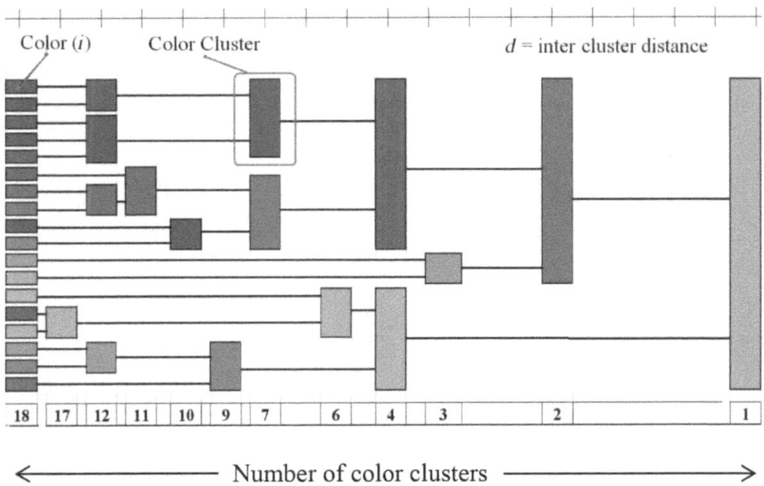

Fig. 37 HAC dendrogram obtained with ripples color marking process (18 ripples orientations). (See the color plate.)

An important question remains: how many reliable colors can be obtained by the method? In our approach, it means the number of colors a human eye or a sensor can recognize without fail. To answer that question, we performed a hierarchical ascendant classification (HAC) (see section 1.5 of Chapter "Transfer of Classical Tools to LIP Context") either in a hue, saturation, value (HSV) color space for its compatibility with human color perception or, for the same reason, using a color LIP metric in RGB space. From the HAC dendrogram (Fig. 37), we observed that the levels corresponding to six and nine classes are the most significant.

Fig. 38 Examples of images from "Asterix" and "Tintin." On the *right*, the real size (here 17 mm). (See the color plate.)

***Remark* 7.** The number of efficient colors cited in this chapter can seem very small. Nevertheless, it is widely sufficient for the application we aim for. In fact, the color mark we apply on an object to authenticate it must possess some properties: it is of very small size and it offers a high combinatory level: with six colors and a 10×10 mark, we obtain 6^{100} different matrices.

To conclude this application on a diverting example of realization, let us present some results obtained by B. Dusser: the reproduction of several cartoons images (Fig. 38).

6. LOCAL LIGHTING CORRECTION

We will conclude this chapter with a significant improvement of the lighting correction we previously proposed. This section refers mainly to the work of Maxime Carré, in particular his PhD thesis (Carré, 2013) and in a paper we published in the Proceedings of the 14th ICSIA (Carré & Jourlin, 2015). In fact, in Section 5, we presented a very fast and simple lighting correction of an image f consisting of assigning a fixed mean value to $f \triangle C$, for example, knowing that it represents a time exposure simulation. The problem is that technique has been presented and illustrated in a global approach: the correction parameter is the same for the whole image. In plenty of situations we need local correction, particularly when the studied image presents a high dynamic range: in such cases, the exposure time of the sensor must be increased to observe a dark region in good conditions and decreased to observe bright regions without saturation.

M. Carré proposed several solutions in Carré (2013) and our goal here is not to explain them in detail. We simply recall the basic approach: the correction applied to each pixel corresponds to a LIP subtraction by a constant computed from the local mean and the reference mean. The idea is first to build an averaging filter. Several approaches are possible: initially, we compute in each pixel x a gray-level value representing the mean on a neighborhood $N(x)$ of arbitrary size, knowing that this mean may be weighted according to the distance to x of each pixel y of $N(x)$. Whatever the chosen averaging filter, let us adopt the following notations: f is the studied image, $A(f)$ the image produced by the averaging filter, A_{ref} the average reference value and f_{res} the resulting image we will obtain. The steps of the proposed algorithm are as follows:

1. Computation, in each pixel x of D, of the constant C_x:

$$\forall x \in D, C_x = A(f)(x) \triangle\!\!\!\!- A_{\text{ref}}$$

2. Application of the LIP subtraction of C_x from $f(x)$:

$$\forall x \in D, f_{\text{res}}(x) = f(x) \triangle\!\!\!\!- C_x$$

Remark 8. The local gray-level mean permits estimating the luminosity, in the chosen neighborhood, for each pixel x of D. Step two performs the adaptation of this luminosity so that it becomes the same in each pixel.

An example of industrial control is given in Fig. 39 where the observed object is a metallic half-cylinder that is impossible to lighten uniformly. The proposed correction considerably facilitates the detection of possible defects.

Such basic correction has been improved by M. Carré: for natural scenes, where it is not possible to model the lighting, the previous simple correction is not always efficient, so he proposed a novel filter and a parameter adjusting the correction strength (Carré, 2013). We present in Fig. 40 the result he obtained on a night image, compared with a global correction and a basic one.

Comment on Fig. 40: The considered initial image is very difficult to process, due to the presence of very bright zones (street lamps) inside a very dark image. The global correction by LIP subtraction of a constant (image B) produces a result of good visual quality, but the saturated zones are enlarged. The basic local correction (image C) is not well adapted here. On the contrary, the improved local correction (image D) realizes a very good compromise between the dark and bright regions. To confirm that, we propose (Fig. 41) a magnification (approximately five times) of the part of image (D) located near the entrance of the laboratory.

Comment on Fig. 41: The balance between dark and bright zones is well done. We observe the presence of a person (probably a smoker with the right hand near the mouth).

Fig. 39 Examples of local lighting correction. (A) Initial image: object without defect. (B) Correction of (A) to get a uniform lighting. (C) Initial image: object with defects. (D) Correction of (C) to get a uniform lighting.

Fig. 40 Comparison between global and local lighting corrections. (A) Initial night image "Laboratory." (B) Global correction of (A): LIP subtraction of a constant. (C) Basic local correction of (A). (D) Improved local correction of (A).

Remark 9. The previously improved local brightness correction also applies to color images (in fact to their luminance). M. Carré tested it in various situations (two examples in Fig. 42).

M. Carré remarks in (Carré, 2013) that the proposed algorithms can be performed at the camera speed with the use of a graphic board.

To conclude this section, let us focus on the similarity between the lighting corrections exposed previously and the tone mapping algorithms. M. Carré's method is similar to that proposed by Reinhard (Reinhard, Stark, Shirley, & Ferwerda, 2002) in both reasoning and results.

Fig. 41 Window extracted from image 40D (Magnification five times).

Fig. 42 Examples of local lighting correction on color images. (See the color plate.)

7. COMPARISON OF LIP ENHANCEMENT WITH EXISTING METHODS

We refer to the beginning of Section 1 where we argued about the great number of algorithms dedicated to enhancement of images contrast, dynamic range, and so on.

Most of these methods are based on look-up tables (Log-transformations, power-law transformations, Gamma correction, etc.) or histogram modifications (see, for example, Hummel, 1977; Kim, 1997; Wang, Chen, & Zhang, 1999). The main drawback common to all these techniques is the lack of scientific justification. On the contrary, in the LIP approach we have presented, the laws $\triangle\!\!\!\!\!+$ and $\triangle\!\!\!\!\!\times$ possess a physical interpretation and the LIP model itself has been demonstrated to be consistent with human vision.

In such conditions, we estimate that it is not really fair to propose comparisons. Nevertheless, we decided to conduct some experiments (realized by M. Carré) for our own information, which are summarized here.

7.1 Material and Method

(a) Tested enhancement techniques. Three methods have been selected:
- Hardware approach based on camera gain
- Classical approach based on histogram equalization
- LIP enhancement

(b) Experimentation conditions
- Night scene with street lamps ("Laboratory" image): 20–40 lux under the street lamp, 1.5 lux in the dark zones
- Extreme conditions: "Conference Room" image without window, door partially open on a low-lighted hall (<0.07 lux)

The light intensity was measured with an Illuminance Meter T-10 (Konica Minolta)

(c) Cameras:
- Basler TXG08 (Color sensor 1/3 in.)
- Basler TXG14 (Monochrome sensor 2/3 in.)

The night scene "Laboratory" has been acquired with both cameras.

The "Conference Room" has been acquired with the Monochrome camera.

Images have been acquired under different exposure times, from 30 to 200 ms.

7.2 Results

Night scene (Laboratory) acquired with monochrome camera

The results are presented in Fig. 43.

Night scene (Laboratory) acquired with color camera

The results are presented in Fig. 44.

Extreme conditions (Conference Room)

The results are presented in Fig. 45.

Comments on these results

- The camera gain and the LIP Model produce similar results until a certain limit: the camera gain is bounded by a maximal value, contrary to the LIP subtraction of a constant, which is able to brighten an image until saturation and is then able to enhance very low-light images. Moreover, applying a gain requires more time than a software correction, which processes the image before displaying it.

Fig. 43 Gray-level image "Laboratory": Comparison of enhancement methods. (A) Initial image "Laboratory" acquired with minimal gain. (B) Image "Laboratory" with maximal gain. (C) Histogram equalization applied to (A). (D) LIP subtraction of a constant $C=230$ applied to (A). (E) LIP subtraction of a constant $C=245$ applied to (A). (F) LIP local correction.

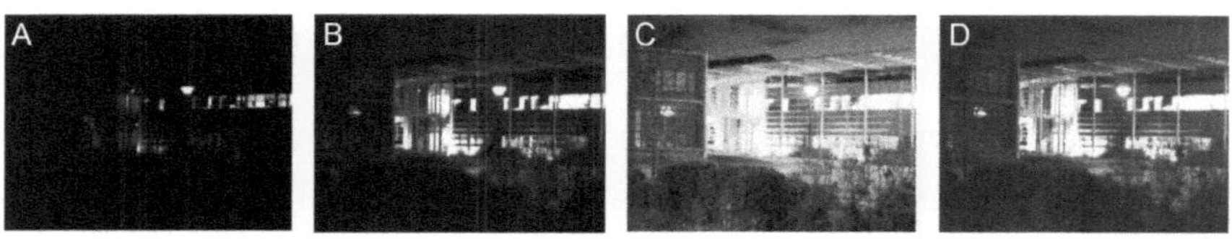

Fig. 44 Color image "Laboratory:" Comparison of enhancement methods. (A) Color image "Laboratory": gain = 1, exposure = 100 ms. (B) Color image "Laboratory": gain = 10, exposure = 100 ms. (C) Histogram equalization applied to image (A). (D) LIP subtraction of a constant $C=245$ applied to (A). (See the color plate.)

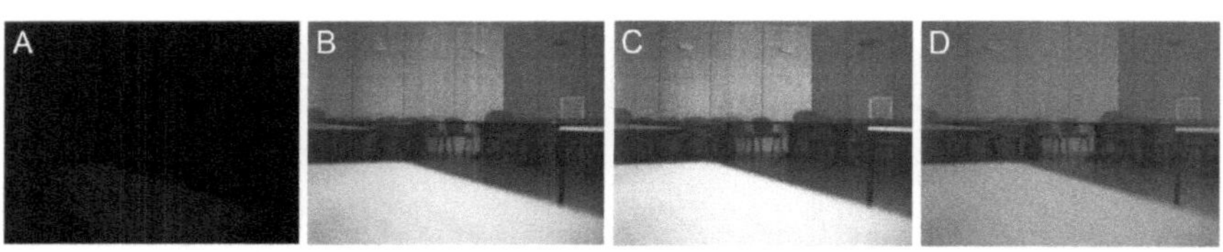

Fig. 45 Gray-level image "Conference Room": Comparison of enhancement methods. (A) Initial image "Conference Room": gain = 10, exposure = 200 ms. (B) Histogram equalization applied to image (A). (C) LIP subtraction of a constant applied to (A). (D) LIP Local correction applied to (A).

- The result of histogram equalization is unique, while the LIP subtraction proposes an adjustable solution.
- The LIP model takes into account the logarithmic character of human vision and then produces images more realistic than gain or histogram equalization.
- The LIP local correction is very promising and requires further developments.

What remains to be done

- The comparison with other enhancement methods must be completed: we have only made it with basic techniques.
- We have seen that in extreme conditions (very low light) our algorithms produce noisy enhanced images. It would be interesting to apply efficient noise filtering algorithms as preprocessing on the studied images.

8. CONCLUSION AND PERSPECTIVES

As we said at the beginning of this chapter, this part of our LIP developments is a key one, due in particular to the use of negative gray levels as light intensifiers. Moreover, this chapter shows clearly the real interest of the vector space structure of the over-space $(F(D,]-\infty, M[, \triangle\!\!\!\!\!+\,, \triangle\!\!\!\!\!\times\,)$. In fact, the physical justifications we have associated to the logarithmic laws $\triangle\!\!\!\!\!+$ and $\triangle\!\!\!\!\!\times$ in the case of images (elements of $I(D, [0, M[)$) find a natural extension to functions of $F(D,]-\infty, M[)$: for example, the addition/subtraction of a constant is interpretable as an exposure time variation or as a source intensity variation, negative gray levels corresponding to sources whose intensity is higher than the reference one (value 0 of $[0, M[$). In the same way, the scalar multiplication $(\lambda \triangle\!\!\!\!\!\times\,)$ is easily interpretable as a *"thickness"* variation of the real (or virtual in case of images acquired in reflection) obstacle generating the studied image. It is thus obvious that a decreasing of λ corresponds to a brightening of the image, until $\lambda = 0$ (image uniformly "white" in the sense of the reference source intensity). Here we highlight the role of negative gray levels: when λ becomes negative, the obstacle no more attenuates the photons emitted by the source: it produces light.

At each step of our presentation, we have illustrated our results by examples (a good image is more convincing that a long speech).

Currently, we are focusing on multispectral LIP and have already obtained some encouraging results in cosmetic applications, such as the evaluation of skin appearance with or without make-up, duration of make-up efficiency, etc. (cf. Corvo, Jourlin, Lépine, Breugnot, & Bordes, 2015).

We defined and tested a multispectral Asplünd's metric based on the LIP scalar multiplication.

Other perspectives are opened by performing images enhancement, stabilization, centering, etc., not only by means of the laws $\mathbin{\triangle\!\!\!\!+}$ and $\mathbin{\triangle\!\!\!\!\times}$ separately, but by combining them. In fact, we have seen that the behaviors of such laws are not similar, so that it would be interesting to mix them to build better algorithms.

REFERENCES

Bertrand, C. (1994). *Développement d'une nouvelle méthode d'imagerie cutanée in-vivo par microscopie confocale tandem.* PhD thesis, France: Saint-Etienne University.

Bertrand, C., & Corcuff, P. (1994). *In-vivo* spatio-temporal visualization of the human skin by real-time confocal microscopy. *Scanning, 16,* 150–154.

Carré, M. (2013). *Interprétation d'images acquises en situation de faible éclairage ou d'éclairage variable.* PhD Thesis, Saint-Etienne, France: Université Jean Monnet. 152 pp.

Carré, M., & Jourlin, M. (2015). Brightness spatial stabilization in the LIP framework. In *14th International congress for stereology and image analysis, Liège (Belgium).* 4 pp.

Corcuff, P., Bertrand, C., & Lévêque, J.-L. (1993). Morphometry of human epidermis by real-time confocal microscopy. *Archives of Dermatological Research, 285,* 475–481.

Corcuff, P., Gremillet, P., Jourlin, M., Duvault, Y., Leroy, F., & Lévêque, J.-L. (1993). 3D reconstruction of human hair by confocal microscopy. *Journal of the Society of Cosmetic Chemists, 44,* 1–12.

Corcuff, P., & Lévêque, J.-L. (1993). *In vivo* vision of the human skin with the tandem scanning microscope. *Dermatology, 186,* 50–54.

Corvo, J., Jourlin, M., Lépine, T., Breugnot, J., & Bordes, S. (2015). A new LIP framework metric applied to analysis of multi-spectral acquisition of in-vivo skin. In *Proceedings of 14th ICSIA, Liège (Belgium).* 3 pp.

de Oliveira Ferreira, E., Yates, E., Jourlin, M., Liu, J., Pilotto Domingues, R., & Goldner, M. (2011). Anaerobe/aerobe environmental flux determines protein expression profiles of *Bacteroides fragilis,* a redox pathogen. *Anaerobe, 17*(1), 4–14.

Dusser, B., Audouard, E., Jourlin, M., Soder, H., Foucou, A. (2009). "Method and device for marking a surface using controlled periodic nanostructures," Patent, WO/2009/090324.

Dusser, B., Sagan, Z., Soder, H., Faure, N., Colombier, J. P., Jourlin, M., et al. (2010). Controlled nanostructrures formation by ultra fast laser pulses for color marking. *Optics Express, 18*(3), 2913–2924.

Hummel, R. (1977). Image enhancement by histogram transformation. *Computer Graphics and Image Processing, 6*(2), 184–195.

Jourlin, M., Breugnot, J., Itthirad, F., Bouabdellah, M., & Closs, B. (2011). Logarithmic image processing for color images. *Advances in Imaging and Electron Physics, 168,* 65–107.

Jourlin, M., & Montard, N. (1998). A logarithmic version of the TOP-HAT transform in connexion with the Asplünd distance. *Acta Stereologica, 16*(3), 201–208.

Jourlin, M., & Pinoli, J.-C. (1995). Image dynamic range enhancement and stabilization in the context of the logarithmic image processing model. *Signal Processing, 41*(2), 225–237.

Kim, Y. T. (1997). Contrast enhancement using brightness preserving bi-histogram equalization. *IEEE Transactions on Consumer Electronics, 43*(1), 1–8.

Lochbihler, H. (2009). Colored images generated by metallic sub-wavelength gratings. *Optics Express, 17,* 12189–12196.

Maini, R., & Aggarwal, H. (2010). A comprehensive review of image enhancement techniques. *Journal of Computing, 2*(3), 8–13.

Pascal, R., Jourlin, M., Moreaud, M. (2007). "Procédé de detection de désordres dans un ouvrage en béton ou en maçonnerie", Patent no 07 58302.

Reinhard, E., Stark, M., Shirley, P., & Ferwerda, J. (2002). Photographic tone reproduction for digital images. *ACM Transactions on Graphics*, *21*(3), 267–276.

Sipe, J. E., Young, J. F., Preston, J. S., & van Driel, H. M. (1983). Laser-induced periodic surface structure. I. Theory. *Physical Review B*, *27*(2), 1141–1154.

Stiles, W. S., & Birch, J. M. (1959). N.P.L. colour-matching investigation: Final report. *Optica Acta*, *6*(1), 1–26.

Vorobyev, A. Y., & Guo, C. (2008). Colorizing metals with femtosecond laser pulses. *Applied Physics Letters*, *92*, 30.

Wang, Y., Chen, Q., & Zhang, B. (1999). Image enhancement based on equal area dualistic sub-image histogram equalization method. *IEEE Transactions on Consumer Electronics*, *45*(1), 68–75.

Young, J. F., Sipe, J. E., & van Driel, H. M. (1984). Laser-induced periodic surface structure. III. Fluence regimes, the role of feedback, and details of the induced topography in germanium. *Physical Review B*, *30*, 2001–2015.

CHAPTER FIVE

Ability of the LIP Model to Simulate Variable Acquisition Conditions

Contents

One of the major problems we encounter at the image acquisition step is linked to the instability of the acquisition conditions:

Lighting: Concerning images acquired in natural conditions (daylight), it is obvious that the lighting of the scene varies: for example, the scenes acquired along highways to perform incident automated detection are subject to this problem. We can also refer to the inspection of the civil engineering structures (see section 5.2 of Chapter "Dynamic Range Expansion, Night Vision. Stabilization, Centering. Industrial, and Biomedical Applications," Mosaïque Software). Some sensors try to counterbalance lighting variations, for example, by an automated diaphragm aperture: unfortunately, this solution is difficult to apply in real time. On the other hand, if we consider software solutions, there are very few image processing tools insensitive to lighting variations. Thus, it remains to imagine postacquisition algorithms aiming at such stabilization, following if possible the camera speed. Throughout the

Advances in Imaging and Electron Physics, Volume 195
ISSN 1076-5670
http://dx.doi.org/10.1016/bs.aiep.2016.04.005

Chapter "Dynamic Range Expansion, Night Vision. Stabilization, Centering. Industrial, and Biomedical Applications," we announced the capability of our enhancement algorithms to work in real time. When the acquisition is performed with a source whose intensity is adjustable (optical microscopy, for example), the problem seems insignificant. Nevertheless, when images are acquired during a long period, in order to build a database, we generally do not have an absolute certainty that the source intensity remains constant throughout the experimentation.

Sensor parameters: Here we principally refer to exposure time and diaphragm aperture.

This chapter is devoted to the possible answers that logarithmic image processing (LIP) tools can bring to these problems.

1. LIGHTING VARIATIONS: SOURCE INTENSITY VARIATIONS

This section constitutes an important part of this chapter, which is why it is cited in the first place. Nevertheless, Chapter "Dynamic Range Expansion, Night Vision. Stabilization, Centering. Industrial, and Biomedical Applications" widely addressed this subject, particularly under the stabilization and centering aspects. For this reason, we will not rediscuss this problem.

2. EXPOSURE TIME VARIATION

The "exposure time" parameter is generally the first that a photographer will modify and adjust according to the lighting of the scene. Considering the importance of this subject and the efficiency of answers provided by the LIP tools, we will discuss it in detail in order to take account of gray-level images as well as color ones.

2.1 Gray-Level Images

In this section, we propose a simulation of increasing the exposure time by subtracting, in the LIP sense, a constant C (representing a homogeneous image) from the studied image f.

For a given source S and a gray-level image f, we have already explained that, in every situation (transmitted or reflected light) it is possible to create a virtual semitransparent obstacle composed of a homogeneous medium such that this obstacle would generate f in transmission. In such conditions, remember that the LIP-subtraction $f \triangle C$ of a constant from f is understandable:

- As the subtraction from the virtual obstacle generating f of a slice of constant thickness or
- As a modification of the source intensity

This last point incited us to experiment with the LIP subtraction as a simulator of changing exposure time (cf. Carré & Jourlin, 2014). In order to show this, the process we propose is (illustrated in Fig. 1):

- Acquisition of a scene at a given exposure time (10 ms in our experimentation) resulting in a dark image f
- Acquisition of the same scene at another exposure time (here: 100 ms) resulting in an image g brighter than f
- Compute the average gray level of f, say $\bar{f}$, and the average gray level of g, say $\bar{g}$
- Knowing that $\bar{f} > \bar{g}$ in the LIP scale, compute the constant C such that $\bar{f} \triangle C = \bar{g}$
- Display $f \triangle C$ and compare it with g

Comment on Fig. 1: The first question we can ask about this result concerns the quality of the simulation. From a visual point of view, the similarity

Fig. 1 Ability of LIP subtraction to simulate exposure time changing. (A) Image f: "Laboratory" acquired with exposure time = 10 ms. (B) Image g: "Laboratory" acquired with exposure time = 100 ms. (C) Starting from (A), simulation of an exposure time of 100 ms.

between g and its simulated image is satisfying, but this evaluation is subjective. In order to make the comparison objective, it is possible to compute the gray-level difference between g and $f \triangle C$ at each pixel $x \in D$ and display it in absolute value. In fact, such a display is not significant, the differences being too small. It remains to compute some global parameters associated to the set of differences:

- The average difference between g and $f \triangle C$ equals 3.3 gray levels
- The standard deviation equals 3.6

The previous results prove the good quality of the method, especially if we consider the large disparity between the two exposure times.

Real time aspects: The method consists of

- Reading a first time the image to compute its average gray-level value
- Computing the constant to be subtracted
- Apply the subtraction.

Note that this subtraction becomes an addition if the initial image is brighter than the target. The previous steps are very simple and fast to perform, which means the method easily applies in real time: we obtained a computation time less than 4 ms for a high-resolution image (1384 × 1036 pixels) with a processor Intel Core i7 (2.8 GHz). Moreover, it is possible to speed up the algorithm by means of a subsampling of considered pixels.

Due to this very short execution time, I decided to consider the same problem for color images.

2.2 Color Images

In order to evaluate the method in good conditions, we selected a color chart as an initial image and acquired it under variable exposure times with a Baumer camera controlled via software development kit (SDK). The range of exposure times is deliberately large to test the quality of simulation images. The initial images are presented in Fig. 2.

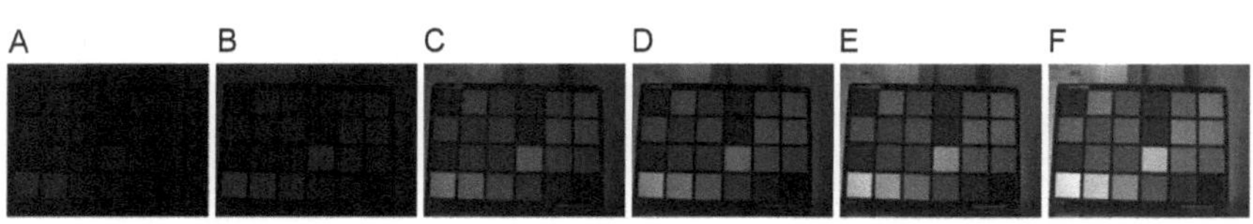

Fig. 2 Color chart acquired at various exposure times: f_3(3 ms), f_6(6 ms), f_{18}(18 ms), f_{24}(33 ms), f_{45}(45 ms). (A) Image f_3: acquired at 3 ms. (B) Image f_6: 6 ms. (C) Image f_{18}: 18 ms. (D) Image f_{24}: 24 ms. (E) Image f_{33}: 33 ms. (F) Image f_{45}: 45 ms. (See the color plate.)

2.2.1 Simulation without Reference

The goal of this section is very ambitious: we aim to simulate exposure time variations for a low-light color image, in real time and without reference.

By "without reference," we mean that, contrary to the example given for gray-level images (Fig. 1), we do not dispose of *two* images corresponding to *two* exposure times. In fact, the question we ask is:

> *Given a color image f acquired at an initial exposure time, is it possible to simulate how f is modified for another exposure time?*

To illustrate this reasoning, we start with the images of the color chart displayed in Fig. 2. We choose one of them as the target, for example, the image f_{33} acquired at 33 ms. This choice is motivated by the fact that we want to test the method in priority to brighten dark images and we can simultaneously keep an image (f_{45}) to test the opposite situation. We are going to simulate this acquisition time, starting from each of the other images $f_3, f_6, f_{18}, f_{24}, f_{45}$, with no a priori knowledge, ie, without using f_{33}, except for a final comparison.

To achieve this goal, some steps will be necessary:

1. We associate to each image $f_3, f_6, f_{18}, f_{24}, f_{45}$, with the enhanced image according to the method expressed in section 3.3 of Chapter "Dynamic Range Expansion, Night Vision. Stabilization, Centering. Industrial, and Biomedical Applications" (subtraction of C_1 according to the formula illustrated in Fig. 20 of this chapter):

$$\sigma\left[\mathrm{GH}_f \triangle C_1\right] = \mathrm{Sup}_{C \in [0, M[}\left\{\sigma\left[\mathrm{GH}_f \triangle C\right]\right\}$$

   The resulting enhanced images are presented in Fig. 3.

   Note that this step permits the association of a stable image to each considered initial exposure time, roughly independent of it. Our objective here is not to evaluate the similarity of these enhanced images. We will perform the comparison between simulated final images.

2. After noting $C_1(f_3)$, $C_1(f_6)$, $C_1(f_{18})$, $C_1(f_{24})$, and $C_1(f_{45})$ the constants associated, respectively, to f_3, f_6, f_{18}, f_{24}, and f_{45} to get the enhanced images of Fig. 3, we decided to study how these constants vary according to

Fig. 3 Enhanced images associated to $f_3, f_6, f_{18}, f_{24}, f_{45}$, from left to right. (See the color plate.)

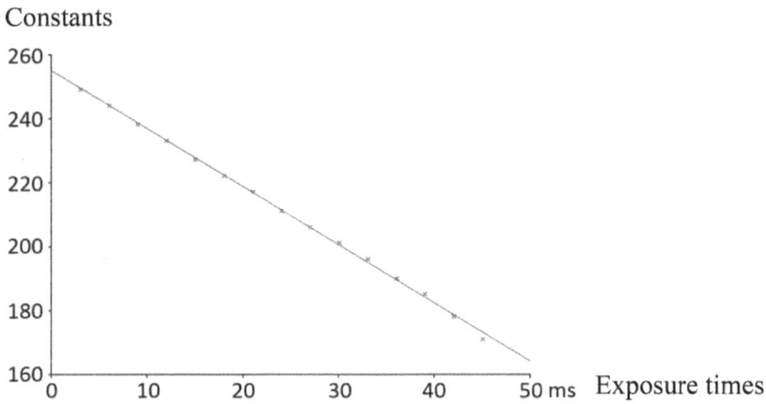

Fig. 4 Proof of the linear relation between the exposure time and the constant subtracted from the acquired image to get the enhanced one.

the corresponding exposure times. Based on a graphic representation, it appeared that the relation was a linear one, as we expected. To strengthen this first observation, we decided to realize more acquisitions of the color chart, namely 15 acquisitions with exposure times from 3 to 45 ms (3, 6, 9, 12, 15, 18, 21, 24, 27, 30, 33, 36, 39, 42, and 45 ms). After computing the corresponding constants according to the formula used in step 1, we confirmed the linearity of the relation (cf. Fig. 4).

3. Given a color image f_{t_1} acquired at the exposure time t_1, this linear relation will permit computing the image f_{t_2} of the same scene we would obtain with an exposure time t_2. In fact, to determine the straight line of Fig. 4, we need two points: the first one is given by $(t_1, C_1(f_{t_1}))$. For the second one, we remark that the straight line intersects the y-axis at the value 256. This is not surprising: if the exposure time tends to 0, the resulting image tends to a black one, corresponding to the limit value of each channel R, G, B. In such a situation, the global histogram is concentrated near M and the constant to be subtracted to get the enhanced histogram tends also to M.

In such conditions, the equation of the straight line is:

$$y = \frac{C_1(f_{t_1}) - 256}{t_1} \cdot x + 256$$

And the constant $C_2(f_{t_2})$ corresponds to the value t_2 of x:

$$C_2(f_{t_2}) = \frac{C_1(f_{t_1}) - 256}{t_1} \cdot t_2 + 256$$

To conclude this exposure time simulation, remember that the enhanced images were similar (Fig. 3), which implies that:

$$\begin{aligned} f_{t_1} \mathbin{\triangle\!\!\!\!\!-} C_1(f_{t_1}) &= f_{t_2} \mathbin{\triangle\!\!\!\!\!-} C_1(f_{t_2}) \\ f_{t_2} &= [f_{t_1} \mathbin{\triangle\!\!\!\!\!-} C_1(f_{t_1})] \mathbin{\triangle\!\!\!\!\!+} C_1(f_{t_2}) \\ f_{t_2} &= f_{t_1} \mathbin{\triangle\!\!\!\!\!-} [C_1(f_{t_1}) \mathbin{\triangle\!\!\!\!\!-} C_1(f_{t_2})] \end{aligned} \quad (1)$$

From Eq. (1), it is possible, considering a scene acquired at an exposure time t_1, to simulate the same scene for another exposure time t_2 by the LIP subtraction of a constant, this constant being positive if the simulated image is brighter $(t_2 > t_1)$ and negative otherwise.

To illustrate this method, let us return to the images of Fig. 2 and consider the image acquired at 33 ms as the target. For each other image $(f_3, f_6, f_{18}, f_{24}, f_{45})$ we compute the new image simulated at 33 ms. The result is presented in Fig. 5.

Comment on Fig. 5: The visual quality appears satisfactory, even for exposure times very different from the target (3 ms, for example). Now it remains to evaluate the quality of such a simulation with objective tools. This will be the subject of Section 2.2.2.

Remark 1. In plenty of situations, we do not know the exposure time of the initial image f, thus the previous method is not available. Nevertheless, it remains possible to subtract (resp. add) from (resp. to) f the same constant to its three components f_R, f_G, f_B in order to simulate larger (resp. shorter) exposure times.

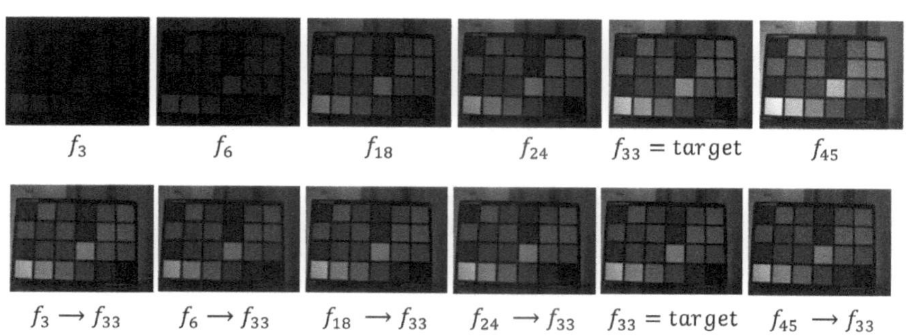

Fig. 5 Simulations of a target, starting from initial images with various exposure times. *First line*: Initial images and the target. *Second line*: Simulations results. (See the color plate.)

Fig. 6 The four colors K_1, K_2, K_3, and K_4 selected in f_{33} after enhancement with C_1-subtraction. (See the color plate.)

2.2.2 Quality Evaluation

To evaluate the simulation quality, we decided to use a metric tool, ie, to compute the distance between the result and the target. Among a large possible choice of metrics, we finally selected the Euclidean one.

We will now consider four colors K_1, K_2, K_3, and K_4 corresponding to four boxes of the color chart acquired at 33 ms. The choice has been made in order to ensure these colors are sufficiently different to not favor a particular channel. In Fig. 6, these four colors are represented after an enhancement by C_1-subtraction (as in Fig. 3).

For each image of the color chart acquired at 3, 6, 18, 24, and 45 ms, we compute the simulated image at 33 ms, namely, $f_3 \longrightarrow f_{33}, f_6 \longrightarrow f_{33}, f_{18} \longrightarrow f_{33}, f_{24} \longrightarrow f_{33}$, and$f_{45} \longrightarrow f_{33}$ and then we compute the Euclidean distance between K_1 and its simulated color in each of the simulated images. The calculus is done on the average color inside each box to avoid problems due to noise.

The results are presented in Table 1.

Comment on Table 1: The four first columns correspond to the four considered colors. The lines represent the simulated images and in each box are given the three R, G, B components of the simulated colors K_1, K_2, K_3, and K_4. Such values must be compared with those associated to the original image (acquired at 33 ms), in bold characters in the table. This comparison is made based on the Euclidean distance (last column).

We remark that these Euclidean distances are generally very small, with a slight increase for very short exposure times (3 ms and to a lesser degree 6 ms).

Section 2.2.2 is devoted to an interesting application of exposure time simulation.

2.2.3 Acquisition of Moving Objects

In various situations and particularly in the field of industrial vision dedicated to control and defects detection, we are confronted with the acquisition of moving objects. This is for example the case for conveyors, rolling mills, traffic control, and also when the sensor is moving: drones, planes, cars equipped with cameras, etc.

Table 1 Euclidean Distances Between Original Colors Acquired at 33 ms and Their Simulations from 3 to 33 ms, 6 to 33 ms, 18 to 33 ms, 24 to 33 ms, and 45 to 33 ms

	K_1	K_2	K_3	K_4	Euclidean Distances
3–33 ms	[19.3;12.7;57.5]	[86.3;36.9;57.0]	[125.2;76.9;9.54]	[90.3;35.1;12.0]	13.14
6–33 ms	[20.9;17.4;56.1]	[89.9;43.6;59.2]	[132.0;86.3;6.61]	[94.7;42.9;7.5]	6.00
18–33 ms	[17.9;19.3;55.0]	[88.0;45.2;59.6]	[132.3;91.3;1.07]	[94.8;45.7;1.2]	1.05
24–33 ms	[17.6;19.3;54.3]	[87.4;45.2;59.0]	[131.6;91.8;1.01]	[93.9;45.8;1.2]	0.96
33 ms	**[17.6;19.9;54.9]**	**[87.5;45.9;59.4]**	**[131.4;91.8;0.04]**	**[94.3;46.5;0.6]**	**0.00**
45–33 ms	[17.4;20.0;55.4]	[89.7;47.0;60.6]	[133.7;93.2;0.01]	[96.1;47.6;0.1]	2.06

The problem to overcome is:

$$\begin{bmatrix} \text{We must find a good compromise between an acquisition time} \\ \text{too large resulting in a motion blur and a very short exposure} \\ \text{time resulting in a dark image.} \end{bmatrix} \quad (2)$$

Motion blur has been the subject of a multitude of papers for a long time, under various points of view:
- Estimating motion blur parameters with the goal of blur removal
- Estimating motion blur parameters with the goal of introducing motion blur inside synthetized images to give them a realistic perception of moving objects,
- Removing blur based on hardware solutions

Let us refer to some of them.
- Potmesil and Chakravarty (1983) described a procedure for modeling motion blur in computer-generated images. The goal was to understand what the characteristics of motion blur are in order to simulate it in synthetic images. Such a simulation is used in static images to give the illusion of speed and is based on a set of system-transfer functions, which are derived from the path and velocity of objects in the scene and the exposure time of a camera.

  Some papers limit their purpose to *linear motion blur*, for example:
- Tanaka, Yoneji, and Okutomi (2007) proposed a method to identify motion blur parameters in linearly blurred images. The method is based on the detection of regular patterns of zeros in the Fourier spectrum of the image degraded by linear motion blur. The authors insist on the robustness of their method against noise.
- Moghaddam and Jamzad (2007) dealt with restoration of motion-blurred images. They presented a novel algorithm estimating two motion blur parameters: direction and length. The first is estimated by Radon transform and the second by means of the bispectrum modeling—thus combining spatial and frequency domains. The paper shows the robustness of the proposed technique for noisy images.

Other publications focusing on "which motion may be recovered from blurred images":

Dai and Wu (2008) exhibit a linear mathematical constraint which applies locally to pixels and permits the solving of difficult problems, such as estimating global affine motion blur, estimating global rotational motion blur, estimating and segmenting multiple motion blur, and estimating nonparametric motion blur field. The proposed approach concerns both blur estimation and image deblurring and can be used for real images or synthetized ones.

Finally, it seems interesting to refer to a hardware solution for motion blur removal: proposed by Cho, Levin, Durand, and Freeman (2010), the technique is based on the capture of the scene with a parabolic motion in two orthogonal directions.

Now let us come back to the problem (2) resulting in real situations of industrial vision and mentioned at the beginning of Section 2.2.2.

In such cases, our exposure time simulation seems to provide an efficient solution.

In fact, it permits us to select a short exposure time to avoid blur and then simulate a larger exposure time to enhance the image. Let us give an example of this approach, presented in the *14th International Congress for Stereology and Image Analysis* held in Liège (Belgium) in July 2015 (Deshayes, Guilbert, & Jourlin, 2015) under the title: "How simulating exposure time variations in the LIP Model. Application: moving objects acquisition."

A reprint of the Congress label is glued on a rotating wheel (3.75 revolutions per second). In Fig. 7A, the wheel is stopped and an acquisition is

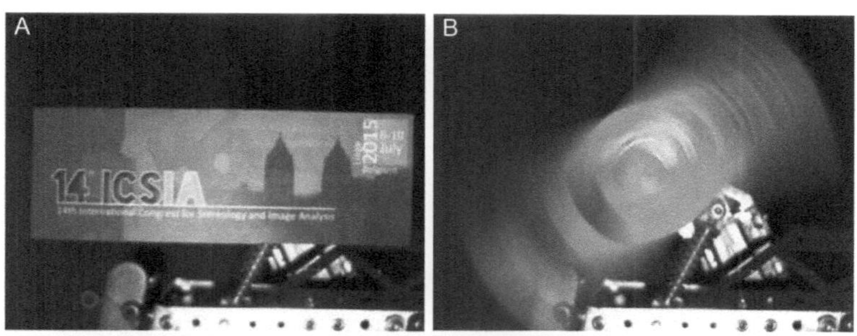

Fig. 7 Acquisition of a moving object through exposure time simulation. (A) Congress label acquired at 30 ms (motionless wheel). (B) Congress label acquired at 30 ms (wheel rotating at 3.75 revolutions per second). (C) Rotating label acquired at 2 ms. (D) Simulation of an exposure time of 30 ms starting from image (C). (See the color plate.)

done with a 30-ms exposure time, which results in a "reference" image. When the wheel rotates, the same exposure time of 30 ms produces a strongly blurred image (Fig. 7B). A drastic reduction of the exposure time at 2 ms gives a very dark image (Fig. 7C). Finally, the image 7C is enhanced by an exposure time simulation of 30 ms, and we get a "readable" image of the rotating label (Fig. 7D).

Comment on Fig. 7: The experimentation conditions were not easy. In fact, the rotation speed corresponds to an instantaneous speed of 1.8 m per second at the external boarder of the label (diameter ~15 cm). Thus the efficiency of the method is established. Nevertheless, the quality of image 7D is better at the center part than at the external one. The final quality of the simulated image greatly depends on the sensor quality, and particularly on the sensor sensitivity in the dark part of the channels.

With the same camera, we have tested a "limit" case with a rotation speed of 6.4 revolutions per second and a reduction to 1 ms of the exposure time. We present the results of this experimentation in Fig. 8. The reference image 8A represents a motionless label acquired at 30 ms. The second

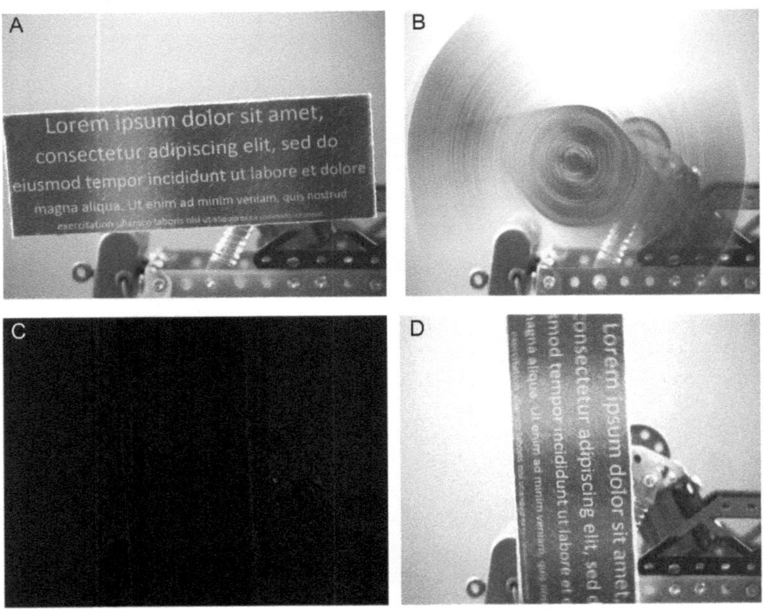

Fig. 8 Acquisition of a moving object through exposure time simulation. (A) Label acquired at 30 ms (motionless wheel). (B) Label acquired at 30 ms (wheel rotating at 6.4 revolutions per second). (C) Rotating label acquired at 1 ms. (D) Simulation of an exposure time of 30 ms starting from image (C). (See the color plate.)

image 8B corresponds to the same exposure time for the rotating label. We can compute the angle described by the label during this acquisition time:

$$\frac{(6.4)\cdot(360\text{ degree})}{s}\cdot\frac{30}{1000}s\sim 70\text{ degree}$$

This rotation produces a strong motion blur.

To remove this motion blur, we made a fast acquisition of the rotating label (exposure time 1 ms) resulting in a very low-light image visualized in Fig. 8C.

Finally, a simulation of exposure time (30 ms) is applied to 8C, producing a "readable" image (Fig. 8D).

3. VARIABLE DIAPHRAGM APERTURE

The promising results obtained for simulating exposure time variations encouraged us to test the ability of LIP subtraction to simulate diaphragm aperture changing.

Remark 2. Before testing this approach, we must point out the limitations of such a simulation. In fact, modifying diaphragm aperture implies not only brightness changing, but also has an effect on other characteristics of the acquired image. For example, the depth of field is transformed: in fact a large aperture such as $\frac{1}{2.8}$ results in a small depth of field. Moreover, large apertures generate a *vignetting* effect (corners of the image darker than the center) and very small ones produce diffraction, implying blur and lower image quality.

In such conditions, we do not expect a simulation regulating all these modifications. We know that the LIP subtraction will mainly act on the image brightness. Nevertheless, it seemed interesting to perform a first experimentation, because in case of lighting changes, it is easier to manually modify diaphragm aperture than program an exposure time modification.

Thus we acquired nine images of the color chart with apertures from $\frac{1}{22}$ to $\frac{1}{1}$, namely: $\frac{1}{22}, \frac{1}{16}, \frac{1}{11}, \frac{1}{8}, \frac{1}{5.6}, \frac{1}{4}, \frac{1}{2.8}, \frac{1}{2}, \frac{1}{1}$ (cf. Fig. 9).

Comment on Fig. 9: In order to test the limits of the method, the proposed experience takes into account extreme values of diaphragm aperture, in particular very small apertures like 1 and 2 producing very low-light images for

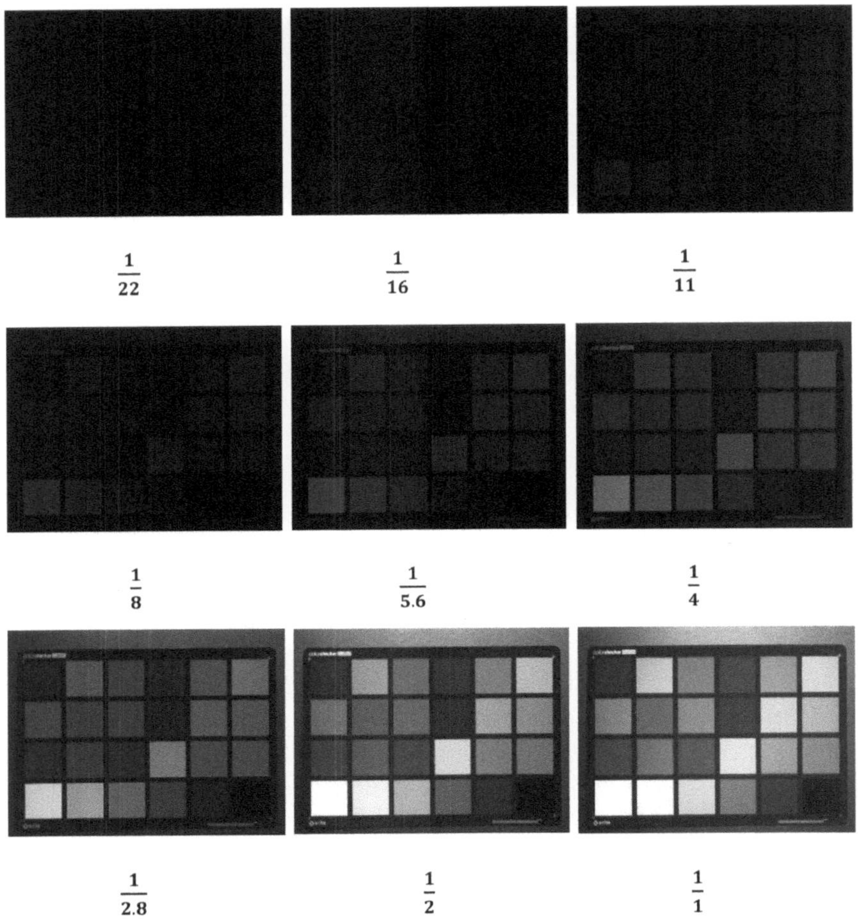

Fig. 9 Color chart acquired with diaphragm aperture varying from $\frac{1}{22}$ to $\frac{1}{1}$. (See the color plate.)

which we cannot expect enhanced images of high quality, knowing that the noise and the signal will be enhanced in the same proportions.

To each image f of Fig. 9, we decided to apply a LIP subtraction of a constant C_1 in order to maximize the standard deviation of the histogram of $f \triangle C$. It means that C_1 is computed according to:

$$\sigma\left[\mathrm{GH}_f \triangle C_1\right] = \mathrm{Sup}_{C\in[0,M[}\left\{\sigma\left[\mathrm{GH}_f \triangle C\right]\right\}$$

where GH_f represents the global histogram of image f, ie, the sum of the red, green, and blue histograms (this method has been explained in section 3.3 of

Chapter "Dynamic Range Expansion, Night Vision. Stabilization, Centering. Industrial, and Biomedical Applications": "Case of C_1").

Let us recall that this method permitted us to perform a stabilization of images acquired under variable exposure times (Fig. 3). Fig. 10 shows the enhanced images of Fig. 9. For each of them, the diaphragm aperture and the value of C_1 are recalled.

Remark 3. As expected, the images of Fig. 10 corresponding to very small diaphragm apertures appear noisy.

Nevertheless, we carry on the method similarly to what has been done for variable exposure time and we represent (Fig. 11) the link between the diaphragm aperture (x-axis) and the constant C_1 to be subtracted (y-axis).

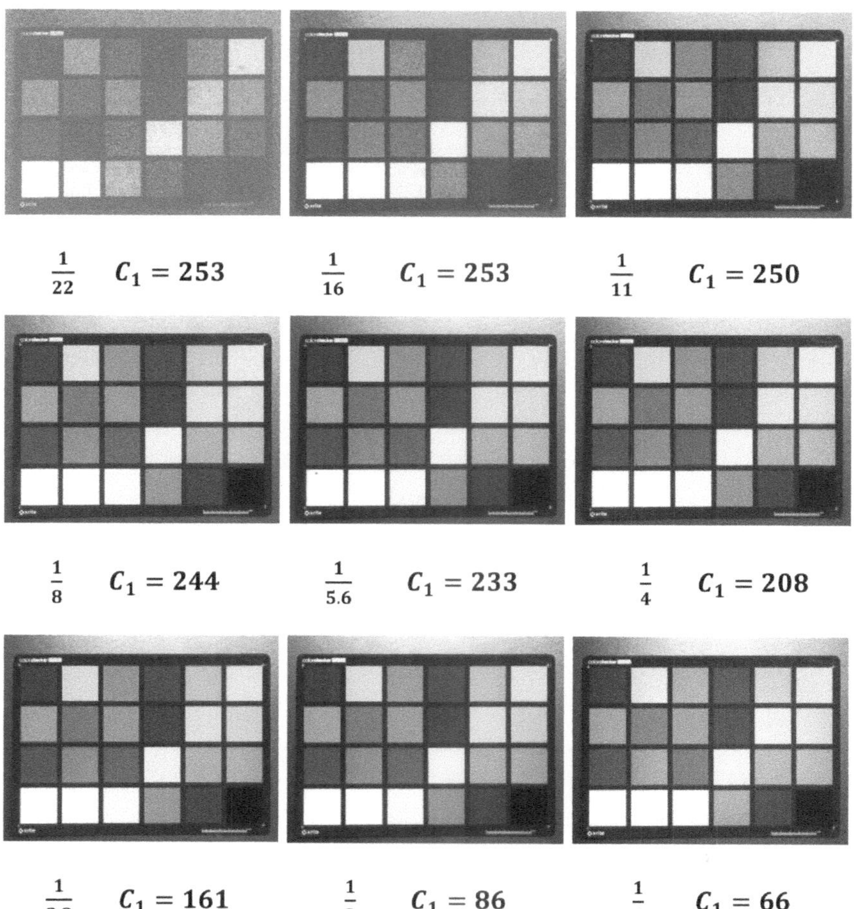

Fig. 10 Enhancement of initial images presented in Fig. 9. (See the color plate.)

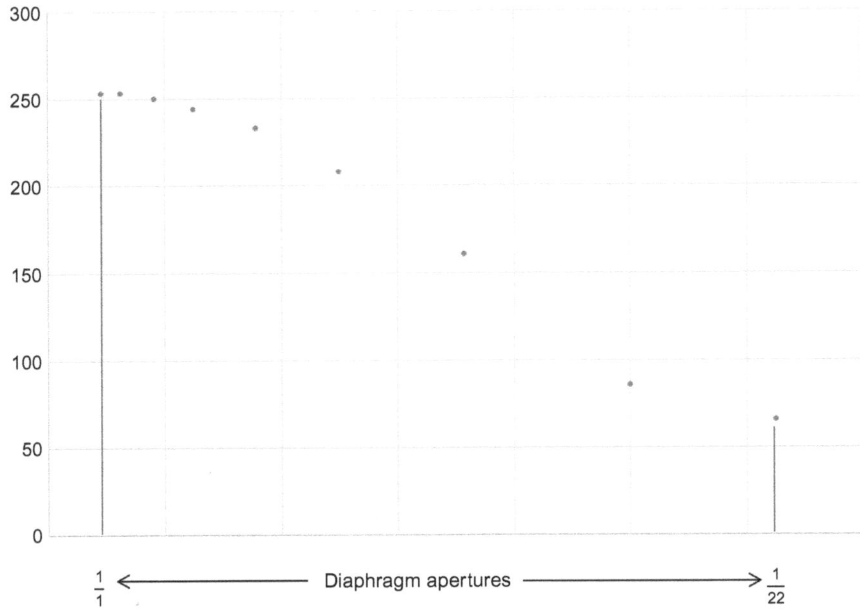

Fig. 11 Relation between diaphragm aperture and constant C_1.

Comment on Fig. 11: It appears clearly that a link exists, but some steps remain to be done:

1. Propose a curve fitting the experimental points. A first observation shows that the answer is not a straight line as for exposure time variations. If we exclude the two extreme points, we remark that a parabola might be suitable. This eventuality seems realistic if we consider that the light received by the sensor is proportional to r^2 where r represents the radius of the diaphragm aperture.
2. Once step 1 realized, one can proceed as for the exposure time simulation: considering an image acquired with a given diaphragm aperture, it becomes possible to simulate the image we would obtain with another aperture.

4. OTHER SIMULATIONS

4.1 Long Exposure Time

This technique is common in photography when we aim to capture a very low-light scene. It consists of increasing the exposure time to get a sufficient lighting of the scene.

We propose here:

- To perform several acquisitions with a shorter exposure time
- To compute the average image of the previous ones
- To apply a logarithmic subtraction of a constant (cf. Section 2) in order to get a brightness similar to a longer exposure time, in fact the sum of exposure times used for the initial acquisitions

An illustration of this method is given in Fig. 12 where we aim to simulate an exposure time of 50 ms, knowing 10 images acquired at 5 ms.

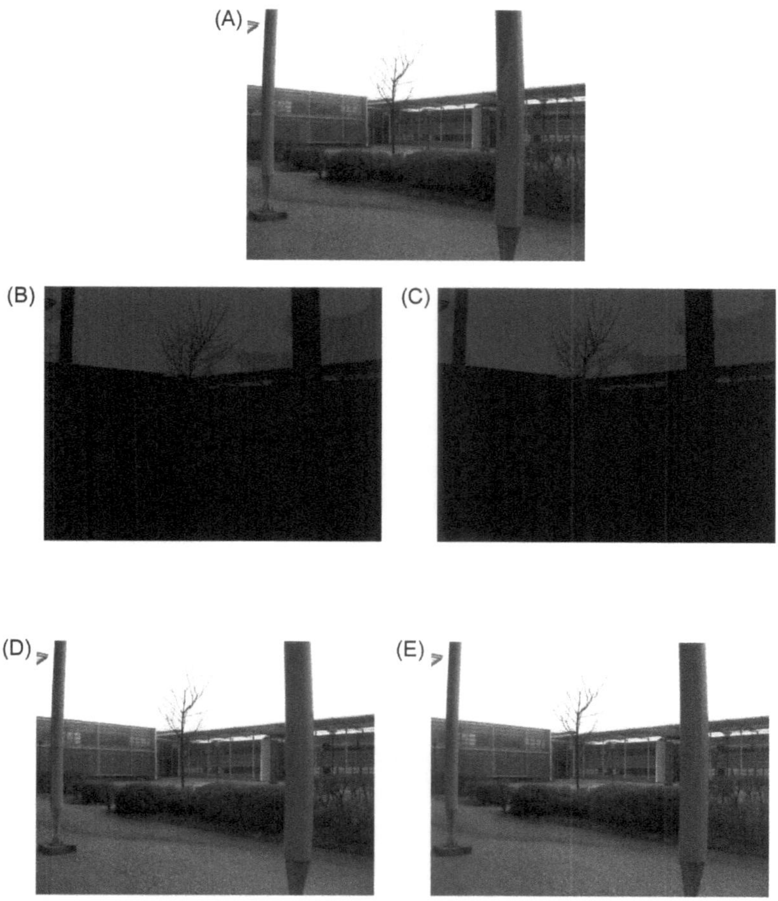

Fig. 12 Simulation of a long exposure time. (A) Image acquired at 50 ms. (B) and (C) Results of short exposure times (5 ms). (D) Average of 10 images acquired at 5 ms, then corrected by LIP subtraction. (E) LIP correction applied on a single image acquired at 5 ms.

(Continued)

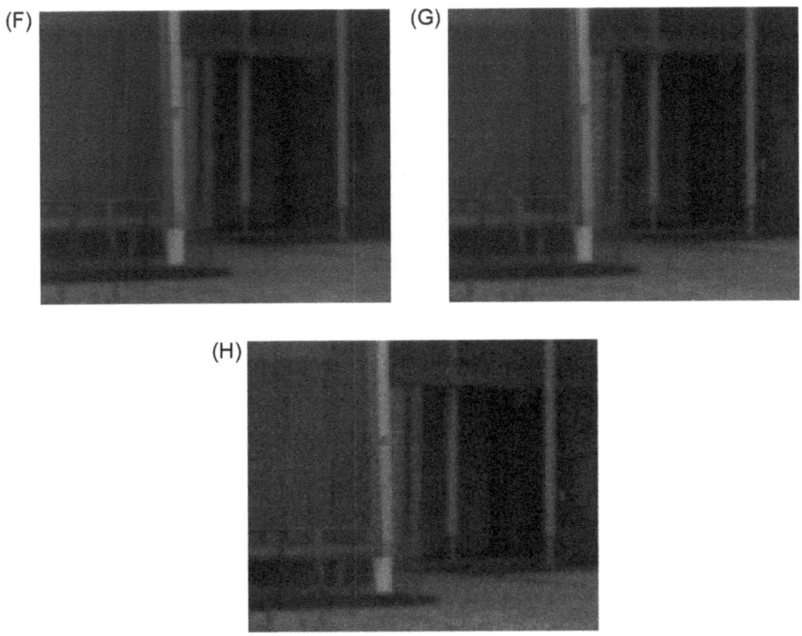

Fig. 12—Cont'd (F) Zoom on image (A). (G) Zoom on image (D). (H) Zoom on image (E).

Comment on Fig. 12: Fig. 12 allows us to compare the result of the LIP correction applied to a single image acquired at 5 ms (noisy image displayed in (H)) with the result obtained by our method (images (D) and (G)). We note also the similarity of (D) with the image acquired at 50 ms (A).

4.2 Binning Simulation

What is *binning*? In case of low-lighting conditions, hardware binning consists of grouping several pixels of the charge-coupled devices (CCD) sensor (for example a set of 2×2 pixels) and cumulating the charges of concerned pixels in the resulting one.

Such a method permits *increasing the brightness while decreasing the resolution*. For a same exposure time, the hardware binning enhances the camera sensitivity without increasing the noise while preserving the same acquisition speed.

The LIP tools permit us to perform a *software binning*. Starting from an image f, it is possible to compute an image g of lower resolution: in g we group a set of neighboring pixels in a single one affected of the average gray-level value and we simulate the enhanced gray level by LIP subtraction.

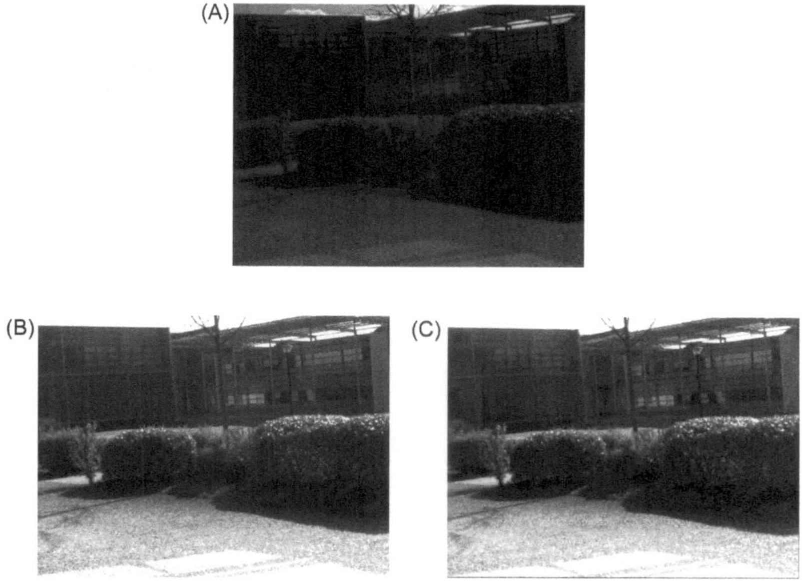

Fig. 13 Comparison of hardware binning and software binning. (A) Initial image acquired at 30 ms. (B) Image acquired at 30 ms using hardware binning. (C) Correction of image (A) by means of software binning, applying LIP subtraction.

The software binning approach offers a lot of advantages, compared with the hardware one:

- It is clearly adaptable in the sense that the number of cumulated pixels can evolve according to the brightness level
- The shape of cumulated pixels is adaptable, permitting performance of local binning on an image presenting various luminosity levels

To verify the ability of LIP tools to realize software binning, Fig. 13 proposes to compare it with hardware binning.

5. HIGH DYNAMIC RANGE IMAGES

An high dynamic range (HDR) image presents a high dynamic range, which means a very large ratio between the brightest and darkest parts of the image. In order to display an HDR image on a normal (low dynamic range) monitor, the huge range of brightness must be compressed to fit into a much smaller range, which results in an overall lack of contrast and thus a very "flat" appearance.

Our objective here is not to expose this problem in detail, but to underline that the tools presented in Chapter "Dynamic Range Expansion, Night Vision. Stabilization, Centering. Industrial, and Biomedical Applications" concerning dynamic range expansion are potentially useful to improve the display visual quality of an HDR image. This can be done in combination with tone mapping techniques, which adjust locally the contrast so that each region (the dark one and the bright one) uses the whole range. For some more information, the interested reader can refer to Carré (2013), where M. Carré explains initial results on the subject.

6. CONCLUSION

In this chapter, we established the ability of the LIP tools to accurately simulate source intensity variations, exposure time changing, and partially simulate (at the luminance level) possible modifications of diaphragm aperture. I successfully used such properties in plenty of industrial or biomedical problems and I encourage the reader to do the same.

REFERENCES

Carré, M. (2013). *Interprétation d'images acquises en situation de faible éclairage ou d'éclairage variable"*. PhD thesis, France: Saint-Etienne University.

Carré, M., & Jourlin, M. (2014). LIP operators: Simulating exposure variations to perform algorithms independent of lighting conditions. In *Multimedia computing and systems (ICMCS), international conference on, Marrakech* (pp. 122–126).

Cho, T., Levin, A., Durand, F., & Freeman, W. (2010). Motion blur removal with orthogonal parabolic exposures. In *Proceeding of the international conference on computational photography* (pp. 1–8).

Dai, S., & Wu, Y. (2008). Motion from blur. In *IEEE conference on computer vision and pattern recognition (CVPR), Anchorage (USA)* (pp. 1–8).

Deshayes, V., Guilbert, P., & Jourlin, M. (2015). How simulating exposure time variations in the LIP Model. Application: Moving objects acquisition. In *14th international congress for stereology and image analysis, Liège (Belgium)*. 4 pp.

Moghaddam, M., & Jamzad, M. (2007). Motion blur identification in noisy images using mathematical models and statistical measures. *Pattern Recognition, 40*(7), 1946–1957.

Potmesil, M., & Chakravarty, I. (1983). Modeling motion blur in computer-generated images. In *SIGGRAPH'83, Proceedings of the 10th annual conference on computer graphics and interactive techniques* (pp. 389–399).

Tanaka, M., Yoneji, K., & Okutomi, M. (2007). Motion blur parameter identification from a linearly blurred image. In *Consumers electronic, international conference on (ICCE), Las Vegas (USA)* (pp. 1–2).

CHAPTER SIX

Transfer of Classical Tools to the LIP Context

Contents

We have already seen some examples where it is quite easy to define a logarithmic image processing (LIP) version of classical tools. In the same vein, a lot of image processing tools are expressed as convolution-based operations: this means that, considering a pixel x and a window neighboring x, the output pixel value is the weighted sum of the input pixels within the window where the weights are the values of the filter assigned to every pixel of the window itself. The same thing is said in slightly different words (Ref. Wikipedia): "In image processing, a *kernel*, *convolution matrix*, or *mask*, is a small matrix useful for *blurring*, *sharpening*, *embossing*, *edge-detection*, and more. This is accomplished by means of *convolution* between a kernel and an image."

Advances in Imaging and Electron Physics, Volume 195
ISSN 1076-5670
http://dx.doi.org/10.1016/bs.aiep.2016.04.006

To perform this kind of convolutions, the only necessary operations are addition, subtraction, and scalar multiplication, whose LIP versions have been defined from the beginning of the model. Thus anyone is able to produce the corresponding LIP operators in order to benefit from a better sensitivity in the dark-gray levels.

Such tools transfers (Deng, 1993; Deng & Pinoli, 1998; Palomares, Gonzalez, Ros, & Prieto, 2006; Pinoli, 1992, and many others) are obviously justified in the case of transmitted signal and in the case of images acquired in reflection by the LIP consistency with human vision.

In this chapter, we will present some examples of classical image processing tools transfer, particularly those we have successfully used for industrial or biomedical applications. Section 1 is widely inspired by a previous extended paper (Jourlin et al., 2013).

1. APPLICATION TO SEGMENTATION METHODS

1.1 Introduction, Context, and Aim of the Section

At the step of acquisition, digitized images require a large number of gray levels in order to account for possible significant details and avoid quantization effects producing coarse transitions at the objects boundaries, due in particular to the "*partial volume effect*," which means that transition pixels overlap two or more regions (anatomical elements, for example), implying that their gray levels are not characteristic of a region of interest.

Classically, the gray-scale digitization is performed on 8-bits, producing 256 possible classes.

In such conditions, it is generally impossible to associate a meaning with each class, for example, to what kind of soil it corresponds (cultures, forest, desert, water, etc.) for remote sensing images. Note that the problem is worse for color images, offering three channels of 256 levels, namely 16.7 mega classes.

This situation requires decreasing the number of classes before image analysis: this is the role of image segmentation, which constitutes one of the most crucial phases of image processing because it determines the quality of image interpretation. Most of the time, image segmentation consists of grouping pixels of different gray levels in the same class if they satisfy various conditions of homogeneity (gray levels similarity).

The *similarity* of two gray levels is generally estimated based on the absolute value of their difference: if this difference is less than a fixed tolerance ε, they are considered "similar."

Remark 1. Such an approach of similarity is rather basic. Moreover, in many situations, it is incorrect: suppose that the studied objects (cells preparation acquired in transmission, for example) are not observed on a homogeneous background. We have discussed this case in Chapter "Various Contrast Concepts" (Fig. 12). In such a situation, pixels of the same object (cell) will not present the same similarity according to the background darkness. I hope that the attentive reader has already understood the interest in replacing a classical gray-level difference with a logarithmic one [logarithmic additive contrast (LAC)].

For more complex images, similarity between two pixels can mean that their neighborhoods present analogous texture, entropy, etc. The possible techniques to evaluate homogeneity or texture are countless and the object of this section is not to propose a survey of segmentation techniques, for which the interested reader can refer, for example, to Freixenet, Muñoz, Raba, Marti, and Cufi (2002), Fu and Mui (1981), and more recently, Raut, Raghuvanshi, Dharaskar, and Raut (2009) and Thilagamani and Shanthi (2011). We also decided to focus on some segmentation algorithms due to their efficiency or novelty.

More precisely, we are first going to shortly recall the simplest approaches comprising multithresholding and *k*-means (Sections 1.2 and 1.3), but the main drawback of these methods is the determination a priori of the desired number of classes. We propose a solution to overcome this problem for multithresholding.

Then we will turn our attention toward region growing (Section 1.4) because it is one of the most popular. A particular case of region growing will be the *systolic algorithm*.

Finally, the heart of this Section 1 dedicated to segmentation methods will focus on the *hierarchical ascendant classification* (HAC) (Sections 1.4 and 1.5). This technique, classical in Statistics, has been widely studied by Benzécri (1970, 1972). It presents the major interest of associating to a given data set all the possible classifications, starting from the initial step where each data is considered as a class, until the final step where all the data are regrouped in a unique class.

Such an algorithm is performed without any a priori.

Moreover, a cost function is naturally generated which designs the most significant steps.

In Section 1.5, we introduce a novel HAC technique, based on the Newton's universal attraction law: the *Gravitational Clustering*.

All these selected methods will be presented in a classical way and also in the LIP framework.

Various applications will illustrate the efficiency of each of the described algorithms.

1.2 Multithresholding

Although it has been presented in Chapter Various Contrast Concepts, Section 1.6.1, this subject is recalled here as a basic segmentation method. We have seen in Chapter Various Contrast Concepts, that most of automated thresholding methods require a prohibitive execution time when they are generalized to multithresholding. This is particularly the case for interclass variance maximization, entropy maximization, and moment preserving techniques. On the opposite side, we have shown the efficiency of Köhler's approach, which presents two major advantages: it takes into account the spatial information and does not require the prior knowledge of the expected number of classes. Moreover, in Section 1.6.1 of Chapter Various Contrast Concepts, we proposed to replace the contrast defined by Köhler with novel ones, namely the LAC or the logarithmic multiplicative contrast (LMC). As a kind reminder, let us illustrate these methods (cf. Fig. 1).

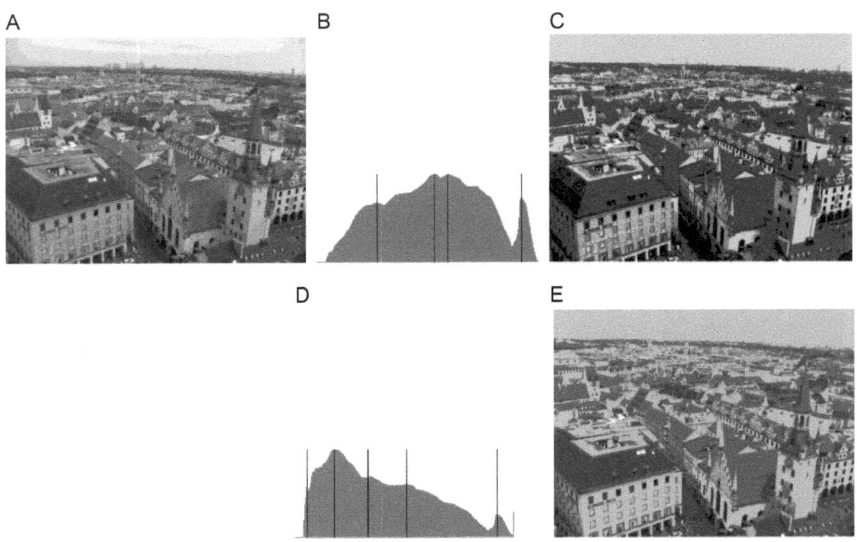

Fig. 1 Multithresholding thanks to Köhler's method. (A) Initial image. (B) Classical contrast curve. (C) Multithresholded image (five gray levels, four peaks). (D) Logarithmic contrast curve. (E) Multithresholded image (six gray levels, five peaks).

1.3 *k*-Means and "Nuées Dynamiques"

The *k*-means clustering (cf. for example, Forgy, 1965; Jain & Dubes, 1988; Kanungo et al., 2002; Mac Queen, 1967) is a simplified version of the expectation-maximization (EM) algorithm in a Gaussian approach. It means that each cluster is supposed to obey a Gaussian law, which permits to represent it by its mean value.

The method hereafter shortly presented ("Nuées dynamiques" in French) is the brainchild of Diday (1971) and constitutes an improvement of *k*-means. All these algorithms are not initially multithresholding methods: they were not conceived for image processing, but to classify a population into subpopulations in general. When applied to gray-level images, they result in a multithresholding of the studied image, and moreover permit to classify color images, for example.

The principle of the Diday's method is extremely simple, and is based on the employment of a "resemblance" criterion or, better put, of a distance between elements of the population. For our purposes, these elements will be gray levels. Therefore, we choose at random n starting points $x_1, x_2, \ldots, x_n$ within the population (elements of the gray scale) if we want to obtain n classes. Each element of the population is allocated to the nearest x_i in terms of the resemblance criterion, thus creating n classes. The corresponding barycenters are calculated, which in turn serve as references to aggregate the nearest points.The process continues until we reach convergence. This is generally quick, even for multithresholding.

To illustrate this process, we have selected (Fig. 2A) a thermal infrared image representing "La Rochelle port," already studied in Fig. 15 of Chapter Various Contrast Concepts. In this image, the very cold regions, such as clouds, are black. The Atlantic Ocean is rather cold, thus dark but does not present a uniform temperature. Such variations can be due to shallow zones or polluted ones. We also remark that two curves (marked by arrows) are darker than their environment: they correspond to ships' wakes, which pull cold water to the surface. I want also to point out an entertaining remark: at the extremities of the wakes, a zoom will show the reader a very small hot zone corresponding to the boat motor.

Shown here (Fig. 2B) is the result of Diday's segmentation in 16 classes obtained with the LIP-distance $d_1^{\triangle\!\!\!\!-}$. So that the results are visually interpretable, the classes have been assigned false colors. We can observe that the information lost by reducing the 256 gray levels to 16 is not negligible: the details we have previously described on the boats wakes are no more perceptible.

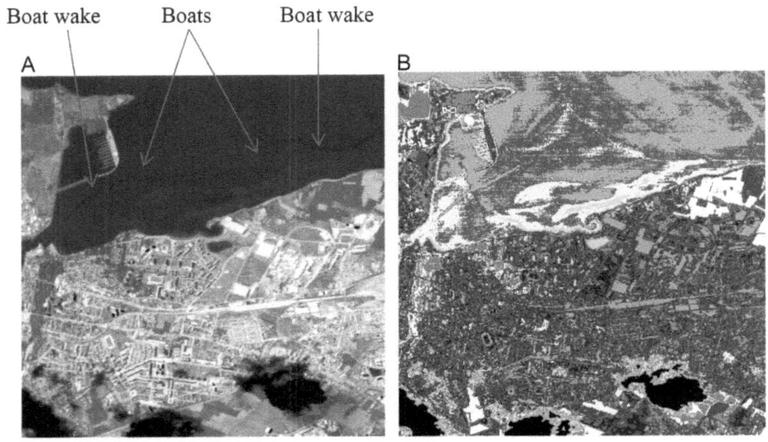

Fig. 2 Diday's algorithm applied to "La Rochelle port." (A) Initial image: "La Rochelle port." (B) Segmented image (16 classes). (See the color plate.)

General comments:

- Obviously, this method is heavily dependent on the distance chosen to evaluate the resemblance criterion. This point is an advantage because it gives rise to a multitude of possibilities, particularly for color images.
- Even for a large number of classes and with randomly chosen initial points, the computation time to reach convergence is acceptable. In order to speed up the algorithm execution, it is possible to work in a monitored fashion: for example, an a priori knowledge concerning the characteristic color of a desired object permits choosing of a well-adapted initial point. This is particularly the case for satellite images, for which we dispose generally of a "ground information:" what zones are cultivated lands, what zones are forests, what zones correspond to water (rivers, lakes), and so on.
- It is often useful to apply the *Mahalanobis correction* because the classes obtained by the algorithm may present rather different standard deviations. Such a correction has been introduced by Mahalanobis (cf. Mahalanobis, 1936) and consists of dividing the Euclidean distance between a point and a barycenter by the standard deviation of this barycenter class. Thus, the hypothesis: "each class is supposed to obey a Gaussian law."
- Let us note that classifications algorithms can be used in image compression. In fact, the efficiency of such methods may be evaluated from two opposite points of view: their ability to preserve the image information on

one hand, while drastically decreasing the number of classes on the other. Historically speaking, let us recall that the *k*-means method (in its "Nuées dynamiques" version) was in competition with the cosine transform to become the standard in compression of television images. G.E. Lowitz was the leader of *k*-means approach and he demonstrated that the 16.7 million possible colors of a RGB image could be limited to approximately 40 without visual significant disturbing. Nevertheless, such a reduction must be executed in real time, which is impossible with most classification algorithms, but reachable through *k*-means.

Partial conclusion: The previous very simple methods, in their initial version, are sufficient to solve many typical situations corresponding to bimodal or multimodal images. It is, however, necessary that the images in question show little light drift and be fairly free of noise. The LIP tools, such as contrasts and various associated metrics, permit conceiving more efficient versions of these segmentations algorithms, particularly in presence of lighting drift and/or lighting variations of different origins. We shall now put forward some complementary methods, each one holding specific interest.

1.4 Region Growing Algorithms

1.4.1 Classical Region Growing

This subject has been already addressed in Chapter "Various Contrast Concepts," Section 1.6.3 and in Chapter "Metrics Based on Logarithmic Laws," Section 5.3.1. In the first case, the objective was to propose a LIP version of the similarity criterion and in the second one to show the insensitivity of region growing to exposure time variations through the Additive Asplünd's metric. Now let us return to this segmentation technique in detail.

A region growing algorithm starts with a seed location and attempts to merge neighboring pixels until no more pixels can be aggregated to the region (cf. for example, Sekiguchi, Sano, & Yokoyama, 1994; Sivewright & Elliot, 1994 for 3D case). The goal is to obtain a final region corresponding to a whole object or a meaningful part of it. The aggregation process is generally governed by a similarity criterion able to evaluate the resemblance between two pixels (at the beginning, the seed is compared with each of its neighboring pixels) and then between a pixel and a region. Various homogeneity criterions have been investigated for region growing (cf. for example, Baraldi & Parmiggiani, 1996; Copty, Ranka, Fox, & Shankar, 1994; Wu, 1993).

The most current region growing method and the simplest one is the *single linkage*. Given a gray-level image *f*, its pixels are considered vertices of a graph, and an arc connects neighboring pixels with similar properties. Maximal sets of pixels belonging to the same connected component constitute the image segments. In the single linkage approach, similar is defined: two pixels x and y are considered similar (noted xSy), or mathematically speaking, related if the absolute difference of their gray levels $f(x)$ and $f(y)$ is lower than a fixed bound α:

$$xSy \Leftrightarrow |f(x) - f(y)| \leq \alpha \tag{1}$$

At the end of the first step dedicated to aggregate neighboring pixels to the seed, we get a region and it becomes necessary to decide if a pixel x neighboring the region is similar to it. In the single linkage approach, this similarity is a new time evaluated based on Eq. (1): x is *similar* to the region if there exists, inside the region, a pixel y neighboring x and satisfying (1).

So defined, S does not satisfy all the properties of an *equivalence relation*. More precisely, it verifies:

- *Reflexivity* (every pixel is similar to itself): $\forall x \in D, xSx$
- *Symmetry*: $\forall x \in D, \forall y \in D, xSy \Leftrightarrow ySx$ but not
- *Transitivity*: xSy and $ySz \Rightarrow xSz$

Nevertheless, it is possible to transform the relation S into an equivalence relation R through transitivity saturation:

$xSy \Leftrightarrow$ there exists a path of pixels $p_1, \ldots, p_n$ such that

$$\begin{cases} p_1 = x,\ p_n = y \\ \text{and } \forall i = 1, .., n-1 : pi\, S\, pi + 1 \end{cases}$$

Remark 2. We have just established that the single linkage is expressed as an equivalence relation. Thus, the associated region growing satisfies a strong property: the image segmentation resulting of this method becomes *independent of the seed*. In fact, it is a well-known mathematical characteristic of an equivalence relation to produce "equivalence classes," which are disjointed and then independent of the chosen points to represent them.

In such conditions, the main weakness of the region growing approach has been overcome. Nevertheless, we get another drawback: the transitivity property results in a *chaining effect*, which often implies segmentation errors, in the sense that the obtained segments are not always interpretable.

A possible answer to this *chaining effect* consists of applying the image, before segmentation, an enhancement (*crispening*) of the boundaries in order to get better contrasts, which permits stopping the transitivity paths.

In Fig. 3, we first show the result of a simple linkage according to Eq. (1) on a well-contrasted region of an image (Fig. 3A and B). Then we consider a radiographic image of one hand (cf. Fig. 3C), where the bones' boundaries present weak contrasts with the background. In such a situation, the simple linkage generates a chaining effect (Fig. 3D–G) and it becomes impossible to determine (even interactively) an efficient value of the "resemblance" parameter α. Nevertheless, the bad quality of the considered image implies a real difficulty to achieve an acceptable segmentation.

Comment on Fig. 3: Two situations are presented. First, a simple one (Fig. 3A), where the region we look for presents regular and sufficient contrasts all along its boundary. This means that contrasts between points of the inside boundary and points of the outside boundary are greater than the threshold α selected in Eq. (1). In such case, the single linkage region growing produces the expected result. Second, the radiographic image of the hand (Fig. 3C) shows a much more difficult case: according to the values of α, we observe an incomplete detection of the bone (Fig. 3D), where $\alpha = 5$) and for values immediately superior to five, the bone is not detected while the region begins to spread out across the expected object (Fig. 3E) and quickly fills the background (Fig. 3F and G).

To conclude this section on classical region growing, let us focus on some other "similarity" evaluations. Opposite from the single linkage, we can point out the complete linkage, which will aggregate a pixel x to a region if it satisfies relation (1) for every pixel y of the region.

Note that such a method presents the interest of strongly limiting the chaining effect, but will *depend on the seeds locations*. Moreover, the aggregation condition being very restrictive, the resulting segments size is generally too small, which implies a merging step. This last process depends on an a priori similarity criterion between regions, adding a new subjectivity aspect to the other ones. Some authors have proposed methods allowing seeds locations independence. Let us refer to Wan and Higgins (2003) who define a set of theoretical criteria that a region growing algorithm must satisfy in order to become insensitive to the selection of initial growing points. Another paper by Lin, Jin, and Talbot (2000) proposes "unseeded region growing" with application to 3D image segmentation.

Concerning the "subjectivity" aspects of region growing, a number of papers try to limit them. Asano and Yokoya (1981) provided a very interesting example that imagined an automated computation of the similarity parameter α (Formula 1). In fact, the gray-levels distribution inside

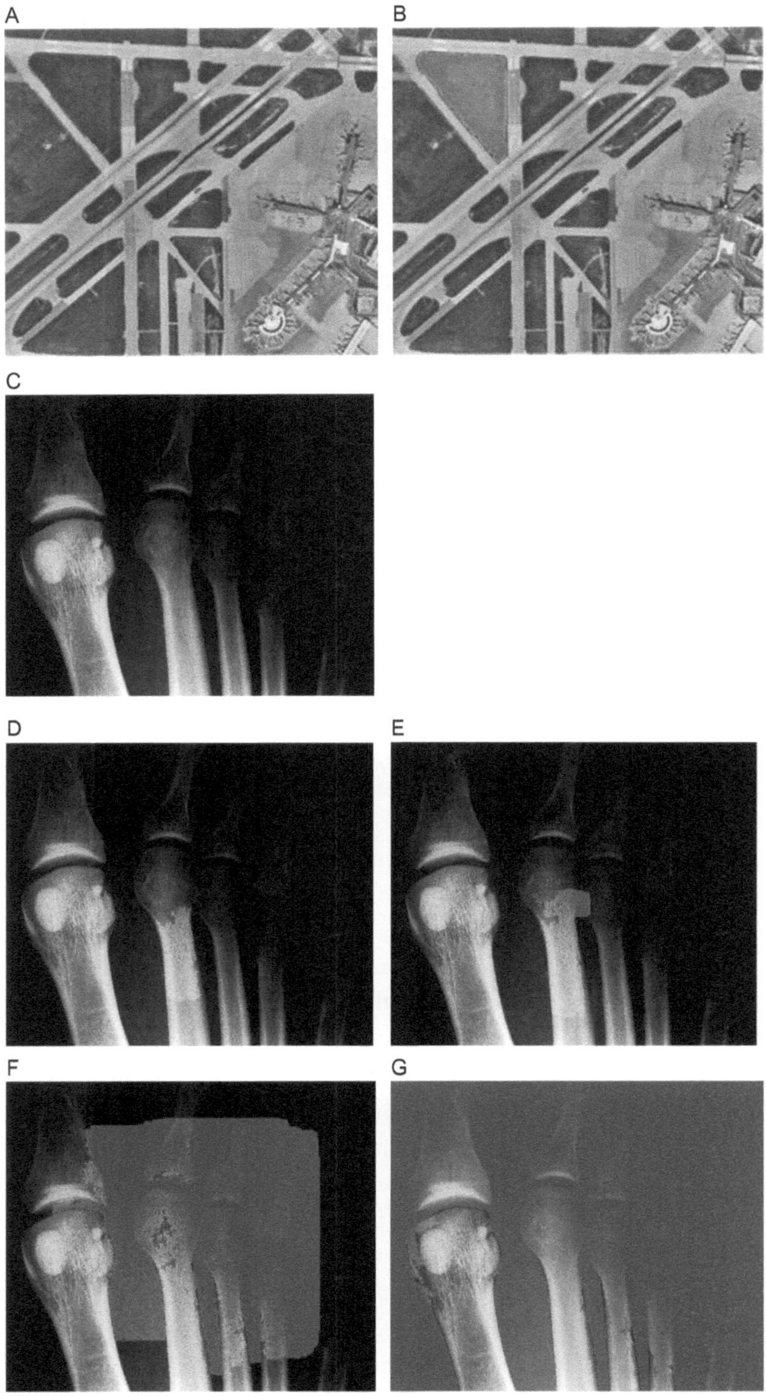

Fig. 3 Two examples of single linkage effects. (A) Well-contrasted initial image. (B) Single linkage region growing on a part of (A). (C) Initial image. (D) Single linkage region growing on (C) for $\alpha = 5$. (E) Single linkage region growing on (C) for $\alpha > 5$. (F) and (G) Single linkage region growing for increasing values of the "resemblance" parameter α. (See the color plate.)

the studied part of the image generates the "best" value of α in order to limit the number of *bridge points* wrongly connecting adjacent, but distinct regions.

1.4.2 LIP Version of Region Growing

The efficiency of LIP operators for creating and then segmenting 3D images has already been proved (cf. Jourlin & Pinoli, 2001) and presented in Chapter "Gray-Level LIP Model. Notations, Recalls, and First Applications, Section 3" and more precisely in Section 3.1.2 of the Chapter "Gray-Level LIP Model. Notations, Recalls, and First Applications." We have shown how, starting from a limited number of x-ray scanner cuts of a human jaw, we realized a 3D visualization of excellent quality using two LIP tools: a logarithmic interpolation and a logarithmic region growing. Let us simply recall that the similarity condition given in Eq. (1) is replaced by its LIP version:

$$xSy \Leftrightarrow \mathrm{Max}\,(f(x),f(y))\triangle\mathrm{Min}\,(f(x),f(y)) \leq \alpha$$

Besides the application of the LIP single linkage to the previous image (Fig. 3C) does not produce a perfect result (Fig. 4), but one of better quality than the classical single linkage. This better quality is mainly due to the over-estimation of the logarithmic contrast in dark regions.

The region growing method has generated plenty of algorithms, and it is not our objective to present a survey of them. We will conclude Section 1.4 by recalling an original approach proposed by Revol and Jourlin (1997).

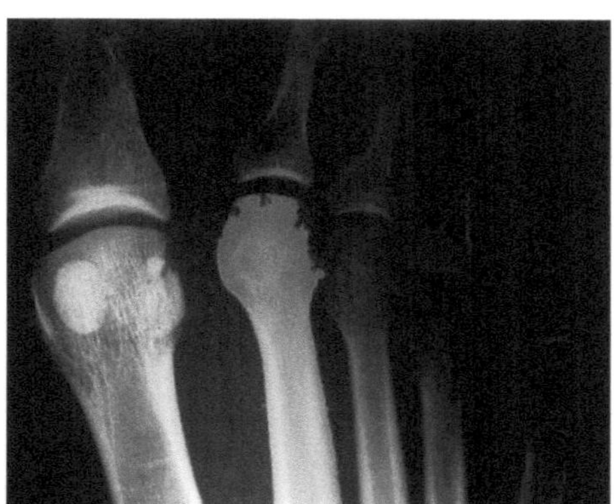

Fig. 4 Logarithmic single linkage on the initial image (Fig. 3C). (See the color plate.)

1.4.3 Systolic Region Growing

The principle of Revol's approach consists mainly of two innovations:

- Replace the resemblance criterion with a homogeneity parameter of the region in construction. Revol proposed to use the simplest homogeneity parameter: the standard deviation of the region.
- At a current step of the growing process, first aggregate every candidate pixels and if the homogeneity criterion is no more satisfied, remove the most penalizing pixels.

More precisely:

Suppose a region R_n has been built after n iterations: instead of performing the aggregation pixel per pixel, all the pixels neighboring R_n are in a first step aggregated to it, resulting in the dilation $R_n \oplus N$ of R_n by the considered neighbourhood N (generally N_8 constituted of the eight nearest pixels in a square grid). The question arising then is to determine if $R_n \oplus N$ is considered homogeneous or not. For that, the standard deviation $\sigma(R_n \oplus N)$ is compared with an a priori fixed threshold σ_0:

- If $\sigma(R_n \oplus N) \leq \sigma_0$:

  The region $R_n \oplus N$ is homogeneous and the new region becomes $R_{n+1} = R_n \oplus N$
- If $\sigma(R_n \oplus N) > \sigma_0$:

  The region $R_n \oplus N$ is inhomogeneous and Revol proposed a strategy in order to reduce it until it satisfies the homogeneity condition, resulting in a region R_{n+1}. Roughly speaking, such a reduction consists of considering the histogram of $R_n \oplus N$ and suppressing the classes located at its extremities. Two situations are then observable at the first time where the homogeneity is satisfied: if R_{n+1} is strictly larger than R_n, the process continues, if not it stops.

Comments concerning this method:

- Such an algorithm is not properly region growing in the sense the region can be reduced before growing again. It is the reason why it seems us more informative to call it *systolic* approach.
- In certain situations, the initial seed can be chosen in an inhomogeneous subset of the image, such that it belongs quickly to an extreme class of the histogram and may be suppressed of the constructed region. Nevertheless, if the user is confident in the seed choice, it is always possible to forbid the seed's exclusion.
- For biomedical images difficult to segment, it is possible to introduce an a priori knowledge concerning the statistical homogeneity of each anatomical element, generating different permitted thresholds σ_0.

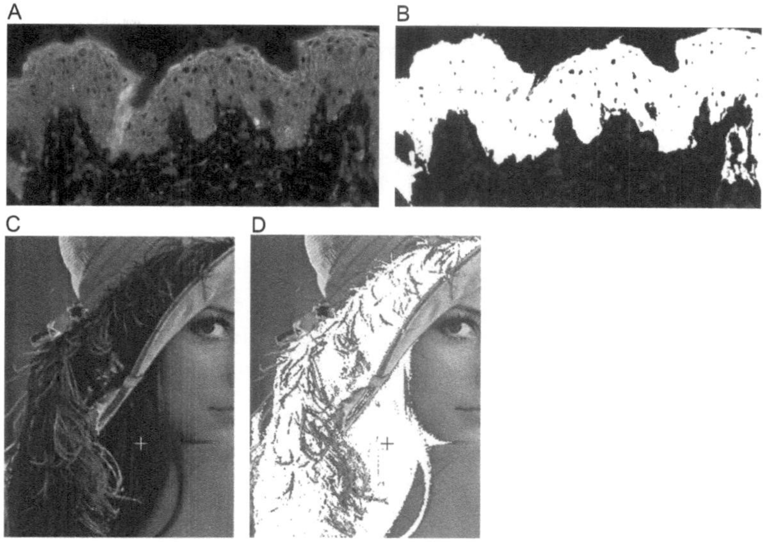

Fig. 5 Examples of Revol region growing. (A) Initial image of fluorescence microscopy on human skin explant with seed (white cross). (B) Systolic segmentation (white region) corresponding to the previous seed (black cross). (C) Initial image (subset of "Lena") with seed (white cross). (D) Systolic segmentation (white region) corresponding to the previous seed (black cross).

In her paper, Revol used the last comment for preparing dental surgical interventions. To perform the segmentation of the 3D image resulting from x-ray scanner cuts, it was necessary to take into account the estimated homogeneity of each anatomical element constituting the image. This technique permitted producing efficient segmentations in this very difficult case. Results are shown in Revol and Jourlin (1997).

Two other examples are given in Fig. 5. The first one represents an image of fluorescence microscopy on human skin explant (Fig. 5A). The second is a subset of the famous "Lena" (Fig. 5C). The initial seed is located through a white (resp. black) cross on initial (resp. segmented) image. In each case, the final corresponding class is displayed in white (Fig. 5B and D).

1.5 Hierarchical Ascendant Classification

1.5.1 Benzécri HAC

Now let us focus on the hierarchical ascendant classification (HAC) described first by Krasner (1947) and Benzécri (1970, 1972) in French publications. Bertrand and Diday (1990) generalized this method to *pyramidal representations*. These four papers concern the field of statistical analysis.

A mathematicians group (CEREMADE), and especially Arbelaez (2005, 2006), transferred the technique to image processing.

The main interest of this approach is that all possible classification levels are computed without any prior knowledge concerning the final number of classes or the distance between two classes. Thus, to the contrary of some other methods, this technique is not seed-dependent. Furthermore, Benzecri established that the structure of hierarchical classification could be reduced to the definition of an ultrametric distance. Our own approach consists of implementing a fast two-dimensional (2D) algorithm of HAC, using different metrics, especially associated with the LIP model, whose main interest resides in the adaptability to both human vision and efficient processing of low-light images.

Now we will define some structures involved in the construction of HAC.

First, *hierarchy* is described by *parts hierarchy* and *stratified hierarchy* definitions and a *dissimilarity index* for image segmentation. Then, the link with ultrametrics is exposed. All definitions have been previously described (Benzécri, 1970) for statistical applications. Here we focus on the particular case of 2D images.

1.5.1.1 Hierarchy

Parts hierarchy: A parts hierarchy of an image is a set of segmentations of this image. Such segmentations are made of subsets (regions) of the initial image and are called levels. These levels are sequenced through a hierarchical order, resulting from the building steps of the hierarchy.

Let F be the finite set (the image) of n elements (n pixels), and $H_F = (h_1, \ldots, h_n)$ the set of segmentations of F. We will say that H_F is a *parts hierarchy* of F if the following properties are satisfied:

$$F \in H_F \tag{2}$$

$$\forall i \in F, \{i\} \in H_F \tag{3}$$

$$\forall h_j, h_k \in H_F, h_j \cap h_k \neq \varnothing \Rightarrow \begin{cases} h_j \subset h_k & \text{or} \\ h_k \subset h_j \end{cases} \tag{4}$$

Condition (2) shows that an extremity of the hierarchy contains the set itself, ie, all image pixels merged into one large uniform region.

Condition (3) means that the other extremity of the hierarchy contains all the elements of the set, ie, all the pixels, here considered as segments.

The third condition (4) shows that if the intersection between two levels is not null, one is included in the other (cf. Diday, 1983). For mathematicians: note that such a property is characteristic of the balls of an ultrametric space.

Stratified hierarchy: A parts hierarchy is said to be *stratified* when a number (called the *hierarchical index*) representing the hierarchical order can be associated with the hierarchy. A stratified hierarchy is a couple (H_F, f) where f is a function: $H_F \longrightarrow \mathbb{R}^+$ such as:

$$\forall i \leq 0, \forall h_i \in H_F, f(h_i) = 0 \text{ and } h_i = h_0 \tag{5}$$

$$\exists M \in \mathbb{R}^+, \forall i \geq M, f(h_i) = M \text{ and } h_i = \{F\} \tag{6}$$

$$\forall h_i, \forall h_j \in H_F,\ h_i \subset h_j \Rightarrow f(h_i) < f(h_j) \tag{7}$$

Property (5) shows that the index starts at 0, (6) the maximum value the index can reach is M, and (7) index values are growing with set inclusion. It is then necessary to define an index to compare F elements with each other.

Dissimilarity index: The dissimilarity index d_{H_F} is defined from a stratified hierarchy (H_F, f) of F, as an application from $FxF \longrightarrow \mathbb{R}^+$. For two elements of F, i and j, the dissimilarity index $d_{H_F}(i,j)$ is defined as the lowest level containing both i and j:

$$d_{H_F}(i,j) = Min\{f(h_x),\ h_x \in H_F,\ i \in h_x, j \in h_x\} \tag{8}$$

It is actually the value of the lowest level that already contains i and j.
Remark 3. It is generally clearer to assign to the hierarchical index the value of the dissimilarity index corresponding to the level.

1.5.1.2 Ultrametrics

HAC shares a link with the mathematical notion of *ultrametric* distances. They ensure each other that properties are well respected: It is equivalent to have on a finite set (Benzécri):

(1) An ultrametrical metric
(2) A stratified hierarchy

Let us now present the notion of ultrametric:

Definition: An *ultrametric* (cf. Krasner, 1947 and Benzécri, 1972) is a metric respecting the following ultratriangular inequality instead of the triangular inequality:

$$\forall (x, y, z) \in F^3,\ d(x, z) \leq Max(d(x, y), d(y, z)) \tag{9}$$

Due to Eq. (9), some surprising properties hold in ultrametric spaces (cf. Krasner, 1947 and Arbelaez, 2005):

- Each side of a triangle is smaller or equal to the largest of the two others.
- Every triangle is isosceles.
- Every point from a circle is its center.
- Two circles are either concentric or disjointed.
- The circle's diameter is equal to its radius.

For simplification, we will use the term *ultrametric* instead of ultrametric metric.

Link with HAC: We have just seen that a dissimilarity index d_{H_F} may be associated with a stratified hierarchy H_F. Benzécri (1972) has shown that the reciprocal implication is true: a unique stratified hierarchy corresponds to a dissimilarity index s if $1-s$ is ultrametric.

1.5.2 Benzécri HAC Application

1.5.2.1 Bidimensional Images

In the case of Benzecri, iteratively elements that are similar are merged because of the similarity index. For 2D images, the situation requires more precisions to keep interest. The classical framework for images deals with 256 gray-level images located on a grid that delimits areas. Such areas are defined owing to their gray-level value and homogeneity, and will be merged, as quickly as the dissimilarity index permits, to create partitions.

Directly applying the Benzécri algorithm to such images would create some useless results. For example, in a real image where all gray-levels are present, each pixel corresponds to one gray level distant from another. For a Euclidean dissimilarity index, the first step produces an image containing only uniform (same gray-level) areas. At the second step, the entire image is merged into one unique region, which has no interest. It is obvious that for images, this algorithm requires taking into account the pixels spatial distribution.

1.5.2.2 Benzécri Algorithm

There are different ways to build an HAC. Benzécri (1970) explains some of them as the Sneath, Sorensen, or Sokal and Michener algorithms. The main algorithm for building a hierarchy H_F, owing to a dissimilarity index d_{H_F} is shown in Fig. 6 (in pseudo-code due to Diday, 1983):

In our framework, it is common that a large number of regions must be compared at each level. Even for a reasonable image size (for example,

1. $h(0)$ = initial image
2. Iteration count = 0
3. While classes number of h (iteration count) >2
   a. Iteration count +1
   b. Aggregate the nearest classes
   c. If more than two classes to merge, random choice or biggest class first
   d. Give h (iteration count) the merged result

Fig. 6 Pseudo-code: algorithm for building an HAC.

1. $h(0)$ = initial image
2. For s from 1 to Maximum
3. *If number of classes of* $h(s\text{-}1) > 2$
   a. Aggregate $h(0)$ classes keeping $d < s$
   b. Give $h(s)$ the merged result

Fig. 7 Pseudo-code: algorithm for building a 2D HAC.

640×480 pixels) execution time quickly becomes enormous if the algorithm is applied as it is. This is mostly due to:

- The initial number of regions.
- The fact that merging only two regions at each step considerably elongates algorithm iterations.

1.5.2.3 Bidimensional Algorithm

A variant of the algorithm has been developed to speed up execution time and propose easy and quick implementation. The variant uses the bounded property of classical image gray scale, which tells us all possible level values. The corresponding classifications are built by merging elements close to a threshold value d.

For example, if the aggregation index d is the Euclidean distance and if it is applied to an 8-bits image, d can take only entire values from 0 to 255.

One can select a threshold value s such that the partition will contain regions merged from elements whose distance d is less than (or equal to) s. The corresponding pseudo-code is given in Fig. 7.

The interest of such an approach is that when classes are merged, it is unnecessary to update the index of the new area. In fact, with this algorithm, we can build any level starting from the initial set $h(0)$. In most other methods, the previous level must be known to build the next. This solves the problem of assigning a value to the newly merged area and strongly simplifies the time-consuming part of the algorithm.

1.5.2.4 Our Specific Approach

This algorithm can be easily applied to various images. Building an HAC needs to sweep all possible aggregations up to the completion (only one region).

For each threshold index value, the corresponding partition is defined by merging neighbors around pixels, which can be directly connected and distant from a value d lower than the threshold.

The following demonstrates on a gray-level synthesis image with known values that HAC steps are well built. One uses a synthesis image for understanding purpose, in order to limit the influence of the initial number of regions and then the steps number of the algorithm. This image and the values of its regions are shown in Fig. 8.

Two metrics, assimilated to the dissimilarity index, are used for comparison: A Euclidean distance and a logarithmic one (LIP framework–based) (cf. Figs. 9 and 10).

The Euclidean distance d_e is defined by relation (10) to compare pixels values $f(x)$ and $f(y)$ from a gray-level image f:

$$d_e(f(x), f(y)) = |f(x) - f(y)| \tag{10}$$

The same comparison is made with a logarithmic distance (in fact the LAC between $f(x)$ and $f(y)$) according to the following relation (11):

$$C^{\triangle\!\!\!\!\!+}_{(x,\,y)}(f) = \frac{Max(f(x), f(y)) - Min(f(x), f(y))}{1 - \dfrac{Min(f(x), f(y))}{256}} \tag{11}$$

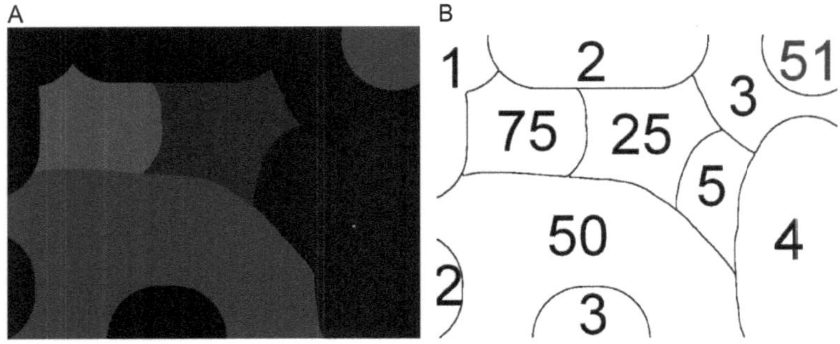

Fig. 8 Synthesis image serving of HAC test. (A) Initial image. (B) Initial image values (standard gray scale: 0 = black).

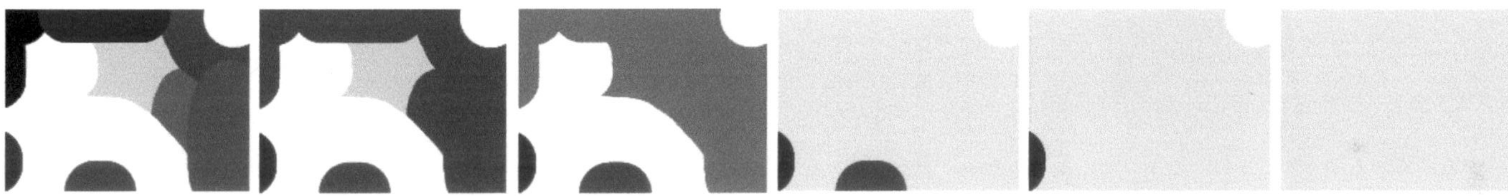

Fig. 9 HAC building steps with Euclidean metric d_e.

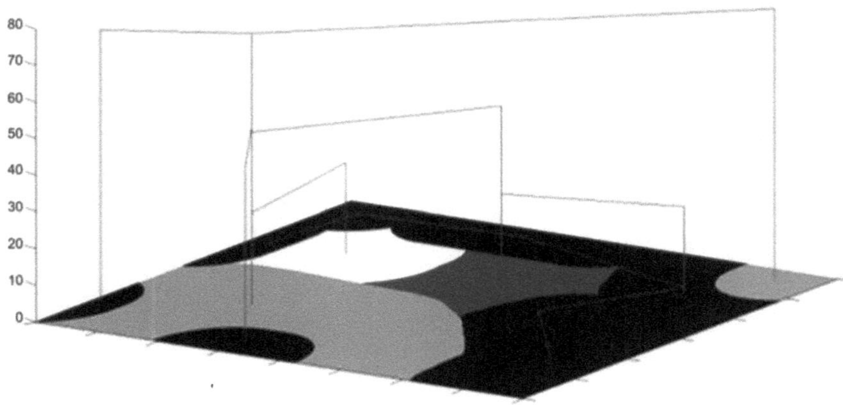

Fig. 10 Dendrogram representation based on HAC of Fig. 9.

Using the previously described algorithm allows us to choose a particular distance to drive merging operations. Some steps of HAC building with Euclidean metric are presented in Fig. 9.

The first regions of the image to be merged correspond to a distance of one gray level. HAC is building itself regularly as expected. Execution on MATLAB with this image size of 800×600 pixels is about 6 s and increases quickly to several minutes as the number of initial regions grows. Another implementation is presented next.

One way to represent results is the dendrogram: The HAC steps are shown on a graph function to the dissimilarity index (Fig. 10).

Fig. 11 illustrates HAC building with the logarithmic distance. This metric tends to ponder low gray-level values, owing to human vision. As expected, regions distant of one gray-level from the brightest pixels are merged first.

Comment on Figs. 8–11: Remember that the gray-level values of Fig. 8B are given in the classical gray scale (with 0 = black). Concerning Fig. 9 obtained with the Euclidean distance d_e, note that the first aggregated zones are those presenting a difference of one gray level (neighboring values: 1, 2, 3, 4, 5). The following step is the aggregation of regions with values 5 and 25, then 25, 50, and 75, then 2 and 50, 3 and 50 and finally 3 and 51 and 2 and 50. The last step produces a homogeneous image.

With the LIP distance based on the LIP additive contrast, the goal is to overweight the distances between low gray levels (here dark ones) so they merge more slowly. The result is consistent with the expected one: among the regions distant of one gray level, the brightest are merged first.

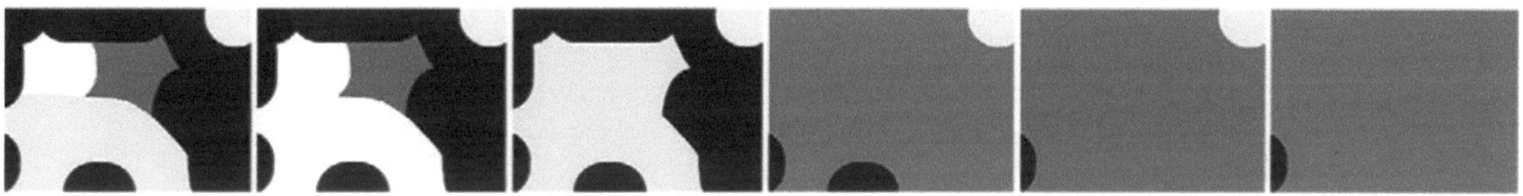

Fig. 11 HAC building steps with the logarithmic metric.

1.5.2.5 HAC Color Version

As in the gray-level version, the color version is defined by processing a Euclidean color distance and a logarithmic color distance. Starting from two color vectors V_1 and V_2, whose components are

$$\begin{cases} R_1 \\ G_1 \\ B_1 \end{cases} \text{and} \begin{cases} R_2 \\ G_2 \\ B_2 \end{cases}$$

The Euclidean color metric is defined by:

$$d_e(V_1, V_2) = \sqrt{(R_1 - R_2)^2 + (G_1 - G_2)^2 + (B_1 - B_2)^2}$$

The logarithmic metric $d^{\triangle_c}$ is defined from the LIP color (LIPC) model (see Jourlin, Breugnot, Itthirad, Bouabdellah, & Closs, 2011). It is based on the search of the semitransparent color layer between two colors, in other words, the layer that must be added to one in order to obtain the other. This is realized as the result of a LIPC subtraction of each channel maxima by each channel minima in order to remain inside the scale. It is also possible to use one of the simplified logarithmic color distances introduced at the end of Chapter "Various Contrast Concepts" (Eqs. (26) and (27)).

1.5.2.6 Results and Examples

The "Peppers" image is used as a test image. Its green channel is extracted for the gray-level algorithm, and the original is used for the color version. Gray-level results for the Euclidean metric are shown in Fig. 12 and for the logarithmic metric (in fact the LAC of Formula (11)) in Fig. 13.

Comments on Figs. 12 and 13: The steps presented in each figure are representative of the hierarchy and show the best results: we search for those corresponding to images containing satisfying information with a minimum number of regions in the hierarchy levels. In Fig. 12, the "best" segmentation is performed at step 2: after, regions are merged which would be important for visual interpretation. The hierarchy produced with the Euclidean metric is not perfect because it includes plenty of different zones. The use of metric $C^{\triangle}_{(x,y)}(f)$ creates a hierarchy converging slowly to stability. The "best" segmentation is given by step 10. In this case, the zones number is lower, with a better uniformity.

Let us now illustrate this property: HAC reaches stability faster with the Euclidean metric than with the logarithmic one. It means that $C^{\triangle}_{(x,y)}(f)$ is

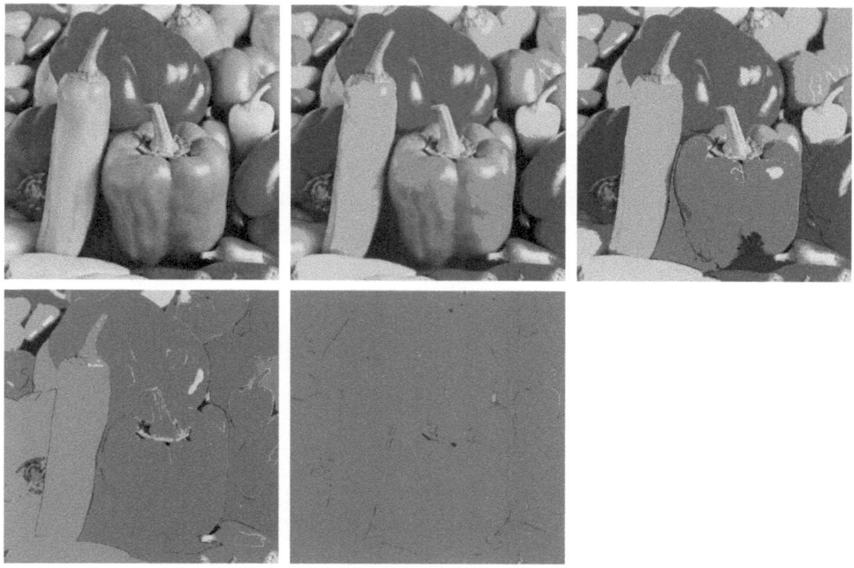

Fig. 12 Gray-level HAC by Euclidean distance d_e: steps 0, 1, 2, 5, and 10.

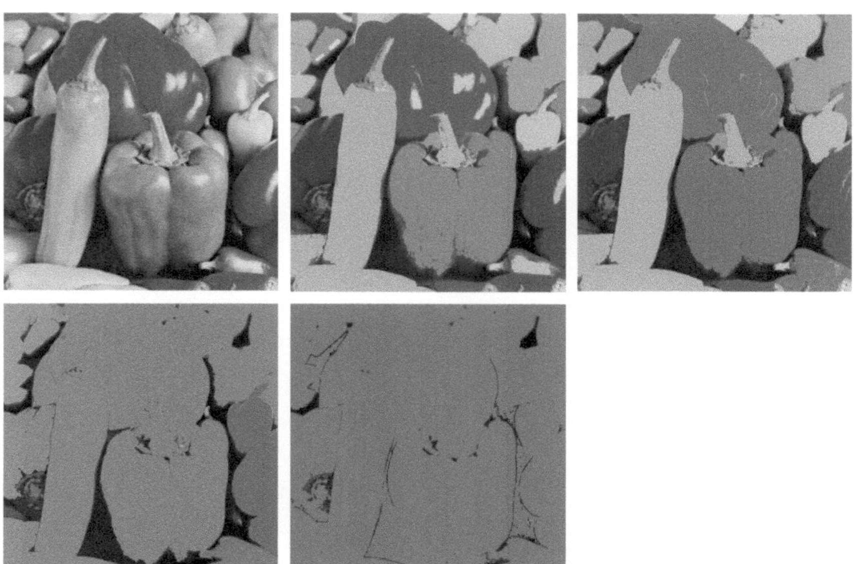

Fig. 13 Gray-level HAC by the Logarithmic distance $C^{\triangle}_{(x,y)}(f)$: steps 0, 7, 10, 15, and 31.

more discriminant than d_e. As can be seen on the "best" segmentation of each HAC, the textured areas generate problems. These areas are not well segmented and stay intact in a high HAC level.

One way to analyze this is to represent images with a random color lookup table (LUT). This allows observation of which areas are really uniform (Fig. 14).

We now apply the same HAC building to color images, with previously defined color metrics: first with the Euclidean metric d_e (Fig. 15), and then with the logarithmic metric $d^{\triangle_c}$ based on the LIPC model (Fig. 16).

Remark 4. One way to compare the results obtained with different metrics is to observe the tree's height, but aggregation speed is different for each metric and some nonsignificant regions can remain far in the tree and disturb this process.

Another solution consists of observing the ratio between the tree's maximum and best segmentation tree's position. The tree's maximum must be adapted due to the problem of small isolated regions (Fig. 17).

To make such an adaptation, we exclude part of the tree. Then, by observing the location of the optimal segmentation, we report the ratio in the other metric's tree and find the corresponding segmentation.

As an example, consider Fig. 15 with the Euclidean metric. First, we define the best segmentation: it corresponds to step 5 (Fig. 18A).

Retaining 92% of the tree allows computation of the proportional distance of the best segmentation from the new maximum. Finally, in the LIP metric tree, an equivalent position is obtained for step 32 (Fig. 18B).

A random color LUT is applied on each best segmentation (Fig. 19) to evaluate the results.

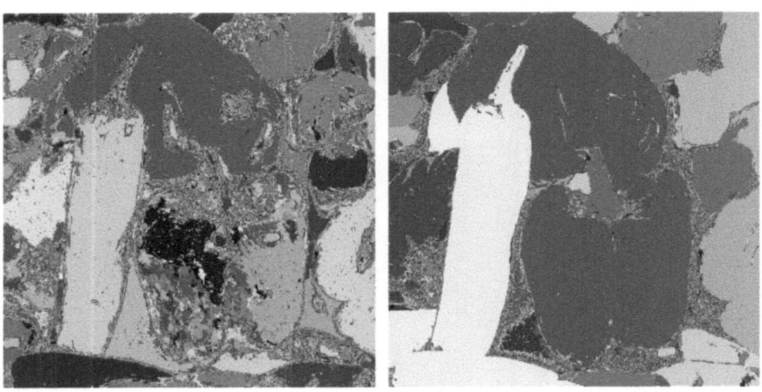

Fig. 14 Random color LUT on HAC steps. (A) Step 2 of HAC by Euclidean distance d_e. (B) Step 10 of HAC by logarithmic distance $C^{\triangle}_{(x,y)}(f)$. (See the color plate.)

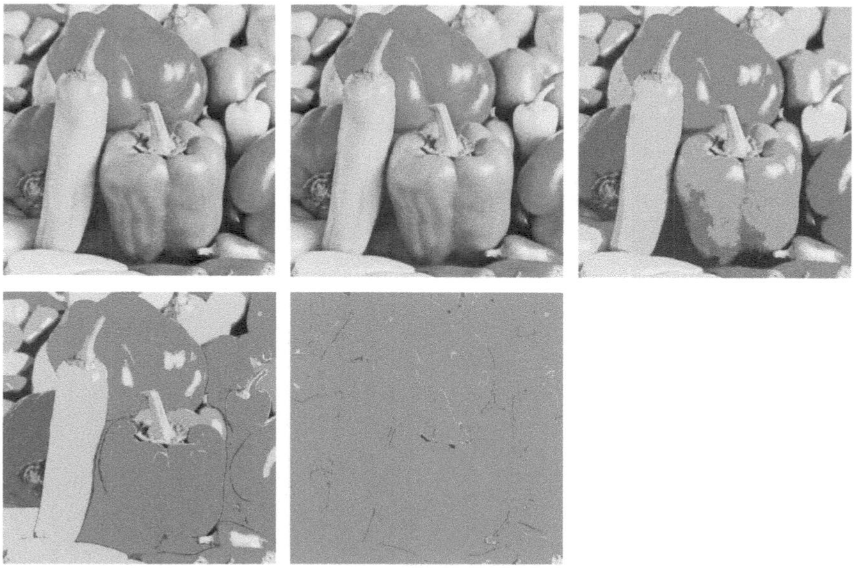

Fig. 15 Color HAC by Euclidean metric d_e: steps 0, 1, 5, 8, and 17. (See the color plate.)

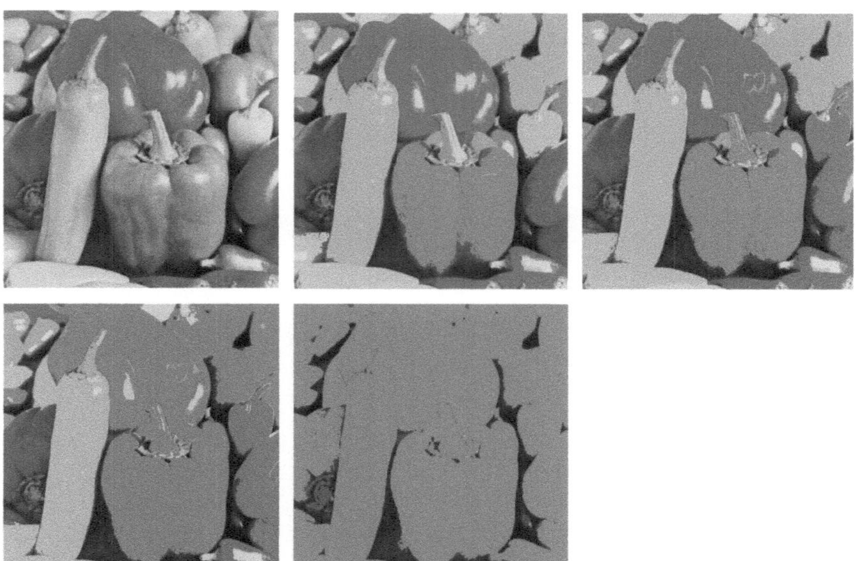

Fig. 16 Color HAC by logarithmic distance $d^{\triangle_c}$: steps 0, 29, 32, 36, and 57. (See the color plate.)

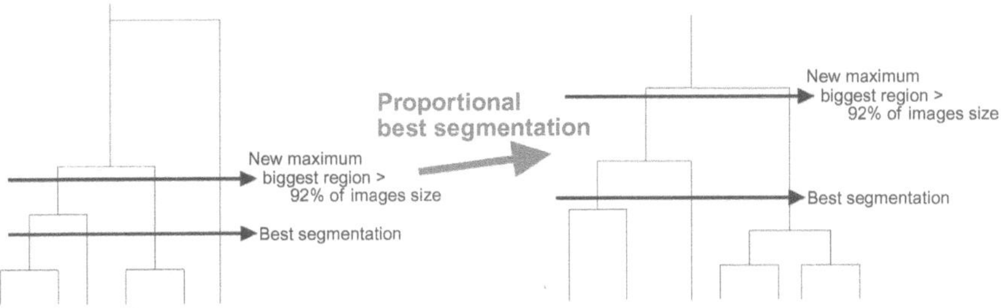

Fig. 17 Tree's observation for intermetric comparison. (See the color plate.)

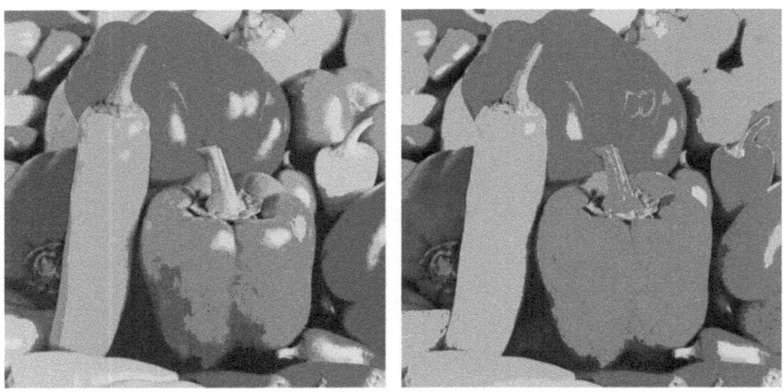

Fig. 18 HAC comparison between different metrics. (A) Optimal step (5) of HAC for the metric d_e. (B) Corresponding step (32) in the tree for the metric $d^{\triangle_c}$. (See the color plate.)

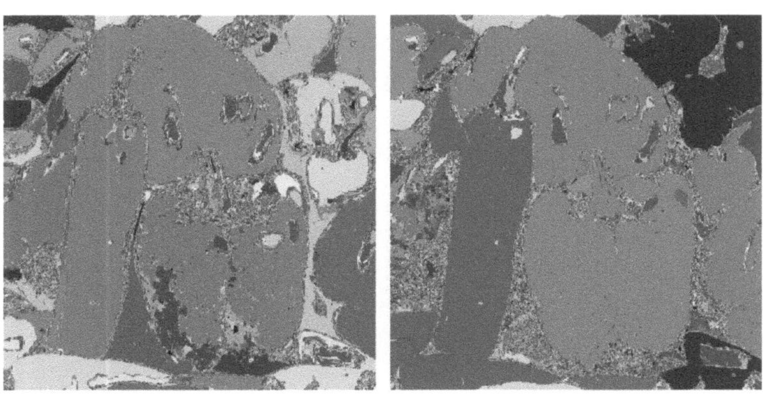

Fig. 19 Random color LUT on HAC steps. (A) Step (5) of HAC by d_e. (B) Step (32) of HAC by $d^{\triangle_c}$. (See the color plate.)

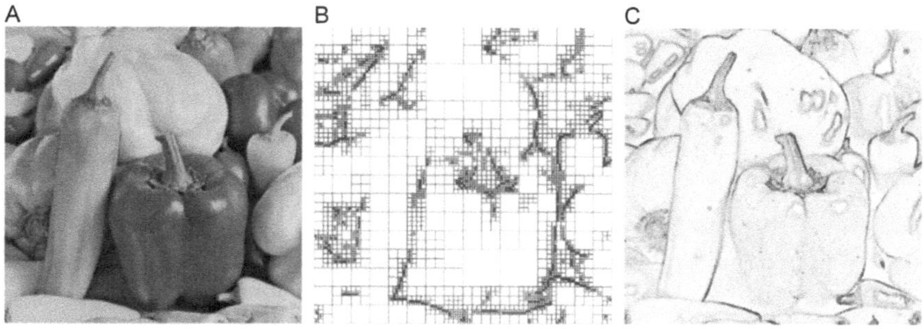

Fig. 20 Pyramidal representation of an image adapted to its information. (A) Initial image. (B) Classical quad tree decomposition. (C) A step in HAC tree.

1.5.2.7 Section Conclusion

We have described a hierarchical classification method based on logarithmic aggregation metrics. It is important to note that it results in a pyramidal representation of an image adapted to its information (Fig. 20C) instead of being an arbitrary grouping of pixels arranged 4×4, 16×16, and so on (Fig. 20B). This representation paves the way to transfer the algorithms and techniques useful to pyramidal image processing to this novel situation.

The following section is dedicated to gravitational clustering, which takes place among HAC algorithms. Considering its interest for image processing, a special focus is done on this approach.

1.6 Gravitational Clustering (Classical and LIP)

Following the paper of Wright, 1977 introducing the concept of gravitational clustering, many authors proposed improvements of the method. Let us refer to Gomez, Dasgupta and Nasraoui (2003), Long and Jin (2006), Rashedi, Nezamabadi-pour and Saryazdi (2009), and recently Rashedi and Nezamabadi-pour (2010). This last publication directly concerns color image processing and will be discussed further. Strangely, we have not found publications adapting gravitational clustering to gray-level images classification: this section achieves such an adaptation and the novelty also resides in the use of logarithmic notions, such as contrasts and metrics issued of the LIP model to define the attraction law. Comparisons will be made between "classical" and "logarithmic" approaches.

For interested reader, more detailed information and/or ideas on gravitational clustering are available, for example, in Sylos Labini (2008).

1.6.1 Principle and Notations

Introduced by Wright (1977), the notion of gravitational clustering performs the classification of statistical data based on an attraction law. The data are considered as a set of moving bodies $(\mathrm{MB}_i)_{i=1\ldots n}$ of mass m_i submitted to the gravitational universal law of Newton. For a pair of bodies MB_i and MB_j, the gravitational interaction results in a force F_{ij} (Fig. 21) according to the Eq.:

$$F_{ij} = -G.\frac{m_i.m_j}{\left(d_{ij}\right)^2}.u_{ij} \tag{12}$$

where

- G represents the universal attraction constant ($G = 6.67384.10^{-11}$)
- d_{ij} is the distance between MB_i and MB_j
- u_{ij} designs the unit vector oriented from MB_i to MB_j

1.6.2 Adaptation to Image Processing

In image processing, the idea will be to consider each element (gray-level value) of the histogram as a moving body whose mass corresponds to the number of pixels with this gray level. The mechanic system will be defined considering only the gravitational forces working on each moving body. When two moving bodies are close enough, we will consider that they merge, resulting in a unique moving body whose mass is the addition of concerned masses.

This technique has already been adapted to color images by Rashedi and Nezamabadi-pour (2010), but in our knowledge, never to gray-level images.

We now present the necessary main notations. We start with a gray-level image f defined on the spatial support D, with values in the gray scale $[0, M[$. For n-bits digitized images, M equals 2^n and we dispose of 2^n gray levels from 0 to $2^n - 1$.

The histogram of f will be denoted h, representing the function which associates with each gray level $i \in [0, 255]$ the number $h(i)$ of pixels $x \in D$ such that $f(x) = i$. The mechanical system is defined by considering together

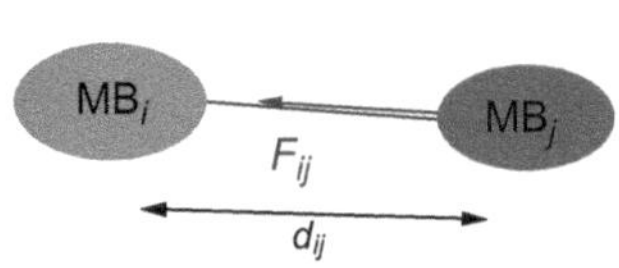

Fig. 21 Gravitational interaction between two bodies.

the gray levels as moving bodies and the gravitational forces that apply to them.

Note that plenty of variants exist for this method, depending of the selected metric d in Eq. (12). In this chapter, we will focus on two particular metrics:

- The absolute value of the difference between two gray levels
- The LAC defined in the LIP framework, cf. (Jourlin, Pinoli & Zeboudj, 1989 and Jourlin, Carré, Breugnot, & Bouabdellah, 2012).

Remark 5. The proposed technique furnishes all the intermediate classifications of an image, starting from the first step, where each gray level corresponds to a class, until the last step, where all the gray levels are grouped in a unique class. In such conditions, the proposed approach clearly enters within the framework of HAC methods.

Now, if we consider a moving body (ie, a gray level) $i \in [0, M-1]$, some complementary notations are needed, as follows:

- the position (respectively, velocity and acceleration) of i at a time t

$$P(i,t)\,(\text{respectively}\,V(i,t)\,\text{and}\,A(i,t)$$

- the ith value $H(i)$ of the normalized histogram of the studied image:

$$H(i) = \frac{h(i)}{Card(D)} \text{for}\, i \in [0, M-1] \tag{13}$$

Note 1: During the execution of the gravitational algorithms, $H(i, t)$ will represent the mass of a moving body i at time t.

- the limiting radius of aggregation between two moving bodies deemed too close:

$$\varepsilon_a \in \left]0, 1\right[$$

- the number of classes that the user wants in the multithresholded image:

$$Nclasses \in \mathbb{Z}$$

Note 2: If $Nclasses \leq 1$ then the number of classes in the given image will be determined automatically

- the number of merges between moving bodies made at each iteration t:

$$NFusions : t \rightarrow NFusions(t)$$
$$\mathbb{R}+ \rightarrow \mathbb{N}$$

- the chosen distance for evaluating the distance between two gray levels

$$dist : [0, 2^n - 1]^2 \to [0, 2^n - 1]$$

1.6.3 Implementation

(A) First Algorithm: Sampling of the Mechanical Model (SM Algorithm)
The SM algorithm, as the name suggests, consists of sampling a mechanical model by considering intervals Δt of infinitesimal times during which the accelerations of the different moving bodies are considered constant. We will see how to manage to forgo choice of an adequate value of Δt (parameter particularly difficult to determine a priori). Therefore, we must define the parameter ε_a and add it to the input of SM algorithm.

The gravitational clustering in our case is limited to one-dimensional (1D) movement along the gray-level axis (see Fig. 22: Normalized histogram in the initial situation for an 8-bits image).

The complete flow chart of SM algorithm is given next (see Fig. 23).

1.6.3.1 Initialization of the Algorithm

When $t = 0$, we have 2^n moving bodies at the integer positions of the gray scale. For each moving body $i \in [0, M-1]$ and standing at the position $P(i, 0) = i$, we assign a mass corresponding to the weight $H(i, 0)$ of the normalized gray-level histogram.

The initial velocity of each moving body i in the initial state is noted $V(i, 0) = 0$.

The LUT we will use later to obtain multithresholded images is initialized as outlined:

For each $i \in [0, M-1], LUT(i, 0) = i$

1.6.3.2 Heart of the Algorithm

The heart of SM algorithm consists of two main steps detailed next:

1. While the number of nonzero mass moving bodies is greater than (or equal to) the maximum between 1 and *Nclasses*:
   - Calculate the accelerations at time t according to the fundamental principle of dynamic adapted to our study, considering only the gravitational forces acting between the moving bodies (Eq. (12)):

Fig. 22 Initial normalized histogram of an 8-bits image. (A) Initial 8-bits image. (B) Normalized histogram of image (A).

$$A(i,t) = -G.\sum_{(j\neq i)}\left(\frac{H(j,t)}{dist\,(P(i,t),P(j,t))^2}\right) \tag{14}$$

Note in this equation that the algorithm result depends on the distance choice (see Sections 1.6.4 and 1.6.5)

- Starting from Eq. (14), we calculate the movement quantity between t and $t+\Delta t$:

$$Mvmt(i,t) = A(i,t).\frac{\Delta t^2}{2} + V(i,t).\Delta t \tag{15}$$

Once this value is computed for each i, we linearly normalize all these values into the interval $[0, \varepsilon_a]$, where ε_a represents the limiting radius of aggregation. Such normalization is performed in order to optimize the algorithm and to overcome a possible setting of G and Δt.

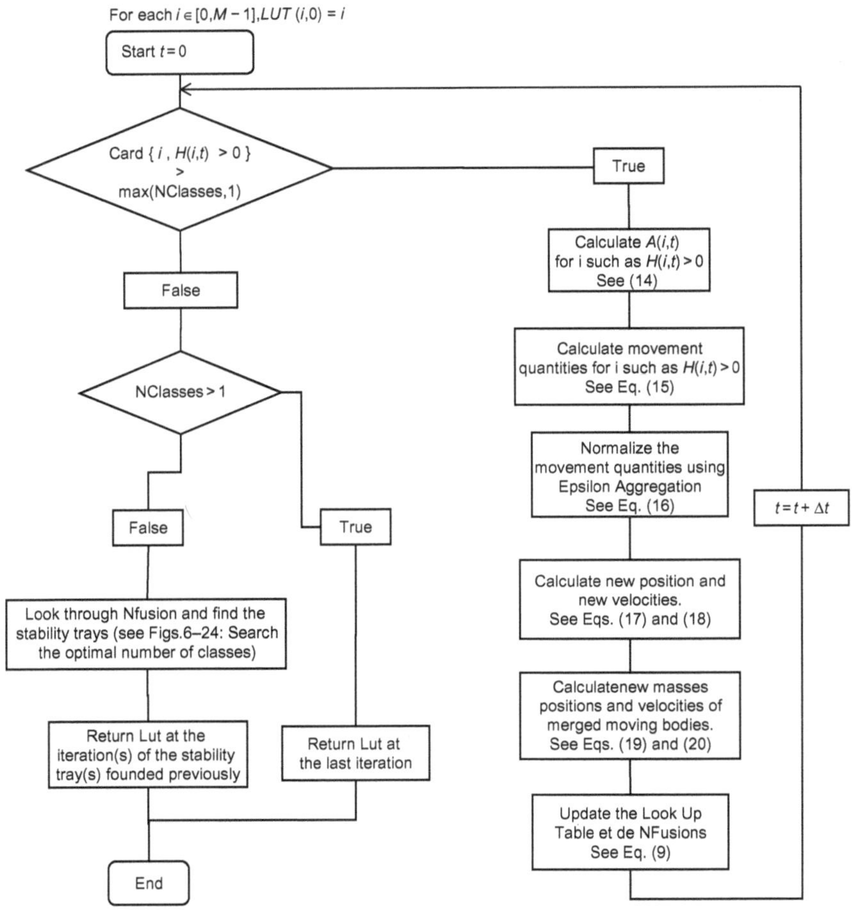

Fig. 23 Flowchart of SM algorithm.

- The normalization of the movement quantity can be written using the equation:

  For each $i \in [0, M-1]$

$$Mvmt(i,t) = \frac{\varepsilon_a}{max_{i \in]0,1[} Mvmt(i,t)} . Mvmt(i,t) \tag{16}$$

- Next comes the calculation of new velocities and positions:

$$P(i, t+\Delta t) = Mvmt(i,t) + P(i,t) \tag{17}$$

$$V(i, t+\Delta t) = A(i,t) . \Delta t + V(i,t) \tag{18}$$

2. Still in the loop defined previously (step 1), for every moving body i of nonzero mass, we look for any moving body j of nonzero mass such as:

$$dist(P(i, t+\Delta t), P(j,t+\Delta t)) < \varepsilon_a \tag{19}$$

In the case where such j exists, i and j are merged (in decreasing proximity order if there are several candidates) according to the following rules:

If $H(i, t+ \Delta t) > H(j, t+ \Delta t)$

Then

$$\begin{aligned}
P(i, t+ \Delta t) &= \frac{P(i, t+ \Delta t).H(i,t+\Delta t) + P(j, t+ \Delta t).H(j,t+\Delta t)}{H(i,t+\Delta t) + H(j,t+\Delta t)} \\
V(i, t+ \Delta t) &= \frac{V(i, t+ \Delta t).H(i,t+\Delta t) + V(j, t+ \Delta t).H(j,t+\Delta t)}{H(i,t+\Delta t) + H(j,t+\Delta t)} \\
H(i, t+ \Delta t) &= H(i, t+ \Delta t) + H(j, t+ \Delta t) \\
H(j, t+ \Delta t) &= 0 \\
Lut(j, t) &= Lut(i, t) \\
NFusions &= NFusions + 1
\end{aligned} \tag{20}$$

Reciprocally otherwise

Note: For each iteration, the number of fusions has also been updated.

1.6.3.3 End of the Algorithm

Once the heart of the algorithm has reached term, two cases can occur:

- If *Nclasses* is strictly greater than 1 (ie, we defined the desired number of classes in the studied image): in such case, it exits the loop defined in Section 1.6.3.2 (2-) with the number of desired classes. It remains only to apply the LUT computed at the last iteration to produce the expected multithresholded image.
- If *Nclasses* is less than (or equal to) 1, it means that we wish to determine automatically the number of classes present in the image. It is in this case that we will use the variable *NFusions* in which we stored the number of mergers based on the number of iterations.

We define a threshold T (for example $=100$ in Fig. 24) for the number of consecutive iterations without mergers during the process. Then we read the

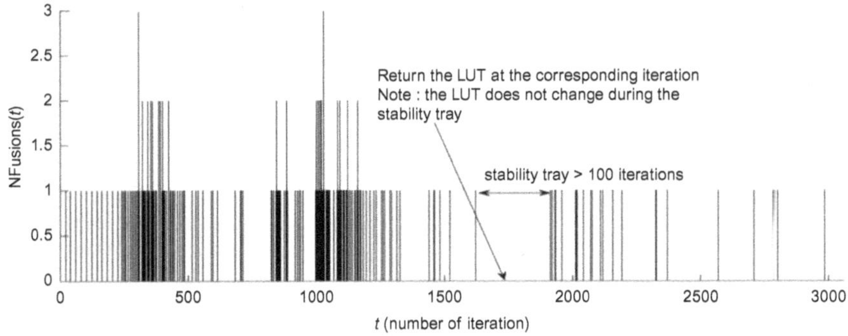

Fig. 24 Search the optimal number of classes.

function *NFusions* and retain the first interval of width greater than or equal to T wherein *NFusions* is null. We remember specifically the first iteration of this plateau as a reference, and the LUT corresponding to this iteration produces the expected multithresholded image.

(B) Second Algorithm: Optimized Automated (OA Algorithm)

This alternative eliminates the choice of any parameter (*Epsilon Aggregation* must be set by the user in the SM algorithm). The objective of this optimization is to aggregate at least two bins of the histogram at each iteration of the algorithm (see complete flowchart in Fig. 25).

From the initial step (see Section 1.6.3.1) to the calculation of acceleration (see Eq. (18), everything is identical to SM algorithm. However, after the acceleration calculation, instead of directly calculating the quantities of motion that flow from it, we will search the pair of moving bodies which will merge first in our set of moving bodies.

In order to find this required time of the first amalgamation between two consecutive neighbors, we use the relationship given by the motion equations (see Eqs. (16), (17), and (22)) and solve the following system (21) with unknown Δt. It will allow us to know the merging time between two moving bodies:

$$\begin{aligned}
&\text{For each } i \in [0, M-1] \text{ such that } H(i,t) > 0 \\
&\text{we find } j = \min(\{j,\ P(j,t) > P(i,t) \text{ and } H(j,t) > 0\}) \\
&\frac{(A(j,t) - A(i,t))}{2}.\Delta t^2 + (V(j,t) - V(i,t)).\Delta t + P(j,t) - P(i,t)
\end{aligned} \tag{21}$$

Solving this system and keeping physically acceptable solutions ($t \in \mathbb{R}^+$), gives us access to the minimal time required for two moving bodies to meet.

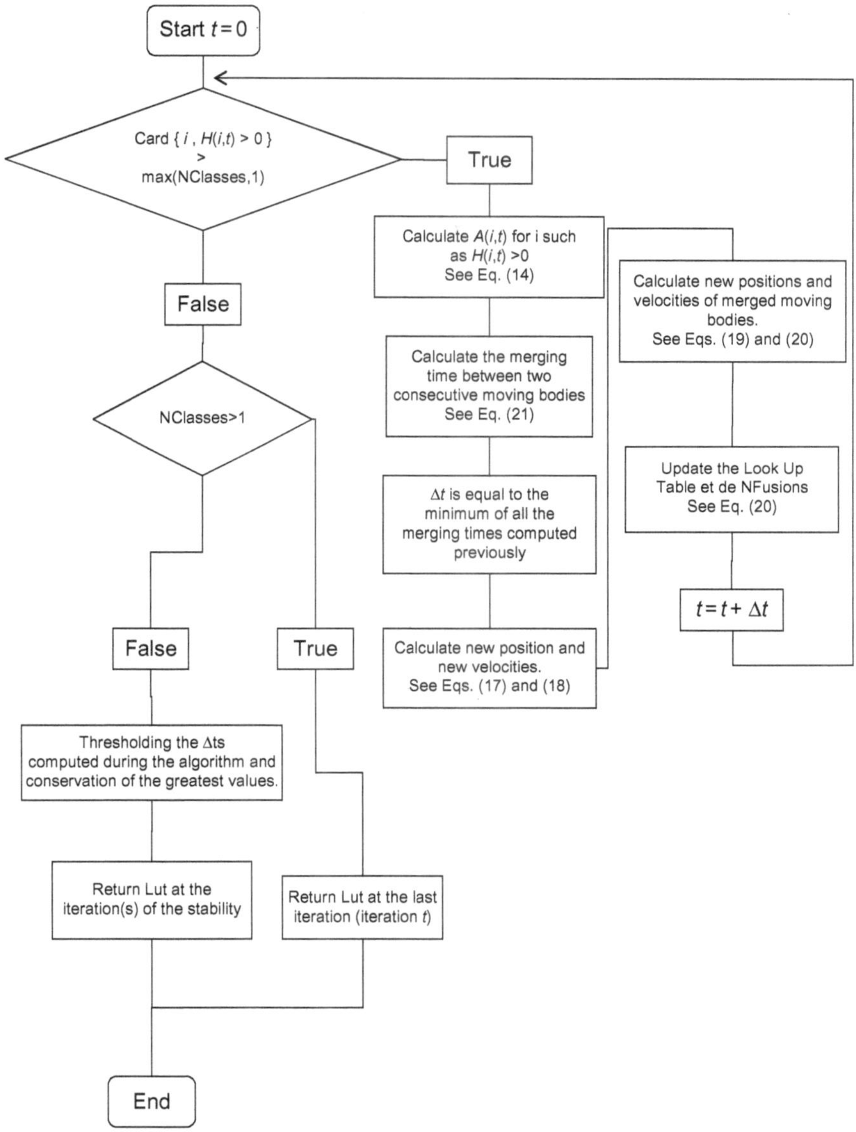

Fig. 25 Flowchart of OA algorithm.

In fact, for each pair of successive moving bodies, we compute the required time needed for them to meet. We then seek the minimum of such times and assign it to Δt. In such conditions, from t to $t+\Delta t$ we guarantee that at least two moving bodies will merge at the exact same position (those for which Δt is the solution of Eq. (21)). Thus we have not to define

Epsilon Aggregation and moreover we speed up the execution time of the algorithm (when compared with the SM algorithm with a small value of *Epsilon Aggregation*).

1.6.4 Metrics Selection

One major interest of the proposed SM and optimized automated (OA) algorithms resides in the various possible manners to compute the distance between two moving bodies. In fact, *the selection of a metric may be determined according to what we desire to emphasize on the studied image.* In this chapter, we have focused on the two following metrics:

Classical Euclidean Metric: If we consider that all the gray levels present on the studied image must be processed in the same manner (ie, if the gray scale is supposed to be approximately linear), it is natural to compute the distance between two moving bodies (gray levels) i and j according to the classical Euclidean metric (standard difference):

$$dist(i,j) = |i-j|$$

Logarithmic Metric: In the case where the considered image has been acquired in transmitted light, or if we aim at interpreting the image in the manner of a human eye, we propose to refer to the LIP framework. Thus we can compute the LAC (logarithmic distance) between two gray levels (moving bodies) according to:

$$\begin{aligned} dist_{log}(i,j) &= Max(i,j) \triangle Min(i,j) \\ &= \frac{|i-j|}{1 - \frac{Min(i,j)}{M}} \end{aligned}$$

In Section 1.6.5, we present the results of the SM and OA algorithms for the two retained metrics.

1.6.5 Results

Figs. 26–29 show the results obtained, respectively, for the SM and OA algorithms, both with the Euclidean distance and LIP metric (logarithmic subtraction). Such results confirm that the logarithmic approach over-evaluates the distance between two dark-gray levels. This allows us to merge moving bodies later in the dark part of the grayscale and to obtain more

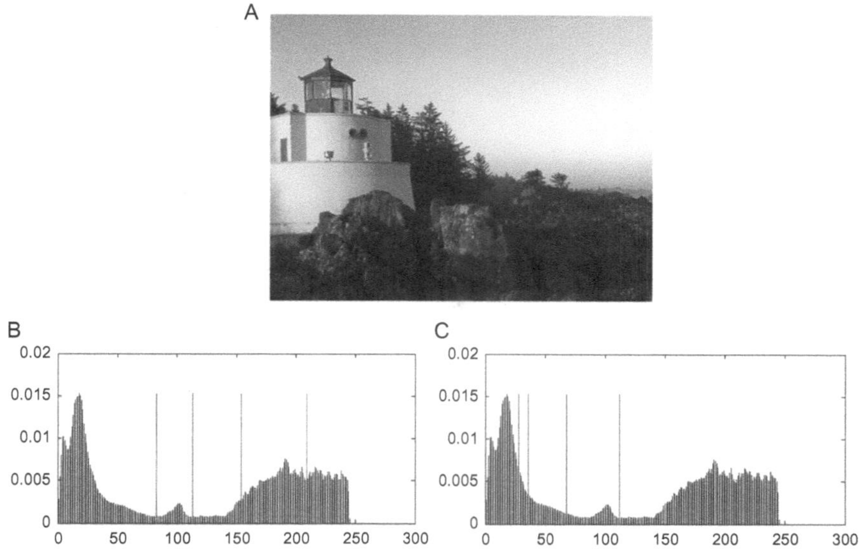

Fig. 26 For SM algorithm: Initial image and its thresholded histogram (for Euclidean and LIP metrics). (A) Initial image. (B) Histogram with four thresholds in the case of Euclidean metric. (C) Histogram with four thresholds in the case of LIP metric.

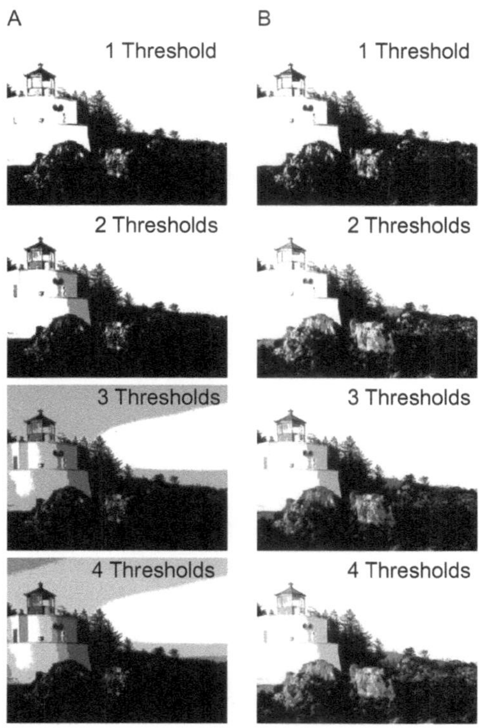

Fig. 27 Results for one to four thresholds and total computing time for the SM algorithm. (A) Results with the Euclidean metric with computing time (MATLAB): 11.26 s. (B) Results with the LIP metric with computing time (MATLAB): 11.65 s.

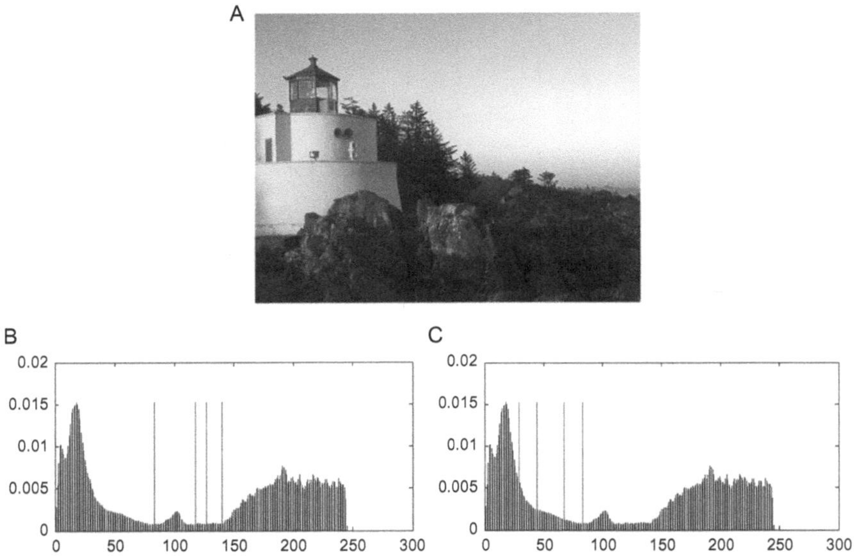

Fig. 28 For OA algorithm: Initial image and its thresholded histogram (for Euclidean and LIP metrics). (A) Initial image. (B) Histogram with four thresholds in the case of the Euclidean metric. (C) Histogram with four thresholds in the case of the LIP metric. (See the color plate.)

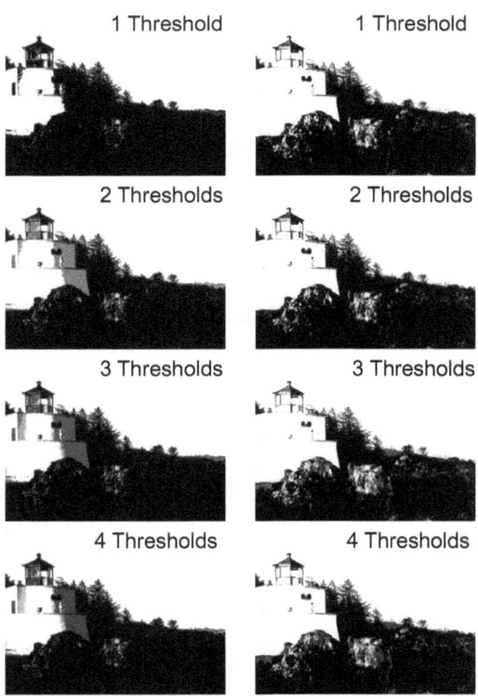

Fig. 29 Results for one to four thresholds and total computing time for the OA algorithm. (A) Results with the Euclidean metric with computing time (MATLAB): 5.22 s. (B) Results with the LIP metric with computing time (MATLAB): 5.31 s.

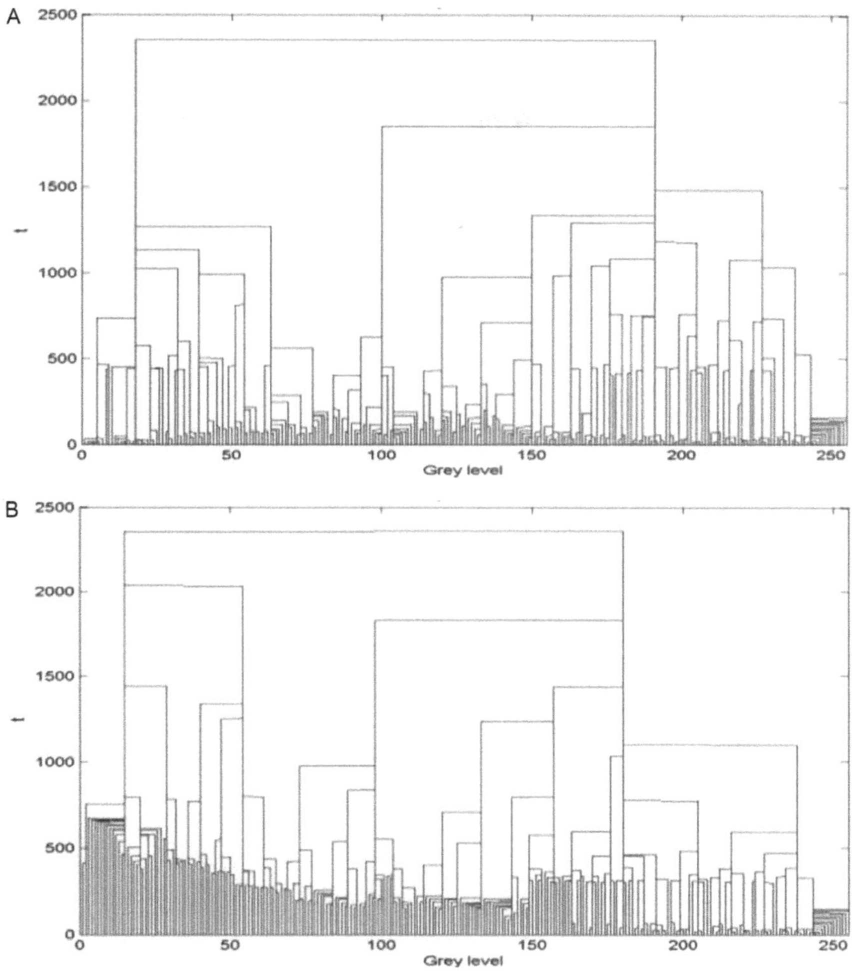

Fig. 30 Merging trees for the SM algorithm. (A) Results with the Euclidean metric. (B) Results with the LIP metric.

details in this part. We can easily observe this phenomenon in the corresponding merging trees (Figs. 30 and 31).

We can observe that the computation time does not depend of the selected distance for a same algorithm. However, we can see how the improvement proposed in the OA algorithm, which consists of achieving at least one aggregation per iteration, allows us to cut down the computation time.

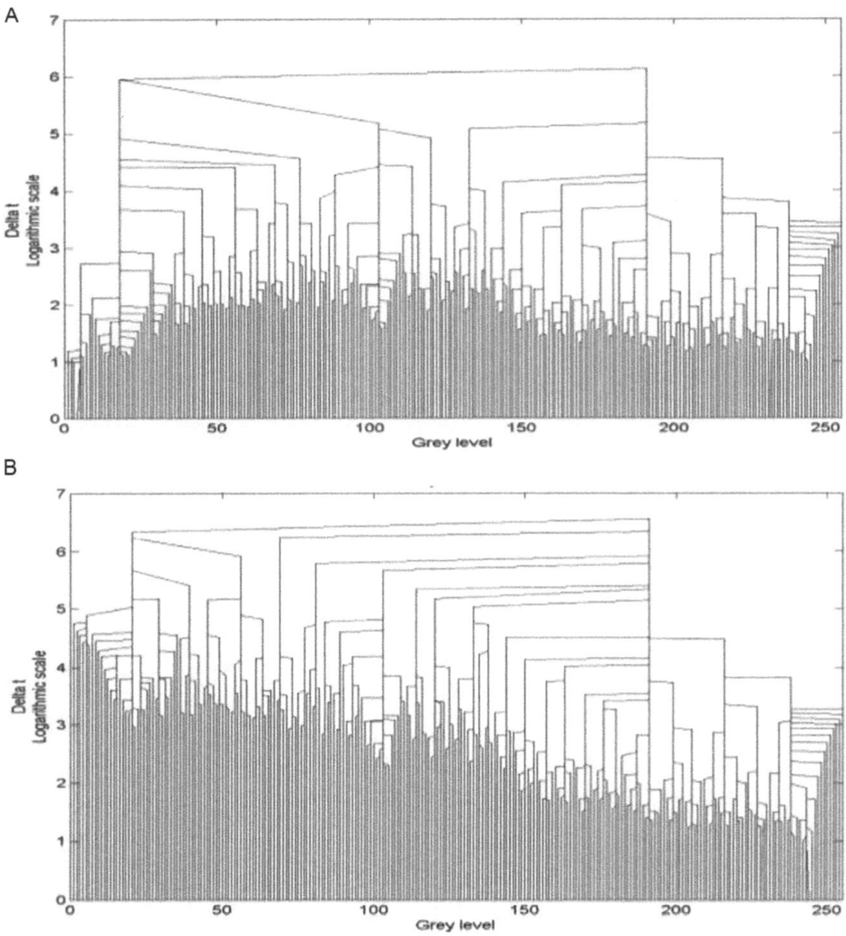

Fig. 31 Merging trees for the OA algorithm. (A) Results with the Euclidean metric. (B) Results with the LIP metric.

1.6.6 Conclusion and Perspectives of this Section

In this Section 1.6, we have presented a new image classification method that belongs to the family of HAC algorithms. It is based on Newton's universal attraction law, which defines the attraction force between moving bodies—here represented by each gray-level class of the image histogram. Two different algorithms have been presented: the sampling of the mechanical model (SM algorithm) and the optimized automated (OA algorithm). Each of them is driven by two metrics: the classical Euclidean metric and a logarithmic metric issued of the LIP model and consistent with human vision. For standard 8-bits images of 512×512 pixels, the execution time of SM algorithm (Matlab) runs approximately 11–12 s and the execution time of the OA algorithm is between 5 and 6 s.

The possibility of selecting various metrics presents the advantage to emphasize details of the image's dark parts (logarithmic approach) and to interpret the image classes as a human eye would.

Perspective: One question concerns the possibility of extending the gravitational clustering to color images.

Such an extension is done without theoretical or practical difficulty, except the computation time which becomes prohibitive. In order to speed up the algorithms (which was not the objective here), it is possible to write them in C language and, if necessary, use graphic cards.

Let us now refer to Rashedi and Nezamabadi-pour (2010), who proposed a gravitational approach to color image segmentation, where each pixel of the studied image is represented by a $\mathbb{R}^5$ vector: the coordinates of the pixel in the spatial domain and its RGB components. The metric then used in $\mathbb{R}^5$ is the Euclidean one. In such a way, spatial coordinates and colors in each channel are taken into account in the same manner. We think that the technique can be significantly improved by the introduction of metrics consistent with human vision, like those issued in the LIP color model (Jourlin et al., 2011) or those we have proposed as a simplified approach (cf. Chapter "Metrics Based on Logarithmic Laws," Section 6).

We present next (Fig. 32) a first example where we use an equivalent of SM algorithm with the Euclidian distance in RGB space. Excluding the problem of execution time, the results appear very hopeful.

Comment on Fig. 32: If we observe carefully the classified image, the quantification effect is visible (in the sky, for example), but we think this is mainly due to the very small number of classes (20) than to a weakness of the method.

Fig. 32 Gravitational clustering (SM algorithm) for a color image driven by Euclidean metric. (A) Initial color image. (B) Classified image (20 colors). (See the color plate.)

2. LOGARITHMIC WAVELETS

On this subject, we proposed a first paper (cf. Courbebaisse, Trunde, & Jourlin, 2002) whose goal was to introduce the concept of logarithmic wavelets. A much more detailed paper has been published by Navarro and Courbebaisse (cf. Navarro, Courbebaisse, & Jourlin, 2014). I recommend the interested reader to refer to this latter paper. Thus, the aim here is not to present its content in depth. Nevertheless, it seemed interesting to recall one application proposed by Navarro and Courbebaisse concerning the well-known ability of wavelets (and in all probability logarithmic wavelets) to denoise images.

2.1 Logarithmic Wavelet Denoising

Image denoising has been widely studied (see Buades, Coll, & Morel, 2005, for example). In our case, the denoising algorithm that we propose is rather simple because the aim of this section is to compare linear and logarithmic wavelets. The denoising is performed by a thresholding of the wavelet coefficients, according to the following steps:

- Computation of the wavelet coefficients of the image to be denoised with orthogonal wavelet transform decomposition on four levels
- Thresholding of the coefficients with a suitable threshold
- Reconstruction of the denoised image by inverse wavelet transform

The condition for thresholding the wavelet's coefficients is:

$$Thresh\,WT = Sup\left\{1 - \frac{T}{|WT|}, 0\right\}.WT \tag{22}$$

where *Thresh WT* represents the thresholded wavelet coefficients, *WT* is the wavelet decomposition of the image to be denoised, and *T* is the chosen threshold.

Logarithmic wavelet denoising stands on the same principle, but the logarithmic wavelet transform is used to decompose the image. The thresholding condition becomes:

$$Thresh\,WT_{\Delta} = Sup\left\{1 - \frac{T}{|WT|_{\Delta}}, 0\right\}.WT_{\Delta} \tag{23}$$

where $Thresh\,WT_{\Delta}$ represents the logarithmic thresholded wavelet coefficients, WT_{Δ} is the logarithmic wavelet decomposition of the image to be denoised and *T* is the chosen threshold.

2.2 Denoising Evaluation

The tests are realized on the classical image "Lena," on which the most common noises have been added, namely, Gaussian, Poisson, Salt and Pepper, and Speckle.

Remember that the Gaussian noise is statistically defined and its probability density function is the normal (Gaussian) distribution. The Poisson noise is also statistically defined and its probability density function is the Poisson distribution.

Here we will limit ourselves to these two noises for which the results of the wavelet logarithmic denoising are *visually* better than the simple wavelet denoising (see Figs. 33 and 34).

Fig. 33 Example of wavelet denoising on the Lena image, with added Gaussian noise. (A) Original image. (B) Noisy image. (C) Linear reconstruction. (D) Logarithmic reconstruction.

Fig. 34 Example of wavelet denoising on the Lena image, with added Poisson noise. (A) Original image. (B) Noisy image. (C) Linear reconstruction. (D) Logarithmic reconstruction.

Remark 6. We said that we compared visually the denoising efficiency of linear and logarithmic wavelets. This observation leads us to propose *objective* tools to *quantify* the superiority of the logarithmic approach.

One of the most classical ways to perform that is the use of signal-to-noise ratio (SNR). The SNR of a noisy image versus the original one can be computed as

$$SNR(I, I_N) = 20.log_{10}\left(\frac{||I||}{||I - I_N||}\right) \tag{24}$$

where I is the studied image and I_N the noisy one.

We already observed in various occasions that the SNR does not evaluate the quality of an image as the human eye would, and it is corroborated by the

Table 1 Comparison of Logarithmic Distances between Noisy Images and Linear and Logarithmic Denoising, with Poisson and Gaussian Noise Added

Added Noise	Logarithmic Distance with Initial Image
Gaussian Noise	
Noisy image	66,72
Linear denoising	49,34
Logarithmic denoising	36,35
Poisson Noise	
Noisy image	45,22
Linear denoising	60,85
Logarithmic denoising	53,58

two previous examples (Gaussian and Poisson noises) where the SNR values of images (B) and (C) are roughly the same and are better that the SNR of images (D).

Remark 7. For the Salt and Pepper and Speckle noises, the results for linear and logarithmic approaches are rather similar, thus we will not present them here. Nevertheless, we invite the interested reader to refer to Navarro et al. (2014) for more information.

Considering the weak information carried by SNR to evaluate the visual image quality after denoising, it seemed interesting to explore another way.

Knowing the consistency of the LIP model with human vision, we decided to compute a logarithmic metric to compare the denoising efficiency of linear and logarithmic wavelets.

The approach we propose consists of computing the distance between a noisy image and its restoration by means of linear and logarithmic wavelets. To do that, we estimate the images mean gray-level difference in the sense of LAC.

The results are presented in Table 1, for Gaussian and Poisson noises added.

3. PERCOLATION

Theory of percolation has been introduced at a mathematical level by Broadbent and Hammersley (1957). It has been studied in depth at a mathematical level [random tiling, percolation in Penrose tiling (Babalievski, 1992)], but is rarely used in image processing, although it is rather natural to adapt the percolation concept to this domain.

In fact, if we define on a gray level or color image a favored direction (for example, from top to bottom like for coffee percolation), it suffices to drive the pixels moving according to a *cost function*. Such a function permits determining a pixel trajectory across the image: at each step of its trajectory, the studied pixel moves to the following location in order to minimize the cost function.

In our opinion, it is possible to innovate in the definition of the cost function: it is possible to favor the choice of bright (resp. dark) pixels or to minimize the contrast (Michelson, or logarithmic additive, or logarithmic multiplicative), and so on.

Percolation can simulate *real* propagations, like heat propagation, forest fire, etc., or virtual propagations aimed at estimating the homogeneity/heterogeneity of a material or more generally of an image.

On our side, we applied successfully the percolation tool in various situations. For example:

- Mayet (2001) used this technique during his Ph.D thesis at the CSTB (Centre Scientifique et Technique du Bâtiment) in order to estimate the wear level of a carpet or the homogeneity of concrete samples: he proposed two methods, one based on the concept of coentropy dimension and one on percolation, knowing that the pixels trajectories are not the same for a worn or homogeneous material and for a new or heterogeneous one.
- For a problem of weaving control (polyester weaving for paper-pulp draining), M. Carré proposed a percolation to detect continuously the weaving front. Unfortunately, for industrial confidential reasons, we are not authorized to present demonstrative images.

Let us now present two particular applications with some more details.

3.1 Application 1: A New Approach of Textures Classification

Here we return to Mayet's idea: the complexity of pixels trajectories through a more or less textured image can contribute to the characterization of the texture itself. Note that this idea has been explored independently by Fricout and Jeulin (2003) under the name of *tortuosity*, in fact trajectories tortuosity. Nevertheless, it seems that a very little literature is dedicated to this idea.

The question we try to answer is:

Is it possible to extract sufficient parameters of the percolation trajectories to characterize and identify a texture?

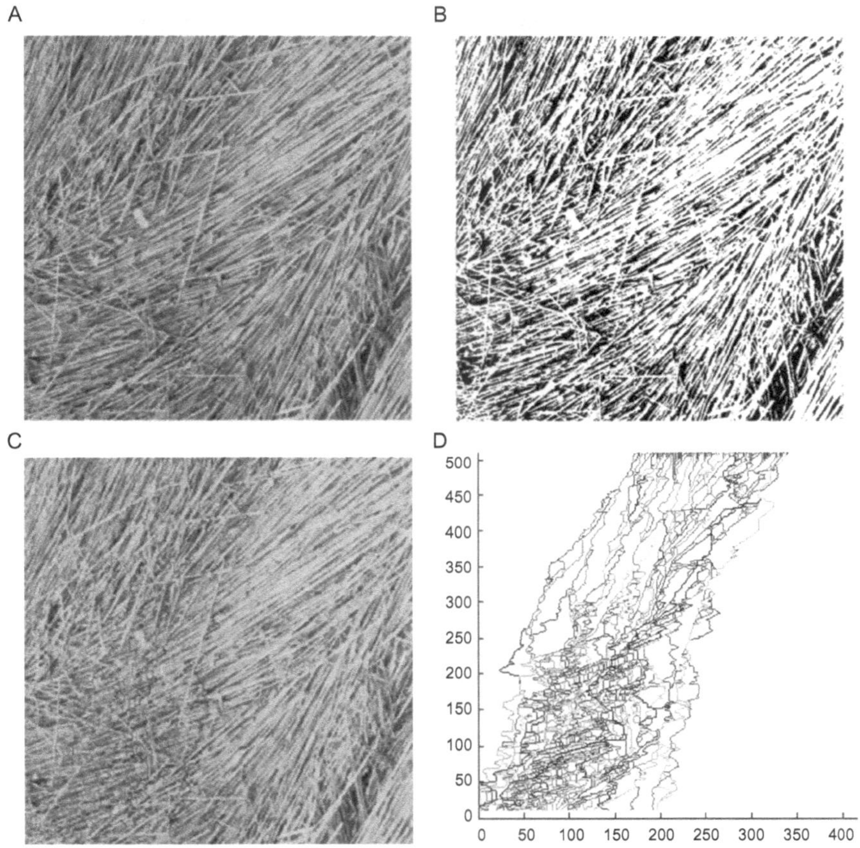

Fig. 35 Percolation trajectories computed on a textured image. (A) Initial image from Brodatz database. (B) Binary image associated to (A). (C) Original image and its percolation trajectories. (D) Percolation trajectories only. (See the color plate.)

3.1.1 Computation of Percolation Trajectories

Fig. 35A–D shows the approach we proposed, considering that the percolation is directed from the top to the bottom of the image (this direction is identical to a gravitational field in the case of physics).

Fig. 35A represents the considered initial image, which is extracted from the very classical Brodatz database of textured images.

We associate a binary image (B) to this image and we apply to (B) a percolation algorithm from top (first line of the image) to bottom (last line of the image).

Thus, each pixel of the first line moves inside the image according to a percolation trajectory. Some of these trajectories are superimposed to the initial image, which is shown on Fig. 35C.

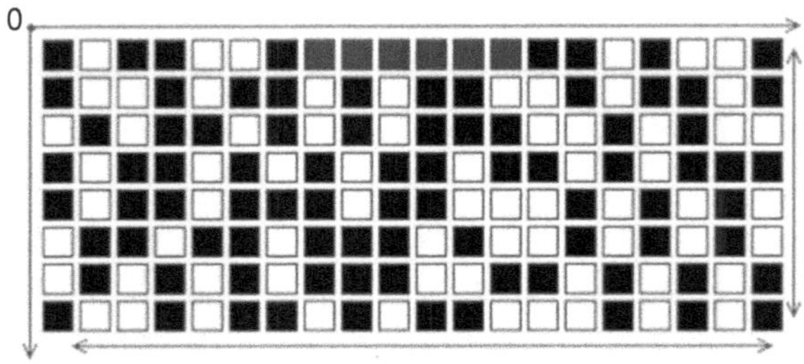

Fig. 36 *Red* (*gray* in the print version) points are selected as starting points for percolation in the binary image.

Finally, these trajectories are represented separately on Fig. 35D to facilitate their readability.

Implementation of percolation: This section has been realized with the collaboration of B. Abdallah [actually completing his Ph.D in Image Processing at the Centre de Morphologie Mathématique (Ecole des Mines de Paris)], and M. Inam Ul Haq during his Ph.D thesis at Saint-Etienne University.

We were particularly interested in the characterization of the so-called random textures, considering that for pseudo-periodic textures, many interesting tools have been developed and are functional at the moment, based on auto-correlation, covariograms, etc.

Although we have defined plenty of cost functions for gray-level and color images, we are going to explain the percolation process for a binary image in order to simplify the comments. This is why we associated a binary image (Fig. 35B) to image 35A, obtained through Otsu's (1979) classical method based on interclass variance maximization.

Initially, we select a number of points on the starting line (in our study the first line of the *binary image*, but any other line can be chosen as starting line). In order to initialize the percolation process, we have chosen to consider points near the middle of the first line to avoid problems of percolation toward unintended edges, which could distort the results (see Fig. 36).

Once the starting points are selected, we perform the process until the last line of the image is reached.

We select each of the current (red) points one by one because we are interested in a portion of its N_8 neighborhood (green in Fig. 37) corresponding to the percolation direction. Since we are working on a binary image, we will try to follow as possible pixel value 1 (white as in Fig. 37).

For each of these neighboring points, a parameter is computed (which could be driven by a gray-scale LUT in case of gray-level image), and which

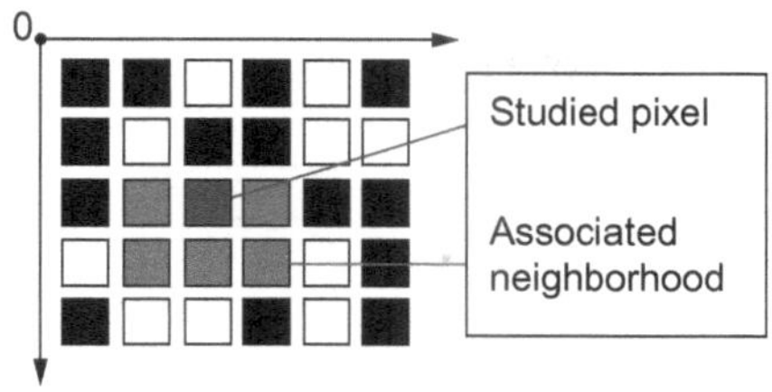

Fig. 37 Useful part (in *green*) of the neighborhood of the studied pixel (in *red*). (See the color plate.)

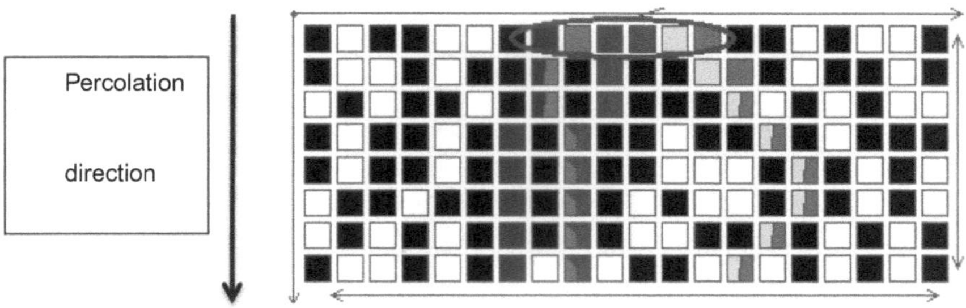

Fig. 38 Result of the percolation of the example presented in Fig. 36. (See the color plate.)

is here simply the value of the pixel. Then we will select the neighboring point to maximize the selected parameter (in our case the lightest point).

Note: In order to overcome the problem of multiple points with the same parameter value, we set up a random selection among the best candidates.

At the end of the process, the trajectory of each initial pixel has been built (cf. Fig. 38).

3.1.2 Percolation Trajectories and Parameters Extraction

This section does not aim to propose a complete and automated method. It constitutes a first attempt to decide if it would be interesting to persist in this way.

Concretely, we tried a multitude of parameters that can be used in the study of these trajectories. After studying the influence of these, we selected the most significant parameters based on principal component analysis, which allowed us to calculate the weights of the various chosen parameters. This study was conducted with the Brodatz base of textured images.

The parameters selected are those producing the best separation of different textures and at the same time bringing together the textures of same nature.

Our approach is exploratory and does not aim to deeply justify the selection of retained parameters. It is the reason why we have limited ourselves to the extraction of nine parameters, listed next:

Histogram: This variable is not used directly as a parameter, but the parameter *MeanHisto* is derived from it. This variable is a vector $(1 \times m)$, where m is the image size, in the direction perpendicular to the percolation axis (in this case the width). Variable *Histo* at each index j represents the number of trajectories arriving in column j (in the case of a vertical percolation).

Mean Histogram: This parameter represents the average offset d (that is to say, on the axis orthogonal to the axis of the percolation) of trajectories. It is calculated by taking the weighted average of the columns (the weight of a column j corresponding to the number of curves approaching this column's end). Thereafter, we subtract the average of the column values initially chosen (variable windows). Thus we obtain the average displacement through our connected curves (percolation candidates) from the departure point to the end one.

In addition, because we want to have a variable independent of the initial image size, these results are normalized, ie, divided by the width m of the image. If the size of initial image I was $n \times m$, the following result holds:

$$MeanHisto = \frac{1}{m}.\sum_{j=1}^{m}(Histo(j).j - Window(j).j)$$

The other selected parameters are:

L: Probability, at a given moment, that a droplet at the position (i, j) arrives at $(i, j-1)$ at next step

R: Probability, at a given moment, that a droplet at the position (i, j) arrives at $(i, j+1)$ at next step

BR: Probability, at a given moment, that a droplet at the position (i, j) arrives at $(i+1, j+1)$ at next step

BL: Probability, at a given moment, that a droplet at the position (i, j) arrives at $(i+1, j-1)$ at next step

B: Probability, at a given moment, that a droplet at the position (i, j) arrives at $(i+1, j)$ at next step

Mean Fractality: Average fractal dimension calculated for each observed droplet trajectory. Note that the fractal dimension of droplets trajectories is computed using the classical method "boxes."

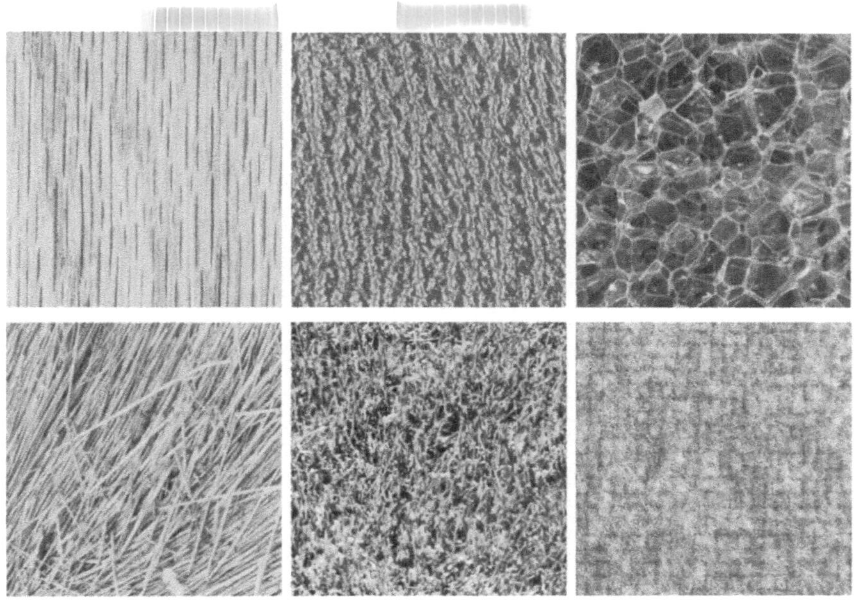

Fig. 39 Images of the six initial textures.

3.1.3 Results

With the previous settings, a $K-NN$ classifier has been programmed on a database of 72 images divided into six different types of random textures. More precisely, we started with six different textures from Brodatz database (Fig. 39) and divided each of them into 12 subimages, resulting in 72 textures.

The percentage of good affectations is approximately 90%, which can be improved by addition of new parameters computed on the trajectories.

3.1.4 A Variant of Percolation Method: Percolation by Propagation Front

We refer here to Mayet (2001). The trajectory of each pixel starting from the top of the image is affected with a "geodesic" time; meaning that the propagation of a point is weighted by two parameters: the gray level of the crossed pixel (for example, a bright pixel is harder to pass through) and the trajectory length—the cost of a diagonal moving is $\sqrt{2}$ time greater than a vertical one. This approach allows us to produce a map of geodesic times (Fig. 40B).

Remark 8. A threshold (time t_O) applied on the map of geodesic times results in a binary image whose boundary represents the *propagation front* at time t_O. If we refer at the previous attempt to classify textures, it would be informative to introduce new parameters describing the length, tortuosity, fractal dimension, etc. of such propagation fronts.

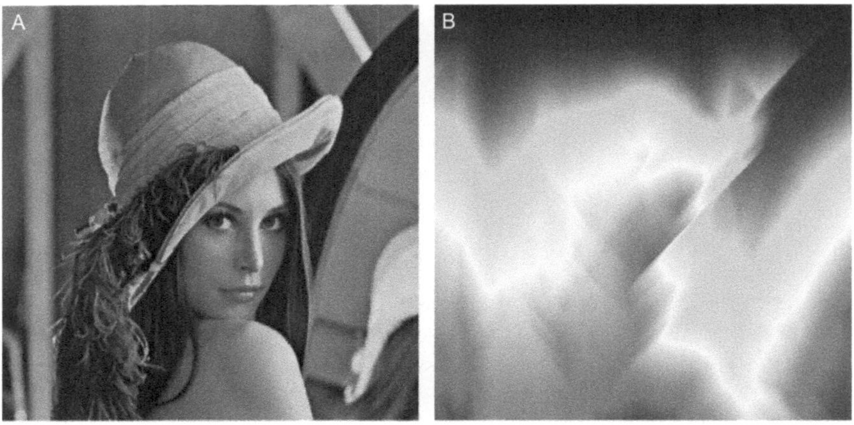

Fig. 40 Geodesic time propagation on a gray-level image. (A) Initial image. (B) Associated map of geodesic times, in false colors. (See the color plate.)

3.2 Application 2: Perception and Automated Detection of Polishing Scratches

This example has been studied in depth by T. Puntous during his Ph.D thesis (Puntous, 2013; Puntous, Rech, Delafosse, & Jourlin, 2011; Puntous, Pavan, Delafosse, Jourlin, & Rech, 2013). The problem addresses quality control of polished metallic surfaces, such as watches. The last step is only done by human controllers, thus there exists a real need for automated and objective control.

Puntous' work covers several steps:

First, he realized scratches of various size and depth on metallic samples.

Then he decided to use LIP tools in order to simulate the human perception of defects.

Finally, the linear scratches seemed adapted to percolation, and more precisely to LIP percolation.

Let us insist on the high difficulty level of the problem, particularly to detect the almost invisible scratches (cf. Fig. 41).

To become aware of the problem to solve, let us examine the behavior of some classical contour detectors applied to image 41 (cf. Fig. 42).

Comment on Fig. 42: The results presented in A–C are clearly not convincing. Even the Canny method (Canny, 1986) produces a noisy image impossible to exploit automatically.

Puntous proposed a solution mixing various tools:

A method without a priori knowledge of the propagation direction proposed by Yamaguchi, Nakamura, Saegusa, and Hashimoto (2008);

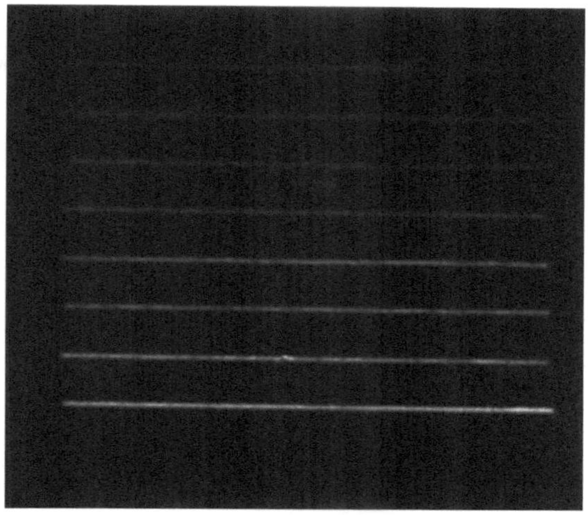

Fig. 41 Initial image: scratches of variable size and depth.

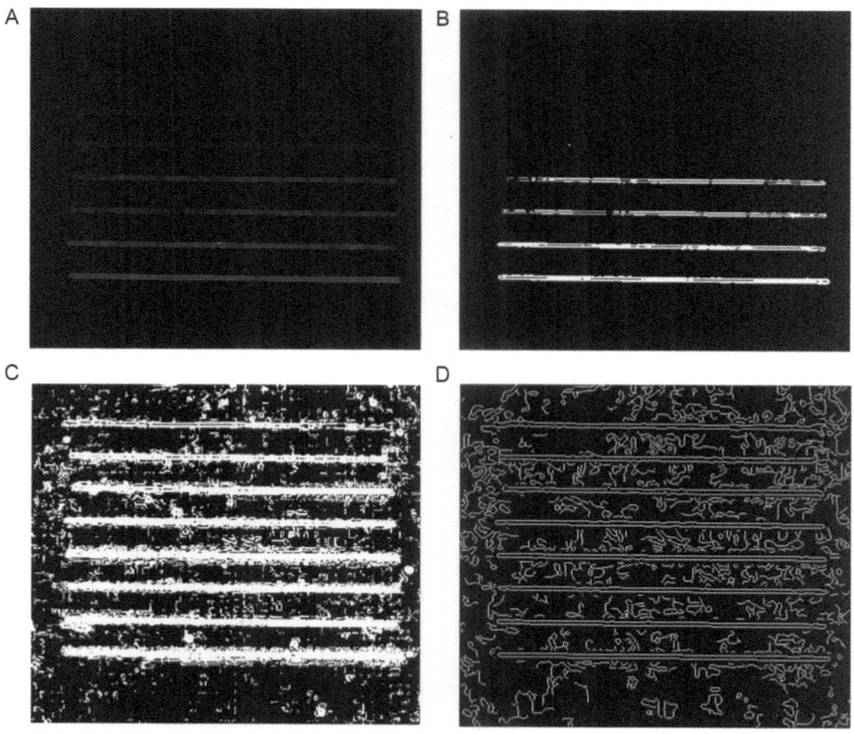

Fig. 42 Examples of scratches detections applied to image 41. (A) Prewitt gradient applied to 41. (B) Automated thresholding of (A) by Otsu method (interclass variance maximization). (C) Manual thresholding of (A) in order to detect the less visible scratches. (D) Canny's contour detection applied to 41.

Yamaguchi and Hashimoto (2010). It consists roughly of a region growing in a homogeneous region and is able to propagate inside a crack (Fig. 43).

A LAC between the percolated zone and its neighborhood.

By this technique, Puntous obtained together a sensitive detection of scratches and a logarithmic contrast between each scratch and its neighboring background (Fig. 44).

Comment on Fig. 44: The width is not precisely estimated for the less visible scratch, which appears normal at this very low initial visibility.

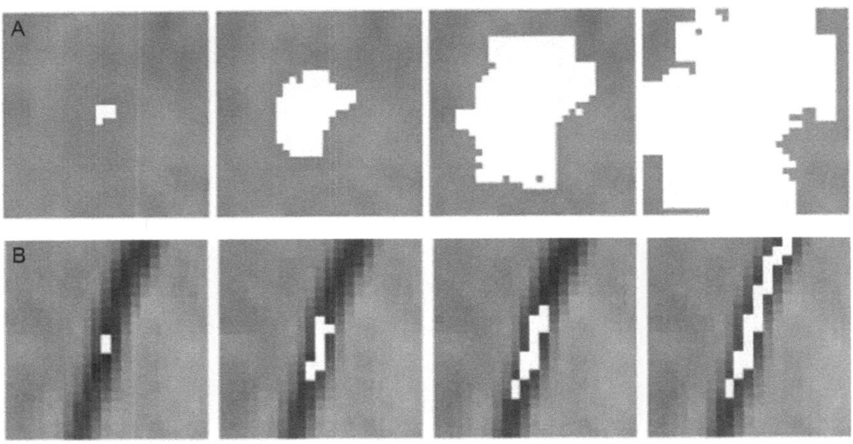

Fig. 43 Examples of Yamaguchi percolation. (A) Percolation in the image background. (B) Percolation in presence of a linear element.

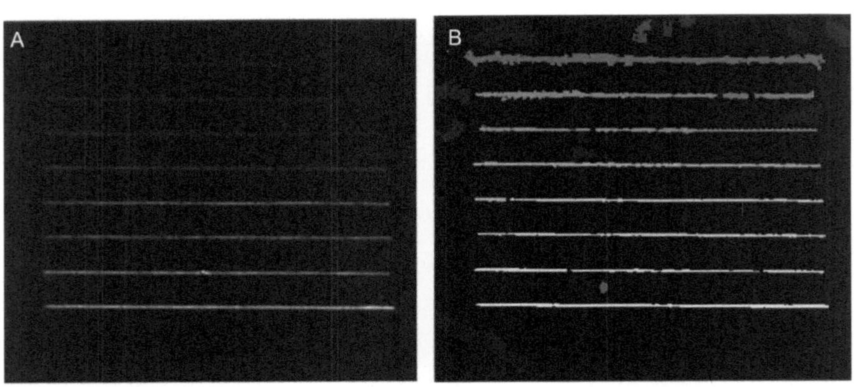

Fig. 44 Scratches detection (percolation) and their LIP contrasts with surrounding background. (A) Recall of initial image. (B) Scratches detection and associated logarithmic contrasts.

4. LIP TOP-HAT

We already evoked the logarithmic version of the Top-Hat. Meyer (1978) introduced the initial Top-Hat transform. It is based on the difference between the studied image and its gray-level opening by some structuring element (white Top-Hat) and dually on the difference between the closing and the input image. Applied according to this definition, the Top-Hat transform will extract too much detail due in particular to noise, thus an a priori condition is required on the height of the extracted features.

Such a transform is particularly efficient to extract peaks and crest lines. Nevertheless, for images acquired in transmission, which are the predilection domain of the LIP model, a remaining problem is due to the attenuation of the contrast generated by a semitransparent obstacle when it is observed on a dark background: such a situation implies that at some moment, the height of detected objects becomes lower than the required one and are thus wrongly rejected.

For this reason, we proposed in Jourlin and Montard (1997) a LIP version of the Top-Hat where the required "height" is replaced by a LAC in order to become insensitive to the background darkness.

A refinement of the LIP Top-Hat has been successfully used for extracting cracks (cf. Chapter "Dynamic Range Expansion, Night Vision. Stabilization, Centering. Industrial and Biomedical Applications," Section 6.2 concerning the software "Mosaïque"). In fact, the simple LIP Top-Hat was not able to perfectly detect cracks: either they were not detected on all their length (high values of the LAC), or we obtained over-detections for low values of the LAC. The solution arose from the concept of Double LIP Top-Hat consisting of applying two LAC values and performing a *marker reconstruction* of the lower detection by the higher one (Rémy, Jourlin, & Moreaud, 2007). An example of crack detection based on this technique is shown in Fig. 45.

5. FUNNEL-SHAPED GROWING IN THE SENSE OF DARSONVILLE

In his Ph.D thesis, Darsonville (1998) suggested for the first time to use a HAC to associate to each point x of an image the funnel composed of its successive growing levels until the whole image. In particular, he associated to x the set of regions $\{R_\varepsilon(x)\}_{\varepsilon \in \mathbb{R}^+}$, each resulting of a region growing

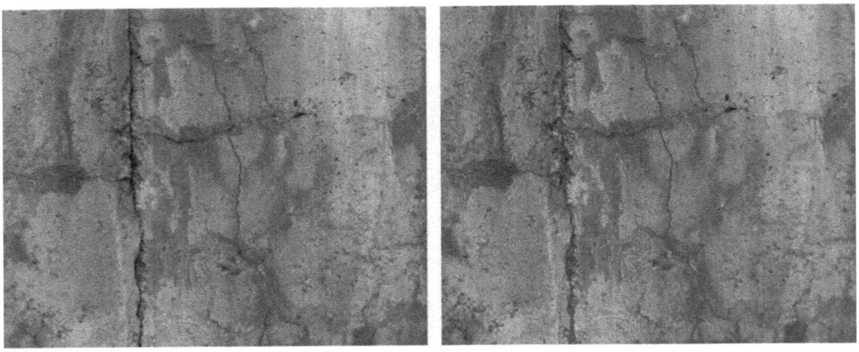

Fig. 45 Crack detection based on double LIP Top-Hat. (See the color plate.)

(single linkage, complete linkage or others) starting from the seed x with an aggregation parameter ε.

For a single linkage, for example, due to the associated equivalence relation (see Chapter "Various Contrast Concepts," *Section* 1.6.3.), the set of regions $\{R_\varepsilon(x)\}_{x\in D}$ for a given value of ε constitutes a *partition* of D, which means it satisfies the following properties:

– $\bigcup_{x\in D} R_\varepsilon(x) = D$

– $\forall(x, y) \in D^2, R_\varepsilon(x) = R_\varepsilon(y) \, or \, R_\varepsilon(x) \cap R_\varepsilon(y) = \varnothing$

Moreover, we have the following inclusion:

– $\forall x \in D, \forall(\varepsilon, \varepsilon') \in \mathbb{R}^{+2}, \varepsilon < \varepsilon' \Rightarrow R_\varepsilon(x) \subset R_{\varepsilon'}(x)$

Remark 9. The main interest of such a stacking of increasing neighborhoods is that it opens the way to novel developments in the field of multiscale image processing. In fact, the common way to perform multiscale algorithms consists of grouping pixels, for example, in growing squares (2×2, 4×4, etc.) without taking into account the possible strong dissimilarities between aggregated pixels. The solution proposed by Darsonville answers this problem: the increasing regions constituting the funnel levels are satisfied by construction similarity properties.

This idea has been widely developed by Debayle and Pinoli (2005), Debayle (2005) and in many papers under the name of logarithmic adaptative neighborhood image processing (LANIP). I only regret that in Debayle (2005), Debayle and Pinoli (2005), and further publications on the same subject, no reference has been made to Darsonville, despite my explicit demand.

6. APPLICATION TO SEMITRANSPARENT MEDIA ACQUIRED IN REFLECTION

The application I chose to illustrate this section has been done years ago for the company National Starck Chemical based near Lyon, France. This company produced auto-adhesive envelopes and desired control over the thickness of the glue spread on the envelope flap. The constraints were rather difficult: the envelopes were conveyed at the speed of 5 m/s and the company refused to work with two cameras to perform stereovision.

After discussion, they opted to mix a fluorescent material with the glue. In such a way, it remained to light the control field with black light to observe the glue as a green/blue region. Moreover, the green level (in the corresponding channel) was logarithmically linked to the glue thickness expressed in g/m^2. Thus the color evaluation permitted precise estimation of the glue grammage in real time and in each pixel. In such conditions, the produced information widely overstepped the required one: it was possible to visualize a 3D thickness map of the glue.

This technique may be adapted with some changing to other situations where we aim to control the thickness of a semitransparent medium, like cosmetic products, varnish, painting, etc.

7. CONCLUSION

The title of this chapter was very ambitious and we limited its content to some important transfers, particularly those on which the transfer was not limited to replace classical addition and scalar multiplication by logarithmic ones.

For example, in the field of segmentation methods, I have chosen to propose *logarithmic* and *systolic* region growing algorithms because they allowed me to solve difficult problems in the biomedical field. I also decided to insist on the gravitational clustering, considering its novelty and the new research it opens, in particular for segmenting color images.

The section dedicated to logarithmic wavelets, mainly based on Navarro and Courbebaisse (cf. Navarro et al., 2014), also appears as a novel field of research and experimentation. We have not, in my opinion, spent sufficient time to dispose to-day of spectacular applications. This remains to be done. The reader will probably have the same opinion concerning percolation. Nevertheless, the situation is very different: we successfully experiment with it in various problems, mainly to detect very badly contrasted boundaries. In such cases, we apply percolation on a contrast (or LIP contrast) map issued from the studied image. Unfortunately, the best examples we have are protected by confidentiality agreements.

More generally, I hope that the present book will engage new researchers to work in the LIP framework.

REFERENCES

Arbelaez, P. (2005). *Une approche métrique pour la segmentation d'images.* Ph.D. thesis, Université Paris Dauphine.

Arbelaez, P. (2006). Boundary extraction in natural images using ultrametric contour Maps. In *Proceedings of the 5th IEEE workshop on perceptual organization in computer vision* (p. 82).

Asano, T., & Yokoya, N. (1981). Image segmentation schema for low-level computer vision. *Pattern Recognition, 14*(1–6), 267–273.

Babalievski, F. (1992). Percolation conductivity of Penrose tiling by the transfer matrix Monte Carlo method. *Physica A: Statistical Mechanics and its Applications, 182*(3), 325–330.

Baraldi, A., & Parmiggiani, F. (1996). Single linkage region growing algorithms based on the vector degree of match. *IEEE Transactions on Geoscience and Remote Sensing, 34*(1), 137–148.

Benzécri, J. (1970). Problèmes et méthodes de la taxinomie. *Revue de Statistique Appliquée, 18*(4), 73–98.

Benzécri, J. (1972). Description mathématique des classifications. *Revue de Statistique Appliquée, 20*(3), 23–56.

Bertrand, B. P., & Diday, E. (1990). Une généralisation des arbres hiérarchiques: les représentations pyramidales. *Revue de Statistique Appliquée, 38*(3), 53–78.

Broadbent, S., & Hammersley, J. (1957). Percolation processes. I. Crystals and mazes. *Proceedings of the Cambridge Philosophical Society, 53*, 629–641.

Buades, A., Coll, B., & Morel, J.-M. (2005). A review of image denoising algorithms, with a new one. *Multiscale Modeling & Simulation, 4*(2), 490–530.

Canny, J. (1986). A computational approach to edge detection. *IEEE Transactions on Pattern Analysis and Machine Intelligence (PAMI), 8*(6), 679–698.

Copty, N., Ranka, S., Fox, G., & Shankar, R. V. (1994). A data parallel algorithm for solving the region growing problem on connection machine. *Journal of Parallel and Distributed Computing, 21*, 160–168.

Courbebaisse, G., Trunde, F., & Jourlin, M. (2002). Wavelet transform and LIP model. *Image Analysis & Stereology, 21*, 121–125.

Darsonville, F. (1998). *Segmentation et Visualisation d'images 3D. Application par croissance de région à l'imagerie médicale dentaire.* Ph.D thesis, France: Saint-Etienne University. 176 p.

Debayle, J. (2005). *Traitement d'image à voisinages adaptatifs généraux.* Ph.D thesis, Ecole Nationale Supérieure des Mines de Saint-Etienne.

Debayle, J., & Pinoli, J.-C. (2005). Adaptive-neighborhood mathematical morphology and its applications to image filtering and segmentation. In *9th European congress on stereology and image analysis, Zakopane (Poland).*

Deng, G. (1993). *Image and signal processing using the logarithmic image processing model.* Ph.D Engineering, La Trobe University.

Deng, G., & Pinoli, J.-C. (1998). Differentiation-based edge detection using the logarithmic image processing model. *Journal of Mathematical Imaging and Vision, 8*(2), 161–180.

Diday, E. (1971). Une nouvelle méthode en classification autpmatique et reconnaissance de formes: la méthode des nuées dynamiques. *Revue de Statistique Appliquée, 19*(2), 19–33.

Diday, E. (1983). Inversions en classification hiérarchique: Application à la construction adaptative d'indices d'agrégation. *Revue de Statistique Appliquée, 31*(1), 45–62.
Forgy, E. W. (1965). Cluster analysis of multivariate data: Efficiency vs interpretability of classifications. *Biometrics, 21*, 768–769.
Freixenet, J., Muñoz, X., Raba, D., Marti, J., & Cufi, X. (2002). Yet another survey on image segmentation: Region and boundary information integration. In *Proceedings of the 7th European conference on computer vision-part III. ECCV '02* (pp. 408–422). London, UK: Springer-Verlag.
Fricout, G., & Jeulin, D. (2003). Textures classification from their morphological properties. In *Transactions of the 9th international conference on metrology and properties of engineering surfaces, Halmstad, Sweden.*
Fu, K., & Mui, J. (1981). A survey on image segmentation. *Pattern Recognition, 13*(1), 3–16.
Gomez, J., Dasgupta, D., & Nasraoui, O. (2003). A new gravitational clustering algorithm. In *Proceedings of the SIAM international conference on data mining (SDM).*
Jain, A. K., & Dubes, R. C. (1988). *Algorithms for clustering data.* New Jersey: Prentice Hall.
Jourlin, M., Breugnot, J., Abdallah, B., Corvo, J., Couka, E., & Carré, M. (2013). Image segmentation in the field of the logarithmic image processing model: Special focus on the hierarchical ascendant classification techniques. In P. W. Hawkes (Ed.), *Advances in Imaging and Electron Physics: Vol. 177.* (pp. 1–44).
Jourlin, M., Breugnot, J., Itthirad, F., Bouabdellah, M., & Closs, B. (2011). Logarithmic image processing for color images. *Advances in Imaging and Electron Physics, 168*(2), 65–107.
Jourlin, M., Carré, M., Breugnot, J., & Bouabdellah, M. (2012). Logarithmic image processing: Additive contrast, multiplicative contrast, and associated metrics. *Advances in Imaging and Electron Physics, 171*, 357–406.
Jourlin, M., & Montard, N. (1997). A logarithmic version of the top-hat transform in connexion with Asplünd's distance. *Acta Stereologica, 16*(3), 201–208.
Jourlin, M., & Pinoli, J.-C. (2001). The mathematical and physical framework for the representation and processing of transmitted images. *Advances in Imaging and Electron Physics, 115*, 129–196.
Jourlin, M., Pinoli, J.-C., & Zeboudj, R. (1989). Contrast definition and contour detection for logarithmic images. *Journal of Microscopy, 156*, 33–40.
Kanungo, T., Mount, D. M., Netanyahu, N. S., Piatko, C. D., Silverman, R., & Wu, A. Y. (2002). An efficient *k*-means clustering algorithm: Analysis and implementation. *IEEE Transactions on Pattern Analysis and Machine Intellegence, 24*(7), 881–892.
Krasner, M. (1947). Espace ultramétrique et valuation. *Séminaire Dubreil, Algebre et théorie des nombres, 1*(1), 1–17.
Lin, Z., Jin, J., & Talbot, H. (2000). Unseeded region growing for 3D image segmentation. In *Visualization 2000, Pan-Sydney workshop on Visual information processing* (pp. 31–37).
Long, T., & Jin, L.-W. (2006). A new simplified gravitational clustering method for multiprototype learning based on minimum classification error training. In *Conference on intelligent computing in pattern analysis/synthesis (IWICPAS' 06)* (pp. 168–175).
Mac Queen, J. B. (1967). Some methods for classification and analysis of multivariate observations. In *Proceedings of 5-th Berkeley symposium on mathematical statistics and probability, Vol. 1.* (pp. 281–297). Berkeley: University of California Press.
Mahalanobis, P. C. (1936). On the generalized distance in statistics. *Proceedings of the National Institute of Sciences of India, 2*(1), 49–55.
Mayet, F. (2001). *Classification d'image par analyse fractale et multi-fractale.* Ph.D thesis, Saint-Etienne, France: University Jean Monnet. 92p.
Meyer, F. (1978). Contrast feature extraction. In J.-L. Chermant (Ed.), *Sonderbände der Praktishen Metallographie: 8.* (pp. 374–380). Stuttgart (Germany): Reiderer Verlag.
Navarro, L., Courbebaisse, G., & Jourlin, M. (2014). Logarithmic wavelets. *Advances in Imaging and Electron Physics, 183*, 41–98.

Otsu, N. (1979). A threshold selection method from grey-level histograms. *IEEE Transactions on Systems, Man, and Cybernetics*, *9*(1), 62–66.
Palomares, J., Gonzalez, J., Ros, E., & Prieto, A. (2006). General logarithmic image processing convolution. *IEEE Transactions on Image Processing*, *15*(11), 3602–3608.
Pinoli, J.-C. (1992). *Modélisation et traitement des images logarithmiques*. Habilitation à Diriger des Recherches(HDR), France: Saint-Etienne University.
Puntous, T. (2013). *Perception et détection automatique des rayures de polissage*. Ph.D thesis, France: University of Saint-Etienne. 141p.
Puntous, T., Pavan, S., Delafosse, D., Jourlin, M., & Rech, J. (2013). Ability of quality controllers to detect standard scratches on polished surfaces. *Precision Engineering*, *37*, 924–928.
Puntous, T., Rech, J., Delafosse, D., & Jourlin, M. (2011). Perceptual percolation based on logarithmic image processing model. In *13th International congress of stereology (ICS 13), Beijing (China)*.
Rashedi, E., & Nezamabadi-pour, H. (2010). A stochastic gravitational approach to color image segmentation by considering spatial information. In *Proceedings of the 1st international conference on communications Engineering* (pp. 87–91).
Rashedi, E., Nezamabadi-pour, H., & Saryazdi, S. (2009). GSA: A gravitational search algorithm. *Information Sciences*, *179*(13), 2232–2248.
Raut, S., Raghuvanshi, M., Dharaskar, R., & Raut, A. (2009). Image segmentation: A state-of-art survey for prediction. In *International conference on advanced computer Control, ICACC'09* (pp. 420–424).
Rémy, P., Jourlin, M., & Moreaud, M. (2007). Procédé de detection de désordres dans un ouvrage en béton ou en maçonnerie, French Patent Number FR 07 58302.
Revol, C., & Jourlin, M. (1997). A new minimum-variance region growing algorithm for image segmentation. *Pattern Recognition Letters*, *18*, 249–258.
Sekiguchi, H., Sano, K., & Yokoyama, T. (1994). Interactive 3-dimensional segmentation method based on region growing method. *Systems and Computers in Japan*, *25*(1), 88–97.
Sivewright, G. J., & Elliot, P. J. (1994). Interactive region and volume growing for segmenting volumes in MR and CT images. *Medical Informatics*, *19*(1), 71–80.
Sylos Labini, F. (2008). Gravitational clustering: an overview, ArXiv e-prints. arXiv:0806.2560 [astro-ph].
Thilagamani, S., & Shanthi, N. (2011). A survey on image segmentation through clustering. *International Journal of Research and Reviews in Information Sciences*, *1*(1), 14–17.
Wan, S.-Y., & Higgins, W. E. (2003). Symmetric region growing. *IEEE Transactions on Image Processing*, *12*(8), 1–9.
Wright, W. (1977). Gravitational clustering. *Pattern Recognition*, *9*(3), 151–166.
Wu, Z. (1993). Homogeneity testing for unlabeled data: A performance evaluation. *CVGIP Graphical Models Image Processing*, *55*(5), 370–380.
Yamaguchi, T., & Hashimoto, S. (2010). Fast crack detection method for large-size concrete surface images using percolation-based image processing. *Machine Vision and Applications*, *21*(5), 797–809.
Yamaguchi, T., Nakamura, S., Saegusa, R., & Hashimoto, S. (2008). Image-based crack detection for real concrete surfaces. *IEEJ Transactions on Electrical and Electronic Engineering*, *3*(1), 128–135.

CHAPTER SEVEN

General Conclusion

Since the results and papers concerning the logarithmic imaging processing (LIP) model are too numerous to be presented in a single volume, I made the following choices:

- I decided to focus on the results I obtained, in collaboration with my students (Masters degree and PhD), because I estimated little interest in making a "survey" of the subject: such an exercise is always ambiguous and requires a high level of precision and diplomacy to assign each result to the right person.
- Even with the previous restricted area, I was forced to limit my presentation to a short summary of the initial results I obtained through the model to favor a detailed presentation of the previous unpublished results.
- I always tried to justify the novel concepts introduced without going too deeply into the mathematical aspects.
- Finally, having successfully applied the LIP tools in a number of industrial and biomedical situations, I proposed real examples throughout the book, despite the confidentiality agreements that prevented me from explaining some other interesting applications.

I hope this book will convince some readers to enter the LIP framework.

Advances in Imaging and Electron Physics, Volume 195
ISSN 1076-5670
http://dx.doi.org/10.1016/bs.aiep.2016.04.007

INDEX

Note: Page numbers followed by "*f*" indicate figures, and "*t*" indicate tables.

CONTENTS OF VOLUMES 151-194

Volume 151[1]

Volume 152

Volume 153

[1] Lists of the contents of volumes 100–149 are to be found in volume 150; the entire series can be searched on ScienceDirect.com

Volume 161

Marian Mankos, Vassil Spasov and Eric Munro, Principles of dual-beam low-energy electron microscopy
Jorge D. Mendiola-Santibañez, Iván R. Terol-Villalobos and Israel M. Santillán-Méndez, Determination of adequate parameters for connected morphological contrast mappings through morphological contrast measures
Ignacio Moreno and Carlos Ferreira, Fractional Fourier transforms and geometrical optics
Vladan Velisavlevic, Martin Vetterli, Baltasar Berufell-Lozano and Pier Luigi Dragotti, Sparse image representation by directionlets
Michael H.F. Wilkinson and Georgios K. Ouzounis, Advances in connectivity and connected attribute filters

Volume 162

Kiyotaka Asakura, Hironobu Niimi and Makoto Kato, Energy-filtered x-ray photoemission electron microscopy (EXPEEM)
Eireann C. Cosgriff, Peter D. Nellist, Adrian J. d'Alfonso, Scott D. Findlay, Gavin Behan, Peng Wang, Leslie J. Allen and Angus I. Kirkland, Image contrast in aberration-corrected scanning confocal electron microscopy
Christopher J. Edgcombe, New dimensions for field emission: effects of structure in the emitting surface
Archontis Giannakidis and Maria Petrou, Conductivity imaging and generalised Radon transform: a review
Olivier Losson, Ludovic Macaire and Yanqin Yang, Comparison of color demosaicing methods

Volume 163

Wolfgang S. Bacsa, Optical interference near surfaces and its application in sub-wavelength microscopy
Ruy H.A. Farias and Erasmo Recami, Introduction of a quantum of time ("chronon"), and its consequences for the electron in quantum and classical physics
Andrew Neice, Methods and limitations of subwavelength imaging
A. Sever Škapin and P. Ropret, Identification of historical pigments in wall layers by combination of optical and scanning electron microscopy coupled with energy-dispersive spectroscopy
Markus E. Testorf and Michael A. Fiddy, Superresolution imaging–revisited

Volume 164

Amos Bardea and Ron Naaman, Magneto-lithography: from the bottom-up route to high throughput
Román Castañeda, The optics of spatial coherence wavelets
Junchang Li, Yanmei Wu and Yan Li, Common diffraction integral calculation based on a fast Fourier transform algorithm
Marcel Teschke and Stefan Sinzinger, A generalized approach to describe the interference contrast and phase contrast method
Dokkyu Yi and Booyong Choi, Nonlinear partial differential equations for noise problems
Henning F. Harmuth, Harmuth corrigenda

Volume 165

Natalie Baddour, Two-dimensional Fourier transforms in polar coordinates
Neil V. Budko, Superluminal, subluminal, and negative velocities in free-space electromagnetic propagation
Rowan Leary and Rik Brydson, Chromatic aberration correction: the next step in electron microscopy
Michele Marrocco, Methods for vectorial analysis and imaging in high-resolution laser microscopy
Tomoya Sakai, Masaki Narita, Takuto Komazaki, Haruhiko Nishiguchi and Atsushi Imiya, Image hierarchy in Gaussian scale space
Yusuf Ziya Umul, The theory of the boundary diffraction wave
Emil Wolf, History and solution of the phase problem in the theory of structure determination of crystals from x-ray diffraction measurements

Volume 166

Valeriy Syrovoy, Theory of Intense Beams of Charged Particles

Volume 167

Emmanuel de Chambost, A history of Cameca (1954–2009)
Johan Debayle and Jean-Charles Pinoli, Theory and applications of general adaptive neighborhood image processing
Mohamed ben Haj Rhouma, Mohamed Ali Khabou and Lotfi Hermi, Shape recognition based on eigenvalues of the Laplacian
Nicolas Loménie and Georges Stamon, Point set analysis
Leonid P. Yaroslavsky, Image recovery from sparse samples, discrete sampling theorem, and sharply bounded band-limited discrete signals

Volume 168

Luca Geretti and Antonio Abramo, The synthesis of a stochastic artificial neural network application using a genetic algorithm approach
Michel Jourlin, Josselin Breugnot, Frédéric Itthirad, Mohammed Bouabdellah and Brigitte Closs, Logarithmic image processing for color images
Rainer A. Leitgeb, Current technologies for high-speed and functional imaging with optical coherence tomography
Sergej A. Nepijko and Gerd Schönhense, Analysis of optical systems, contrast depth, and measurement of electric and magnetic field distribution on the object's surface in mirror electron microscopy
Chad M. Parish, Multivariate statistics applications in scanning transmission electron microscopy
Hidetaka Sawada, Fumio Hosokawa, Takeo Sasaki, Toshikatsu Kaneyama, Yukihito Kondo and Kazutomo Suenaga, Aberration correctors developed under the Triple C project
Tobias Schulz, Martin Albrecht and Klaus Irmscher, Spatially resolved thermoluminescence in a scanning electron microscope

Volume 169

Erchan Aptoula and Sébastien Lefèvre, Morphological texture description of grayscale and color images
Vera Guarrera and Herwig Ott, Electron microscopy of ultracold gases
Konstantinos Konstantinidis, Ioannis Andreadis and Georgios Ch. Sirakoulis, Application of artificial intelligence to content-based image retrieval
Xingwei Yang, Daniel B. Szyld and Longin Jan Latecki, Diffusion on a tensor product graph for semi-supervised learning and interactive image segmentation
S.A. Nepijko and G. Schönhense, Electron holography for electric and magnetic field measurement and its application for nanophysics

Volume 170

Alex S. Eggeman and Paul A. Midgley, Precession electron diffraction
Ray Hill, John A. Notte and Larry Scipione, Scanning helium ion microscopy
Hone-Ene Hwang and Pin Han, Signal reconstruction algorithm based on a single intensity in the Fresnel domain
Kazuhisa Sato, Toyohiko J. Konno and Yoshihiko Hirotsu, Electron microscpy studies on magnetic $L1_0$ FePd nanoparticles
D.A. Zanin, H. Cabrera, L. de Pietro, M. Pikulski, M. Goldmann, U. Ramsperger, D. Pescia and John P. Xanthakis, Fundamental aspects of near-field emission scanning electron microcopy

Volume 171

Gregor Esser, Wolfgang Becken, Werner Müller, Peter Baumbach, Josep Arasa and Dietmar Uttenweiler, Derivation of the reflection equations for higher order aberrations of local wavefronts by oblique incidence
Lila Iznita Izhar and Maria Petrou, Thermal imaging in medicine
Jean-Michel Tualle, Derivation of the radiative transfer equation in a medium with a spatially varying refractive index: a review
Kamlesh Shrivas and Mitsutoshi Setou, Imaging mass spectrometry. Sample preparation, instrumentation and applications
Robert T. Thompson and Steven A. Cummer, Transformation optics

Tobias Klein, Egbert Buhr and Carl Georg Frase, TSEM – a review of scanning electron microscopy in transmission mode and its applications
Michel Jourlin, Maxime Carré, Josselin Breugnot and Mohamed Bouabdellah, Logarithmic image procesing: additive contrast, multiplicative contrast and associated metrics

Volume 172

Jay Theodore Cremer, Neutron and x-ray microscopy, Part 1

Volume 173

Jay Theodore Cremer, Neutron and x-ray microscopy, Part 2

Volume 174

Silicon-based Millimeter-wave Technology. Measurement, Modeling and Applications
M. Jamal Deen and Ognian Marinov, Measurement techniques and issues
Guennadi A. Kouzaev, M. Jamal Deen and Natalia K. Nikolova, Transmission lines and passive components
Mohamed H. Bakr and Mohamed H. Negm, Modeling and design of high-frequency structures using artificial neural networks and space mapping
Oana Moldovan, Antonio Lázaro, François Danneville, Rodrigo Picos, Bogdan Nae, Benjamin Iniguez and M. Jamal Deen, Nanoscale FETs
M. Daneshmand and R.R. Mansour, RF MEMS switches and switch matrices
Natalia K. Nikolova, Maryam Ravan and Reza K. Amineh, Substrate-integrated antennas on silicon

Volume 175

Jay Theodore Cremer, Small angle scatter with correlation, scatter and intermediate functions
Jay Theodore Cremer, Nuclear scatter of neutron spin states
Christian Dwyer, Atomic-resolution core-level spectroscopy in the scanning transmission electron microscope
Partha Pratim Mondal and Alberto Diaspro, Point spread function engineering for super-resolution single-photon and multiphoton fluorescence microscopy
Paul Murray and Stephen Marshall, A review of recent advances in the hit-or-miss transform
Stephen J. Sangwine, Perspectives on color image procesing by linear vector methods using projective geometric transformations

Volume 176

Katsushige Tsuno, Damaschin Ioanoviciu, Early History of Wien Filters
Damaschin Ioanoviciu, Katsushige Tsuno, Aberration Theory of the Wien Filter
Katsushige Tsuno, Damaschin Ioanoviciu, Wien Filter Instrumentation
Katsushige Tsuno, Damaschin Ioanoviciu, Simulation of Multipole Wien Filters
Damaschin Ioanoviciu, Katsushige Tsuno, Wien Filter Applications to Ions
Katsushige Tsuno, Damaschin Ioanoviciu, Application of Wien Filters to Electrons

Volume 177

Michel Jourlin , Josselin Breugnot, Bassam Abdallah, Joris Corvo, Enguerrand Couka , Maxime Carré, Image Segmentation in the Field of the Logarithmic Image Processing Model: Special Focus on the Hierarchical Ascendant Classification Techniques
Petros Maragos, Representations for Morphological Image Operators and Analogies with Linear Operators
Kenneth C. A. Smith Electron Microscopy at Cambridge University with Charles Oatley and Ellis Cosslett: Some Reminiscences and Recollections
Miguel José-Yacamán, Arturo Ponce, Sergio Mejía-Rosales, Francis Leonard Deepak, Advanced Methods of Electron Microscopy in Catalysis Research

Volume 178

Tony Lindeberg, Generalized Axiomatic Scale-Space Theory
Agnieszka Lisowska, Smoothlet Transform: Theory and Applications

Evgeniy M. Yakushev, Theory and Computation of Electron Mirrors: The Central Particle Method

Volume 179

Claude Daviau, Invariant Quantum Wave Equations and Double Space-Time
Niels de Jonge, In-Situ and Correlative Electron Microscopy
Vladimir P. Oleshko, James M. Howe, Electron Tweezers as a Tool for High-Precision Manipulation of Nanoobjects
Pilar Sobrevilla, Eduard Montseny, Aina Barcelo, Robustness Analysis of the Reduced Fuzzy Texture Spectrum and its Performance on Noisy Images
Arturo Tejada, Wouter Van den Broek, Arnold J. den Dekker, Measure-by-Wire (MBW): An Automatic Control Framework for High-Throughput Transmission Electron Microscopy

Volume 180

Mikhail Ya. Schelev, Mikhail A. Monastyrskiy, Nikolai S. Vorobiev, Sergei V. Garnov and Dmitriy E. Greenfield, Aspects of Streak Image Tube Photography

Volume 181

Ying Bai, Xiao Han, Jerry L. Prince, Octree Grid Topology-Preserving Geometric Deformable Model (OTGDM)
Maïtine Bergounioux, Second-order Variational Models for Image Texture Analysis
Victoria Klang, Nadejda B. Matsko, Electron Microscopy of Pharmaceutical Systems

Volume 182

Hans R. Gelderblom, Detlev H. Krüger, Helmut Ruska (1908–1973): His Role in the Evolution of Electron Microscopy in the Life Sciences, and Especially Virology
Hans R. Gelderblom, Detlev H. Krüger, Peter W. Hawkes Publications from the Düsseldorf University Institute for Biophysics and Electron Microscopy (Institut für Biophysikund Elektronenmikroskopie der Universität Düsseldorf) 1958-1973
Nebojsa Neškovič, P. Beličev, I. Telečki, S. Petrovič, Rainbow Lenses
Ben Adcock, Anders Hansen, Bogdan Roman, Gerd Teschke, Generalized Sampling: Stable Reconstructions, Inverse Problems and Compressed Sensing over the Continuum

Volume 183

M.M. El-Gomati, C.G.H. Walker, Toward Quantitative Scanning Electron Microscopy
Laurent Navarro, Guy Courbebaisse, Michel Jourlin, Logarithmic Wavelets
F. Lanusse, J.-L. Starck , A. Woiselle, M. J. Fadili, 3-D Sparse Representations

Volume 184

Anatoli A. Ischenko, Sergei A. Aseyev, Time-Resolved Electron Diffraction: for Chemistry, Biology and Materials Science

Volume 185

Pawel Berczynski, Slawomir Marczynski, Gaussian Beam Propagation in Inhomogeneous Nonlinear Media. Description in Ordinary Differential Equations by Complex Geometrical Optics
David Agard, Yifan Cheng, Robert M. Glaeser, Sriram Subramaniam, Single-Particle Cryo-Electron Microscopy (Cryo-EM): Progress, Challenges, and Perspectives for Further Improvement
Martin Welk, Michael Breuß, Morphological Amoebas and Partial Differential Equations

Volume 186

Niels de Jonge, Marina Pfaff, Diana B. Peckys
Practical Aspects of Transmission
Electron Microscopy in Liquid
Jian-Jiun Ding, Soo-Chang Pei
Linear Canonical Transform
Andrey I. Denisyuk, Alexey V. Krasavin, Filipp E. Komissarenko, Ivan S. Mukhin

Mechanical, Electrostatic, and Electromagnetic Manipulation of Microobjects and Nanoobjects in Electron Microscopes

Volume 187

Ahmed Elgammal, Homeomorphic Manifold Analysis (HMA): Untangling Complex Manifolds
Teruo Kohashi, Spin-Polarized Scanning Electron Microscopy

Volume 188

Allen M. Carroll, Pattern Generators for Reflective Electron-Beam Lithography (REBL)
Frank Gunzer, Jürgen Grotemeyer, Recent Developments in Time-of-Flight Mass Spectrometry
Margit Pap, A Special Voice Transform, Analytic Wavelets, and Zernike Functions
Colin J.R. Sheppard, Shan S. Kou, Jiao Lin, The Hankel Transform in *n*-dimensions and Its Applications in Optical Propagation and Imaging

Volume 189

Georges Lochak, Theory of the Leptonic Monopole, Part 1
Harald Stumpf, Symmetry Breaking by Electric Discharges in Water and Formation of Lochak's Light Magnetic Monopoles in an Extended Standard Model, Part 2

Volume 190

Niels de Jongek, CISCEM 2014: Proceedings of the Second Conference on In situ and Correlative Electron Microscopy, Saarbrücken, Germany, October 14–15, 2014
A. R. Faruqi, Richard Henderson, and Greg McMullan, Progress and Development of Direct Detectors for Electron Cryomicroscopy
Peter W. Hawkes, Electron Optics and Electron Microscopy Conference Proceedings and Abstracts: A Supplement
Grzegorz Wielgoszewski and Teodor Gotszalk, Scanning Thermal Microscopy (SThM): How to Map Temperature and Thermal Properties at the Nanoscale

Volume 191

Martin Berz, Philip M. Duxbury, Kyoko Makino, and Chong-Yu Ruan, Femtosecond Electron Imaging and Spectroscopy
Ronald E. Burge, Imaging with Electrons, X-rays, and Microwaves: Some Scattered Thoughts

Volume 192

Energy Analysis of Charged Particle Flows
Analytical Design Methods
Approximate-Analytical Method of Calculating the Charged Particle Trajectories in Electrostatic Fields
Numerical Methods in the Design of Energy Analyzers

Volume 193

Leanne Delma Duffy and Alex J. Dragt, Utilizing the Eigen-Emittance Concept for Bright Electron Beams
Igor Spivak-Lavrov, Analytical Methods for the Calculation and Simulation of New Schemes of Static and Time-of-Flight Mass Spectrometers

Volume 194

Complements of Geometrical Light Optics
Complements of Wave Optics
Particle Theory of Image Formation
Electromagnetic Lenses
Electron, Particles, or Waves?
The Wavefunction of the Paraxial Electrons
Fourier Optics
Other Interference Experiments
Interpretation of the Experimental Results
Off-Axis Electron Holography: A Short Introduction
Waveoptical Analysis of the Spherical Aberration
Epilogue

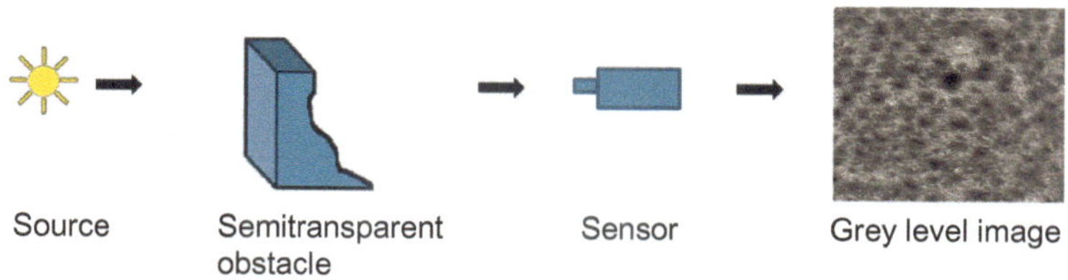

Chapter 1, Fig. 2 Acquisition in transmission.

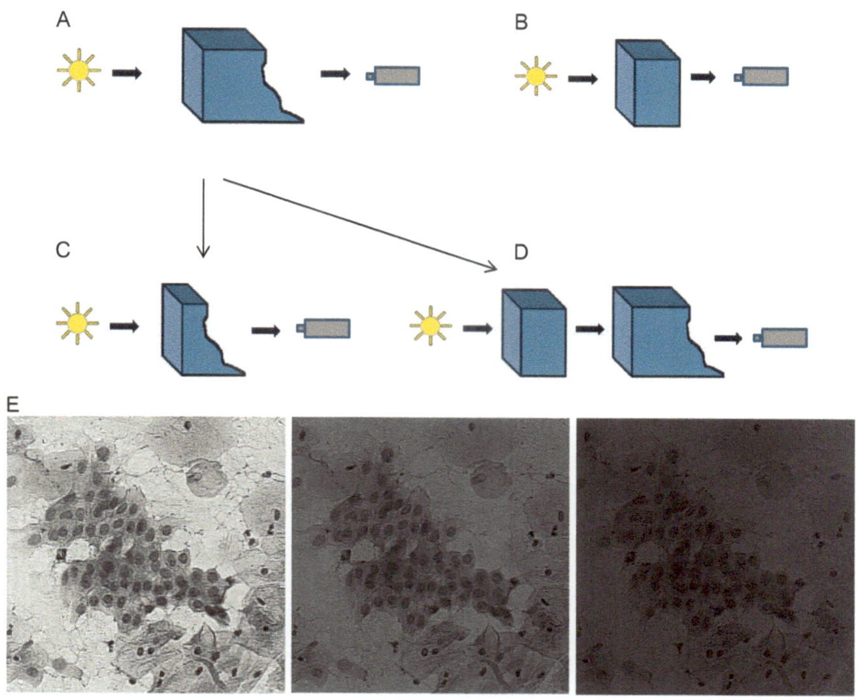

Chapter 1, Fig. 3 Subtracting (resp. adding) a constant from (resp. to) an image. (A) Acquisition of initial image *f*. (B) Uniform image *C*. (C) Acquisition of $f \triangle C$. (D) Acquisition of $f \,{\triangle\kern-0.62em{\scriptstyle +}}\, C$. (E) Visualization of $f \triangle C$, f, and $f \,{\triangle\kern-0.62em{\scriptstyle +}}\, C$ (from *left* to *right*), case $C = 128$.

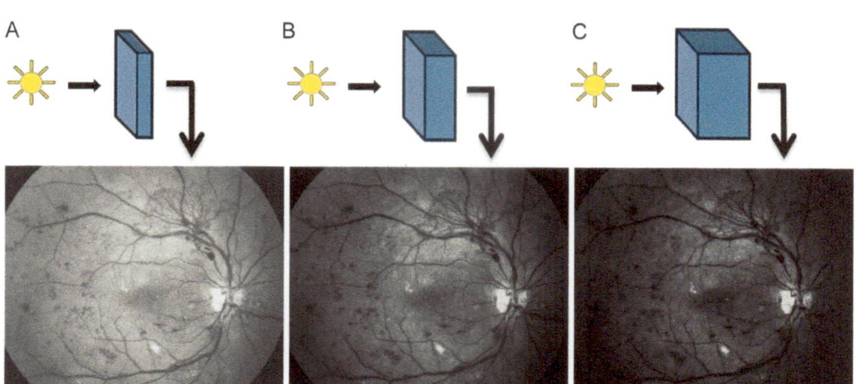

Chapter 1, Fig. 4 Multiplying an image by a real number. (A–C) Visualization of $0.5 \,{\triangle\kern-0.62em{\scriptstyle \times}}\, f$, f, and $2 \,{\triangle\kern-0.62em{\scriptstyle \times}}\, f$ (from *left* to *right*).

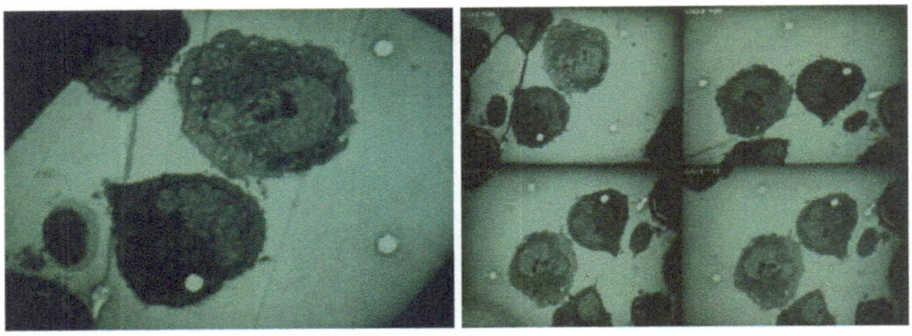

Chapter 1, Fig. 5 Examples of images of T4-lymphocytes cuts. Note the laser marks (*white disks*).

Chapter 1, Fig. 6 Virtual changing of a cut thickness: here +10% (A) and +20% (B). (A) Initial image, (B) Image (A) △1.1, (C) Image (A) △1.2.

Chapter 1, Fig. 7 Vertical cutting along a laser hole.

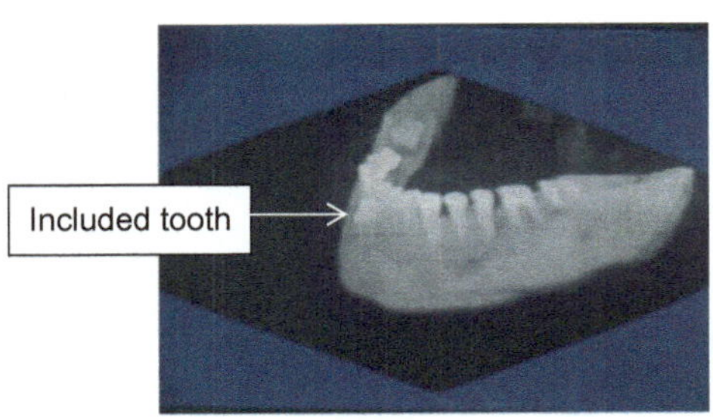

Chapter 1, Fig. 10 Visualization of the tooth to be extracted.

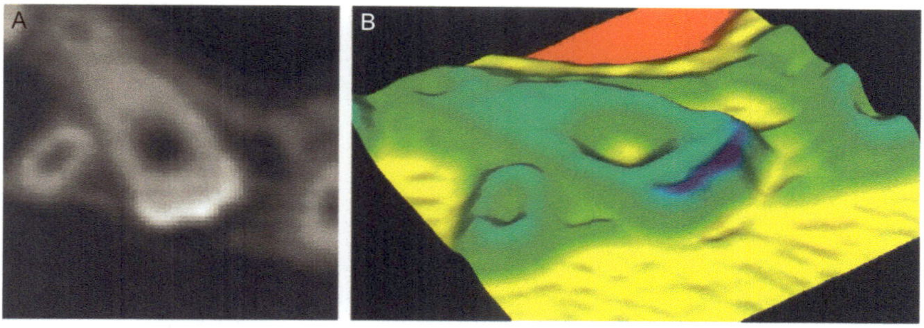

Chapter 1, Fig. 12 Visualization of the blurred transitions due to the partial volume effect. (A) Gray-level image of a subset of an X-ray scanner cut. (B) Representative surface of image (A).

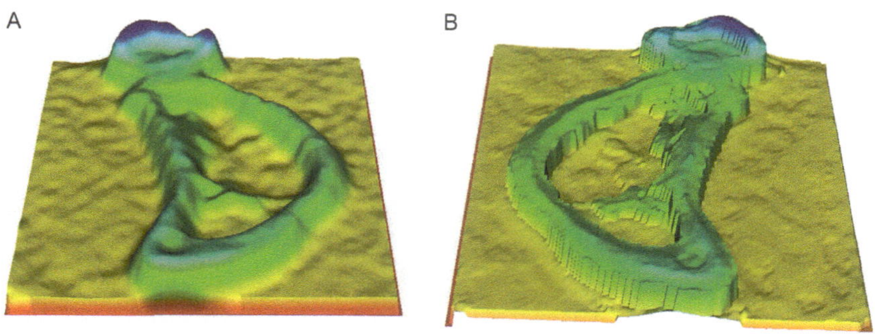

Chapter 1, Fig. 14 Visualization in false colors of the deblurring preprocessing effect. (A) Representative surface of an X-ray scanner cut. (B) Effect of a preprocessing of (A) including noise filtering and deblurring (edge crispening).

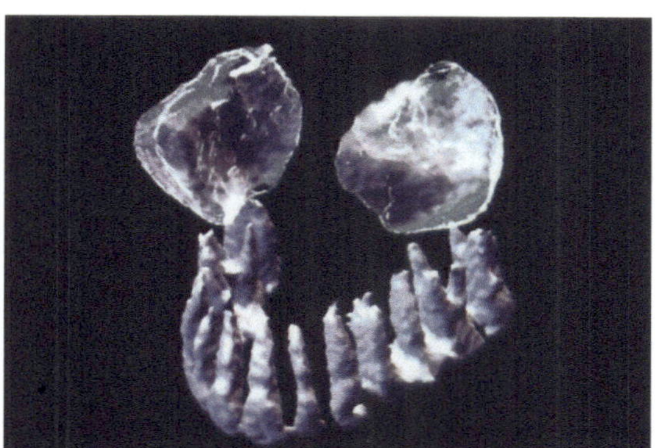

Chapter 1, Fig. 16 3D visualization of two anatomic elements of the superior part of a jaw: the teeth roots and the sinuses.

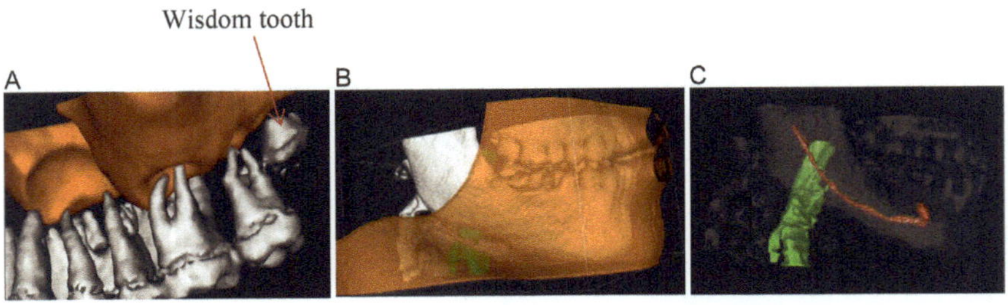

Chapter 1, Fig. 17 Examples of 3D images improvements. (A) Interaction between teeth roots and sinuses. (B) Visualization of the jaw and the flesh (semitransparent rendering of the flesh). (C) Visualization of the jaw without the flesh. Semitransparent rendering of the jaw bone: in *red* the inferior dental nerve, in *green* the trachea.

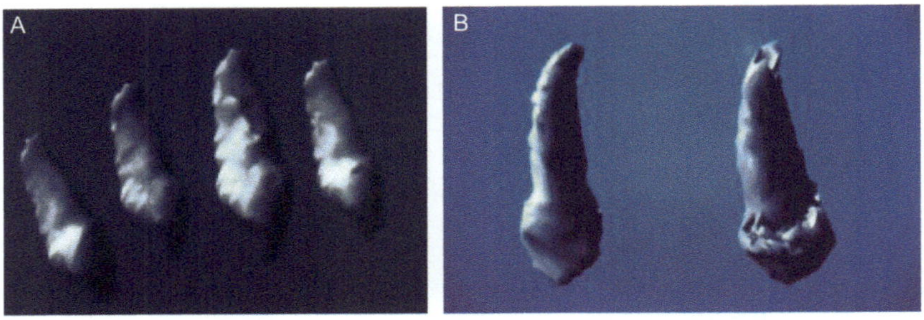

Chapter 1, Fig. 18 (A) and (B) Examples of shapes modification of segmented regions due to the selected algorithm.

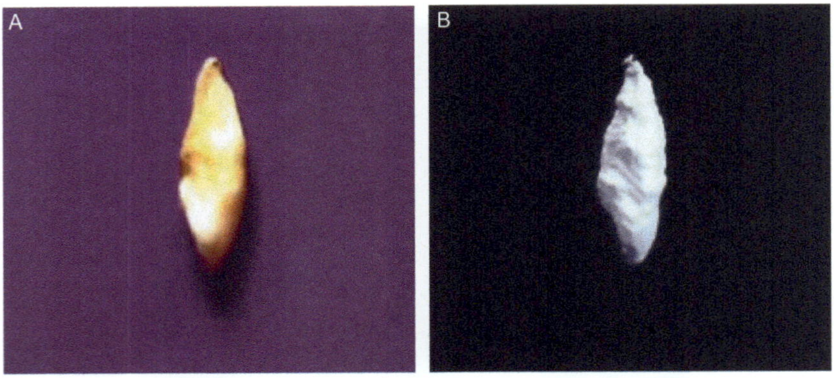

Chapter 1, Fig. 19 Comparison of a real tooth to its 3D image before extraction. (A) Image of the real tooth after extraction. (B) Segmentation of the same tooth obtained from the 3D image before extraction.

Chapter 2, Fig. 6 Comparison between Köhler and LIP-Köhler thresholdings. (A) initial image, (B) and (D) curves representing the Köhler and LIP-Köhler contrasts, (C) and (E) thresholding of image (A) at the maximal value of (B) and (D).

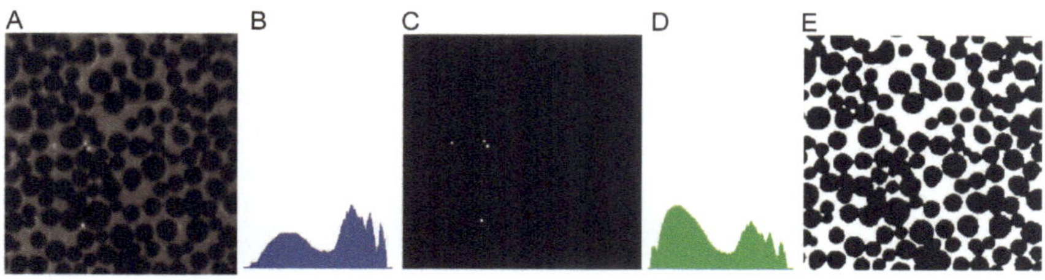

Chapter 2, Fig. 7 Comparison between Köhler and LIP-Köhler thresholdings. (A) Initial image. (B) (resp. (D)) Curves representing the Köhler (resp. LIP-Köhler) contrasts. (C) (resp. (E)) Thresholding of image (A) at the maximal value of (B) (resp. (D)).

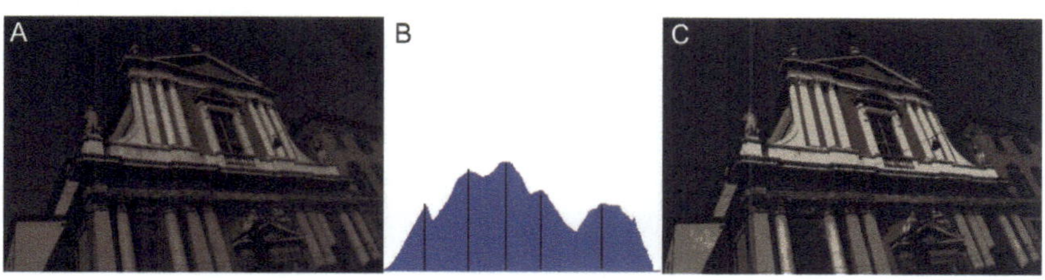

Chapter 2, Fig. 9 Example of multithresholding based on Köhler's method. (A) Initial image. (B) Curve of Köhler's contrasts with local maxima (values 39, 79, 112, 145, 202). (C) Multithresholding of the initial image based on the local maxima of (B). Example of multithresholding based on Köhler's method.

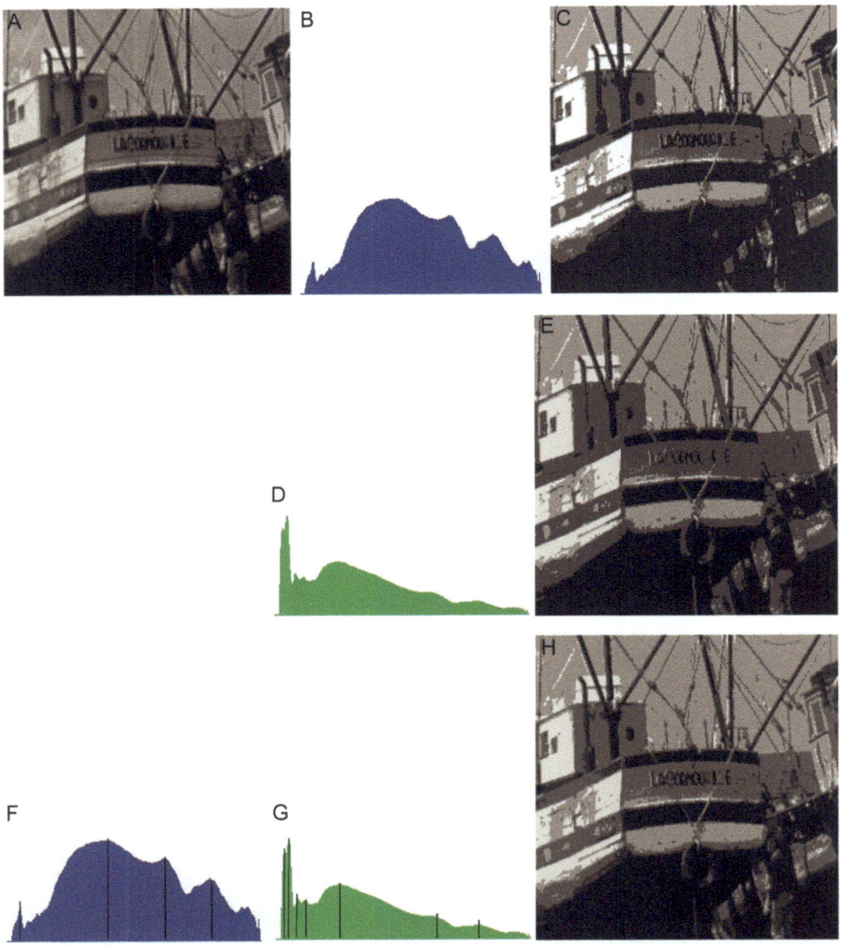

Chapter 2, Fig. 10 Multithresholding gathering Köhler and LIP-Köhler multithresholdings. (A)–(C) Initial image, (resp. Köhler contrast curve), (resp. result: five classes generated by the four thresholds values 13, 100, 160, and 205, displayed in (F)). (D) and (E) LIP-Köhler Multithresholding of (A) (local maxima of (D)): 6, 13, 21, 30, 64, 160, and 205. (F) and (G) Köhler and LIP-Köhler thresholds on the corresponding contrast curves. (H) Multithresholding obtained by gathering the local maxima of Köhler and LIP-Köhler: 6, 13, 21, 30, 64, 100, 160, 205.

Chapter 2, Fig. 12 Robustness to light condition of texture detection using the LIP additive maximal contrast. On the *left*: original image, at the *center*: contour detection using $\mathrm{MC}_x^{\triangle}(f)$, on the *right*: contour detection by Sobel operator.

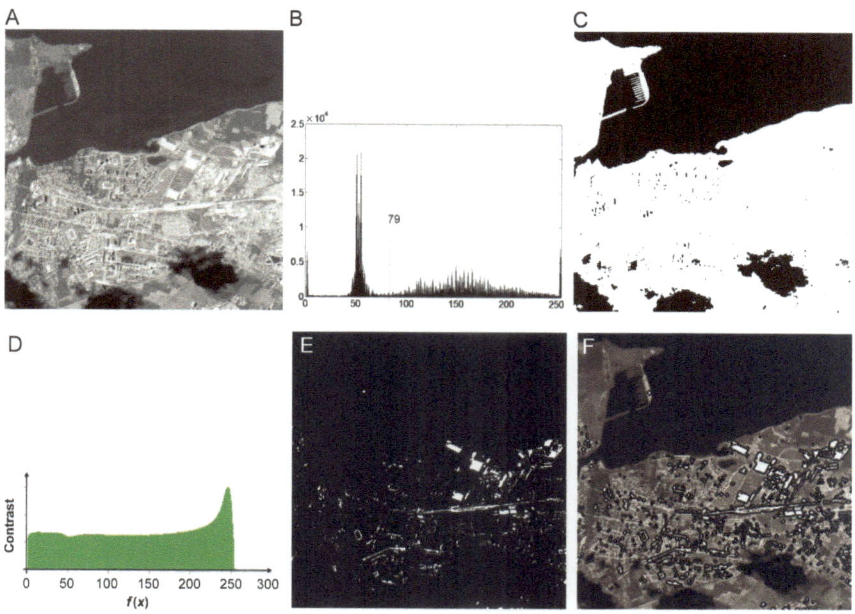

Chapter 2, Fig. 15 Comparison of thresholdings based on interclass variance maximization and on LMC map. (A) Initial image; (B) Histogram of (A) with highlighted value 79 corresponding to the threshold based on interclass variance maximization; (C) Binary image after thresholding (A) at gray level 79; (D) Histogram of the LMC map of (A) with a unique peak; (E) Binary image after thresholding (A) at the peak of (D); (F) Thresholded area highlighted on (A).

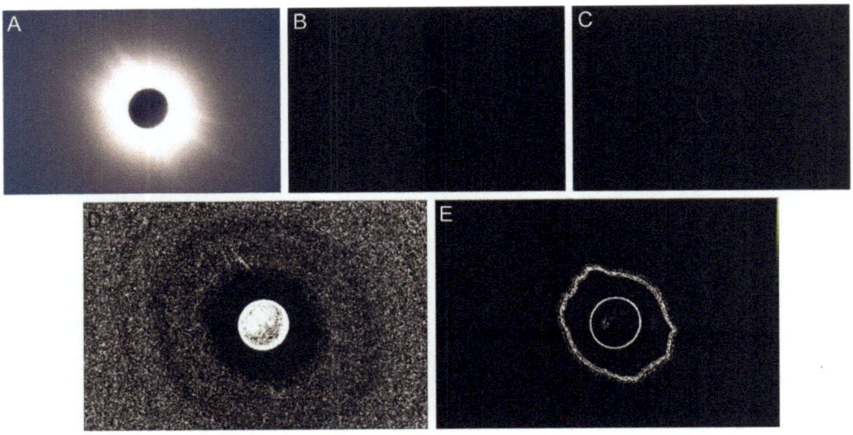

Chapter 2, Fig. 16 Contour detection based on LMC. (A) Initial image "Sun eclipse 2006" (with authorization of the author L. Ferrero); (B) Classical gradient of (A); (C) Contrast map of (A) corresponding to $MC_x^{\triangle}$; (D) Thresholded gradient; (E) Thresholded contrast map.

Chapter 2, Fig. 17 Insensitivity to lighting drift of LIPC color contrast and norm. (A) Initial image "Peppers"; (D) Image "Peppers" with lighting drift; (B) resp. (E)) Color contrast image of (A) (resp. (D)); (C) (resp. (F)) Norm image of (A) (resp. (D)).

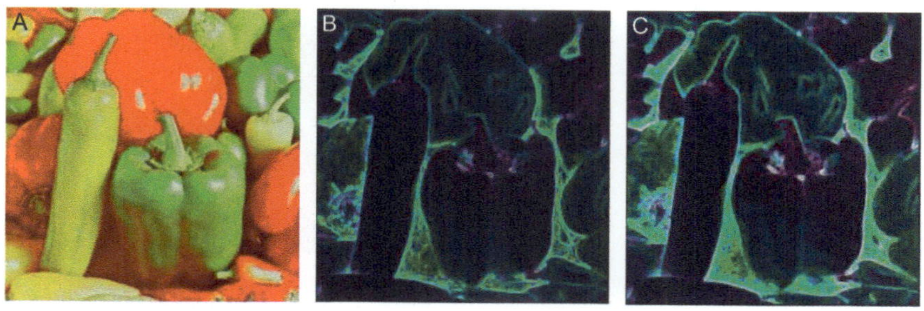

Chapter 2, Fig. 18 Average and maximal LAC as color vectors. (A) Initial image "Peppers". (B) Image (A) processed with the average LAC: each pixel of "Peppers" has been replaced by the vector given in Eq. (24). (C) Image (A) processed with the maximal LAC: each pixel replaced by the vector given in Eq. (25).

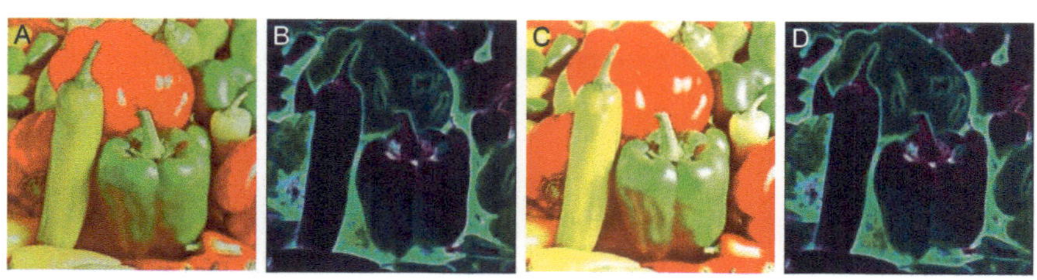

Chapter 2, Fig. 19 Insensitivity of the MLAC vector to brightness variations. (A) Initial "Peppers", (C) Image (A) with higher brightness, (B) and (D) Images (A) and (B) processed by the MLAC vector.

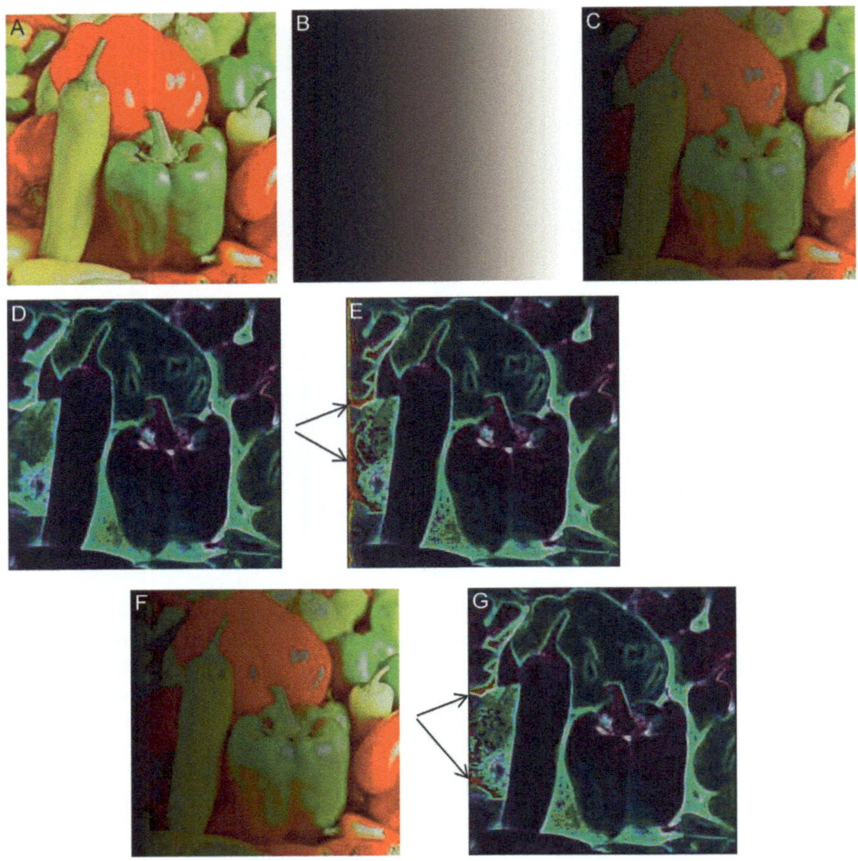

Chapter 2, Fig. 20 Behavior of the MLAC vector in presence of lighting drift. (A) Initial image "Peppers". (B) An example of strong lighting drift. (C) "Peppers" with lighting drift (B). (D) (resp. (E)), MLAC vector applied to (A) (resp. (C)). (F) "Peppers" with a lightly lower lighting drift. (G) MLAC vector applied to (F).

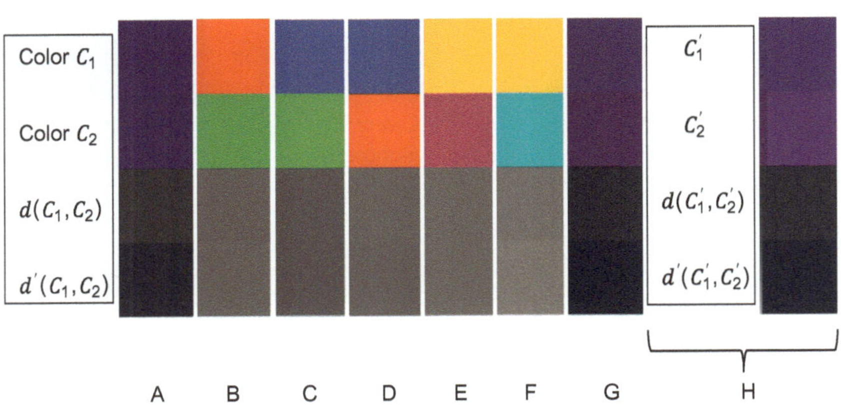

Chapter 2, Fig. 21 Representation, in gray levels, of the distances $d(C_1, C_2)$ and $d'(C_1, C_2)$ for various pairs of colors (C_1, C_2).

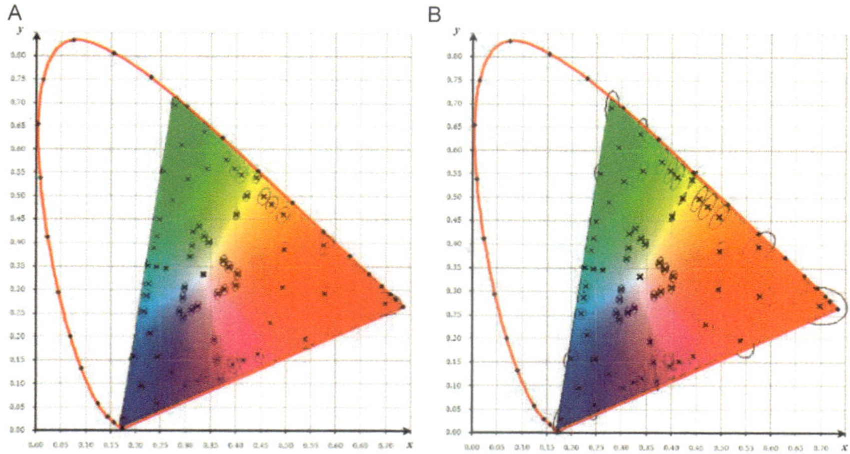

Chapter 2, Fig. 22 Comparison of colors neighborhoods shapes in (x,y)-space. (A) Euclidean neighborhoods, (B) Neighborhoods associated with metric d_1.

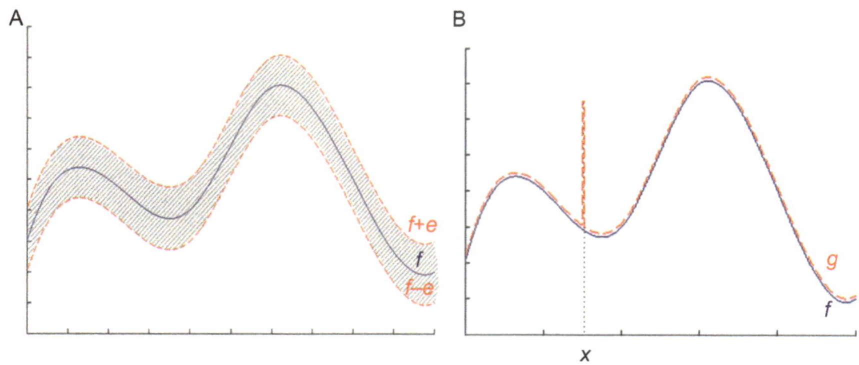

Chapter 3, Fig. 1 Neighborhoods shapes for d_1 and d_∞. (A) The tolerance tube of f is represented by the hatched area. (B) The local difference between a function f and a function g lying in the ε-neighbor of f may be arbitrarily large.

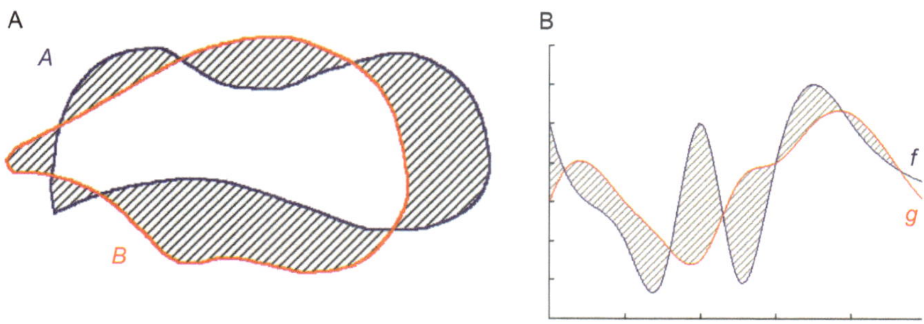

Chapter 3, Fig. 3 Illustration of symmetric difference in the case of binary shapes and in the case of functions. (A) Symmetric difference between A and B (hatched area). (B) Symmetric difference between the subgraphs of functions f and g (hatched area).

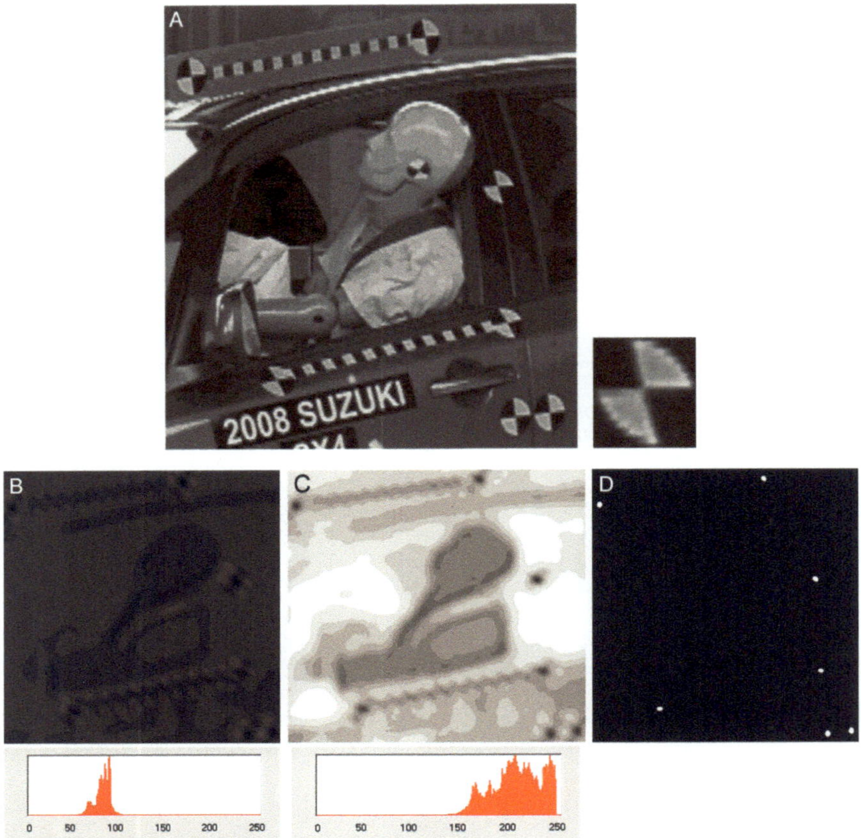

Chapter 3, Fig. 7 Example of target tracking based on the logarithmic metric $d_1^{\triangle}$. (A) Image of car crash test (with the authorization of "Insurance Institute for Highway Safety") with reference target (magnified). (B) Represents the distance map associated to the classical metric d_1. (C) Represents the distance map associated to the logarithmic metric $d_1^{\triangle}$. (D) Location of the targets obtained by an automated thresholding of (C).

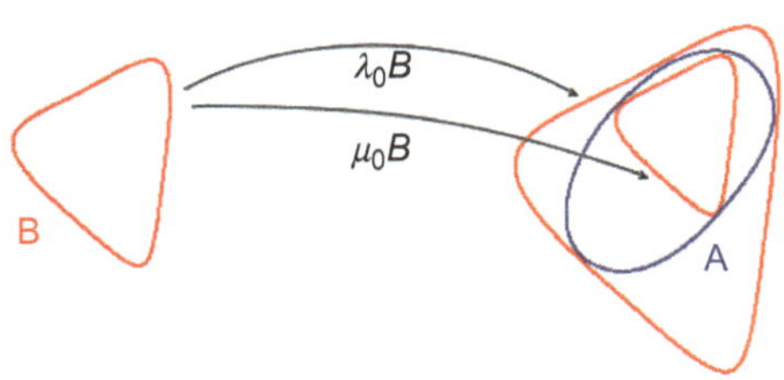

Chapter 3, Fig. 10 Asplünd's metric for binary shapes.

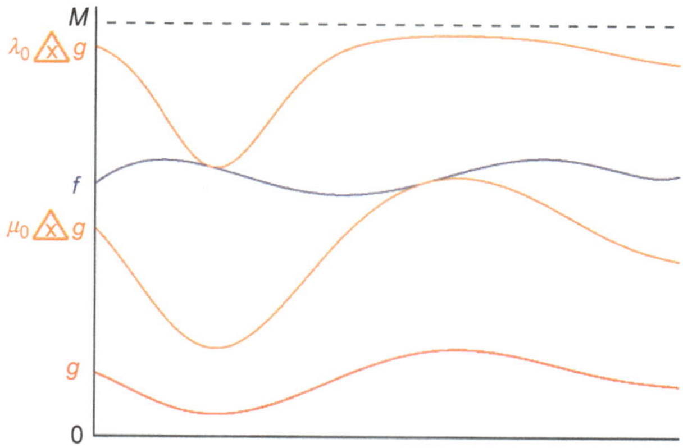

Chapter 3, Fig. 11 Computation of the double-sided probing metric $d^{\triangle}_{As}(f, g)$, when g is the probing function.

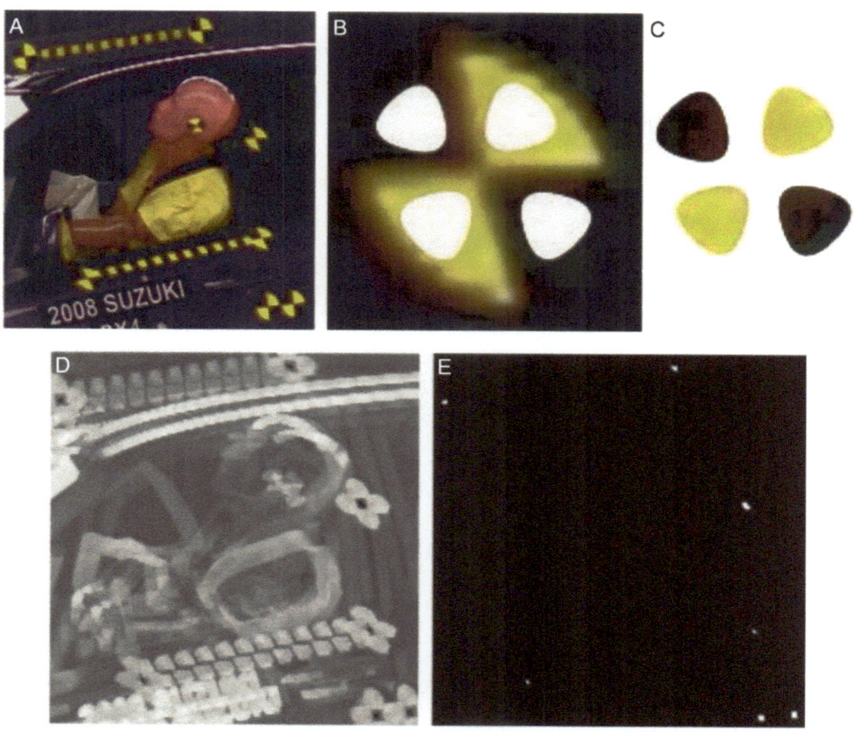

Chapter 3, Fig. 14 Target detection based on multiplicative Asplünd's metric. (A) Crash test image. (B) Probing zones inside the target magnified 10 times. (C) Probing function magnified 10 times. (D) Asplünd's map. (E) Thresholded image of (D).

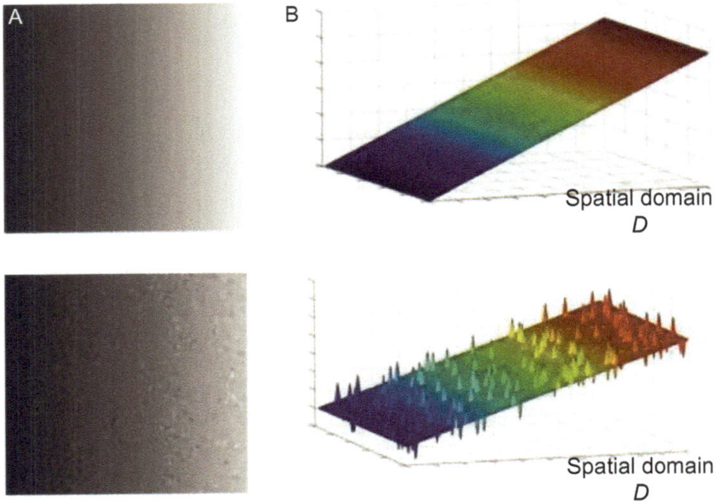

Chapter 3, Fig. 17 A synthesis image *f* and its noised version. (A) Initial image *f* with a linear drift. (B) Representative surface of image *f*. (C) Image *g*: image *f* with salt and pepper noise added. (D) Representative surface of image *g*.

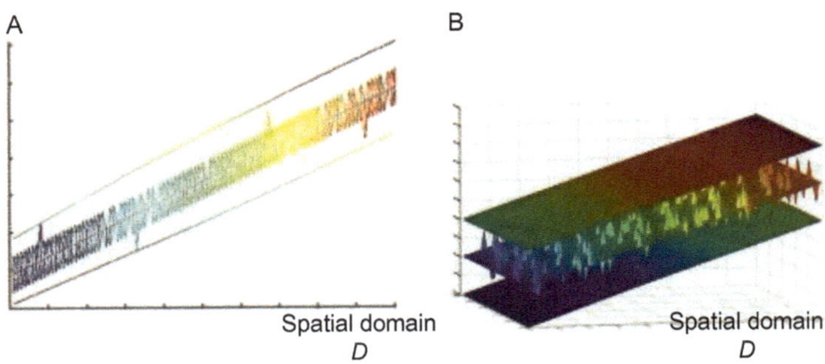

Chapter 3, Fig. 18 Sensitivity to noise of $d_{As}^{\triangle}$. (A) Representation of the probing of *g* by *f* (2-D representation). (B) Representation of the probing of *g* by *f* (3-D representation).

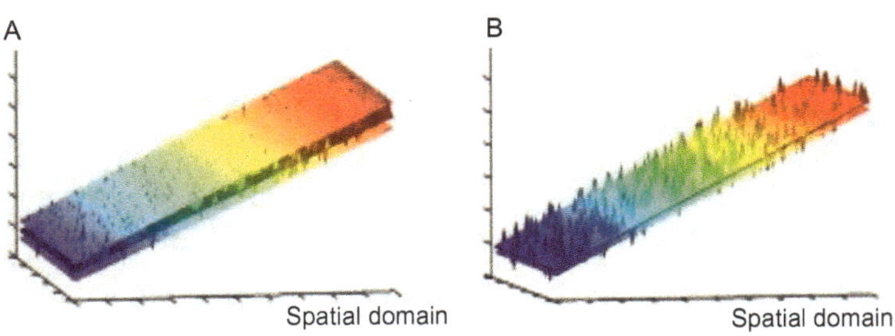

Chapter 3, Fig. 19 Getting probing functions closer to the target. (A) Probing of *g* by *f* with tolerance $p=0.98$. (B) Probing of *g* by *f* with tolerance $p=0.95$.

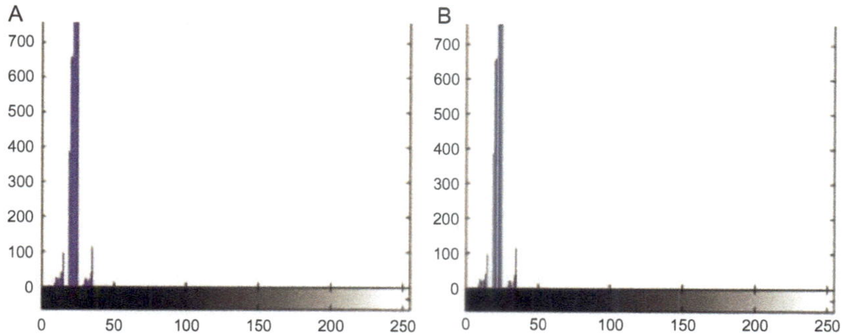

Chapter 3, Fig. 20 Differences between g and its probing functions. (A) Histogram of the differences between λ⨻f and g. (B) Histogram of the differences between g and μ⨻f.

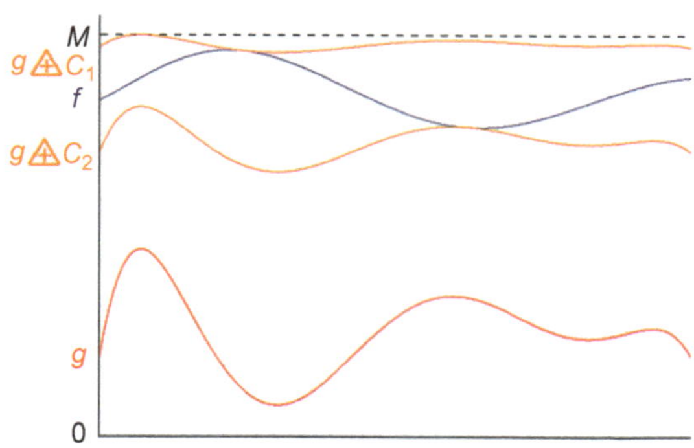

Chapter 3, Fig. 22 Additive Asplünd's distance between f and g, where g is the probing function.

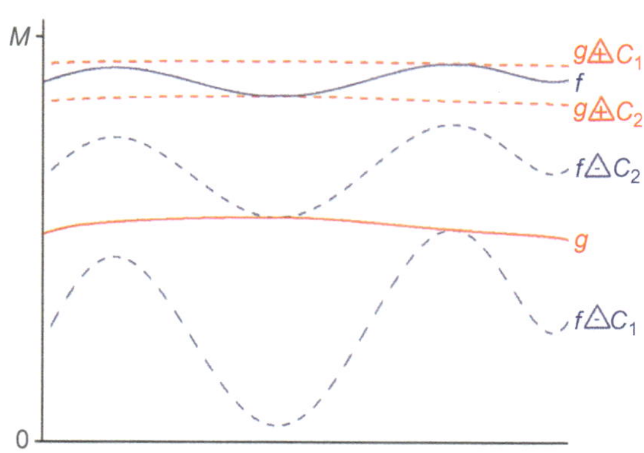

Chapter 3, Fig. 23 Illustration of the symmetry property.

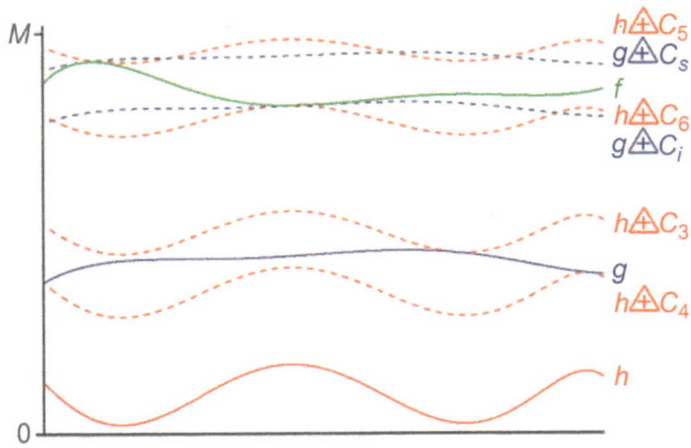

Chapter 3, Fig. 24 Triangular inequality property.

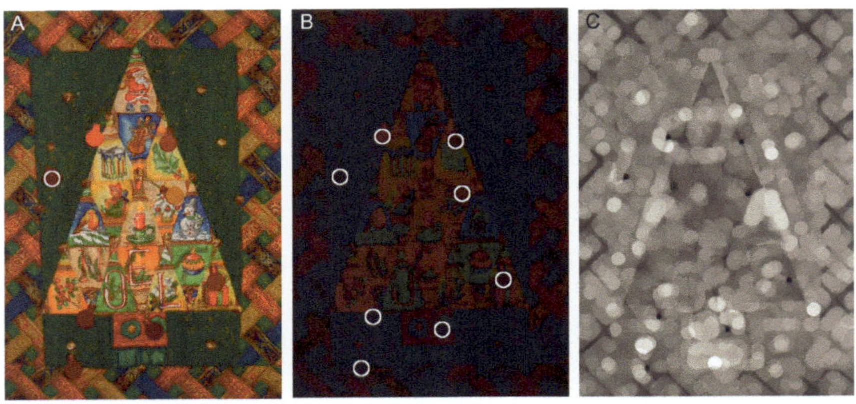

Chapter 3, Fig. 31 Color target tracking on a dark image. (A) Initial color image *f* (Christmas Tree) with selected target *t*. (B) Detected Christmas balls on a low-light version of (A). (C) Asplünd's map of *t* inside the low-light version of (A).

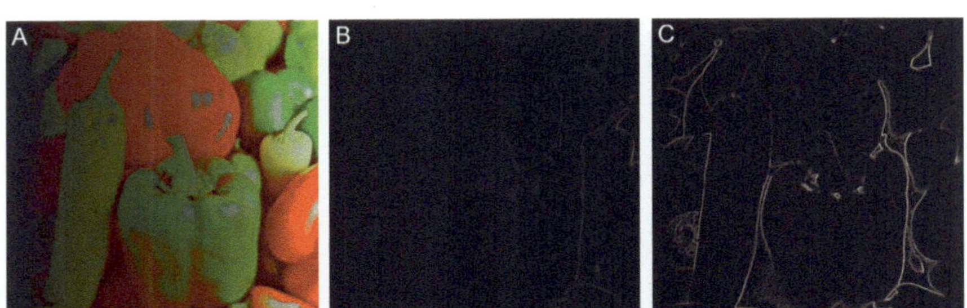

Chapter 3, Fig. 32 Sobel applied to the luminance image compared to color additive Asplünd gradient. (A) Initial image "Peppers" with lighting drift. (B) Sobel filter applied to (A). (C) Result of contour detector defined in Formula (5).

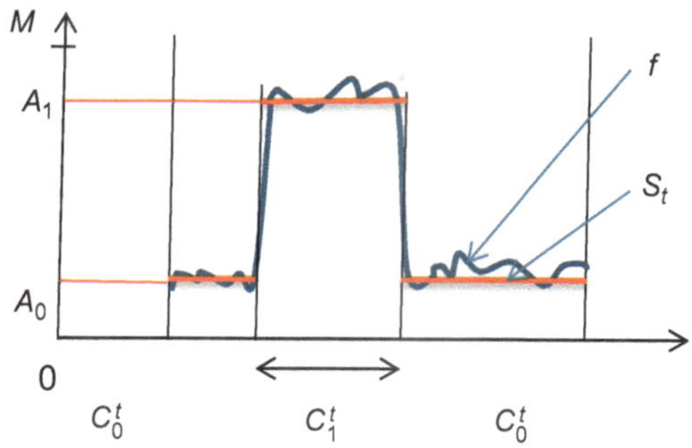

Chapter 3, Fig. 34 Initial image f (in *blue*) and associated step function (in *red*).

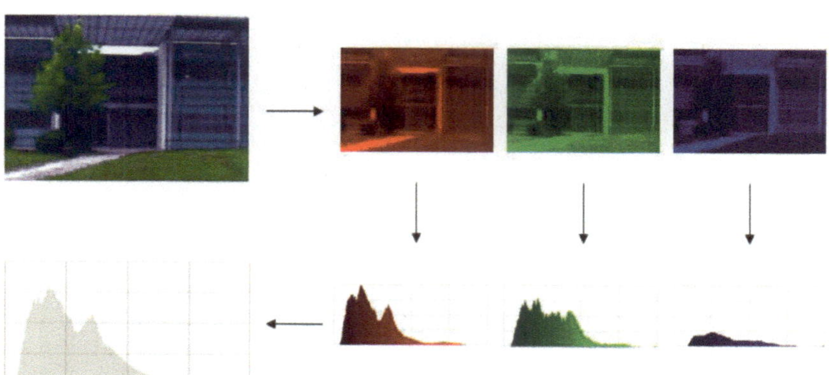

Chapter 4, Fig. 11 Global histogram of a color image. Initial image f (upper *left*) and according to successive arrows: Images f_R, f_G, f_B; Histograms H_{f_R}, H_{f_G}, H_{f_B}; Finally: global gistogram GH_f.

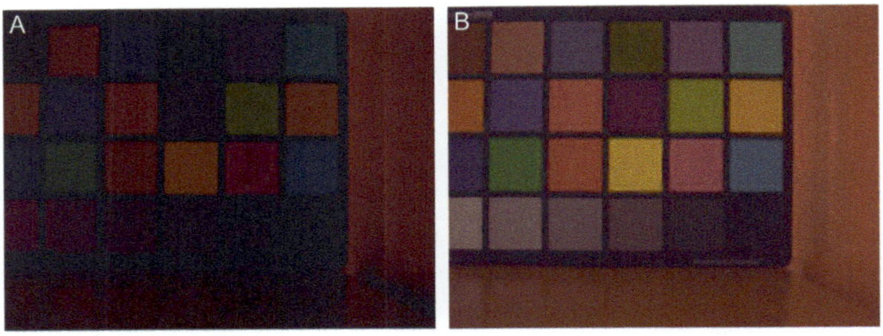

Chapter 4, Fig. 12 λ_0-Enhancement of a color image. (A) Initial low-light image f "color chart." (B) Enhanced image $\lambda_0 \triangle f$.

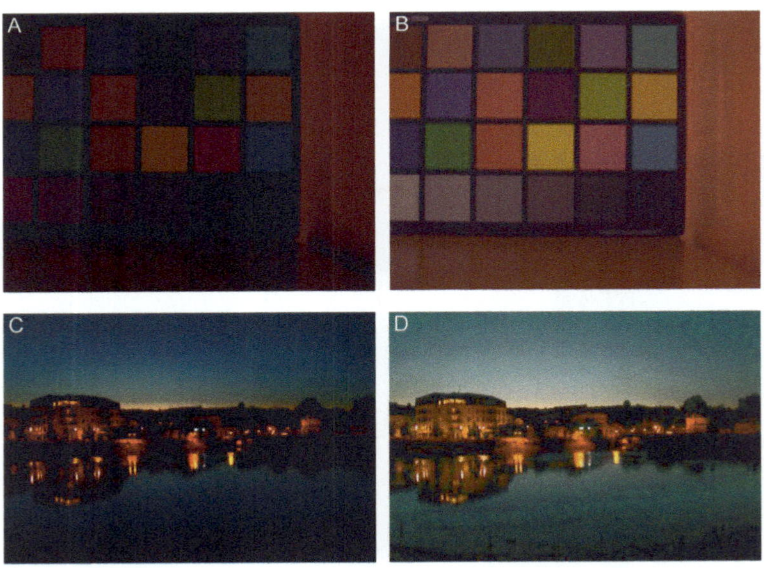

Chapter 4, Fig. 13 λ_1-Enhancement of a color image. (A) and (C) Initial color images. (B) and (D) Enhanced images of (A) and (B) (λ_1 parameter applied on the global histogram).

Chapter 4, Fig. 14 Examples of λ_2-full dynamic range expansion. (A) Low-light color image f "Castle." (B) Enhanced image $\lambda_2 \triangle f$. (C) Color image g "Campus" (back lighted). (D) $\lambda_2 \triangle g$ inside a window. (E) Very low-light color image h "Young Girl." (F) Enhanced image $\lambda_2 \triangle h$.

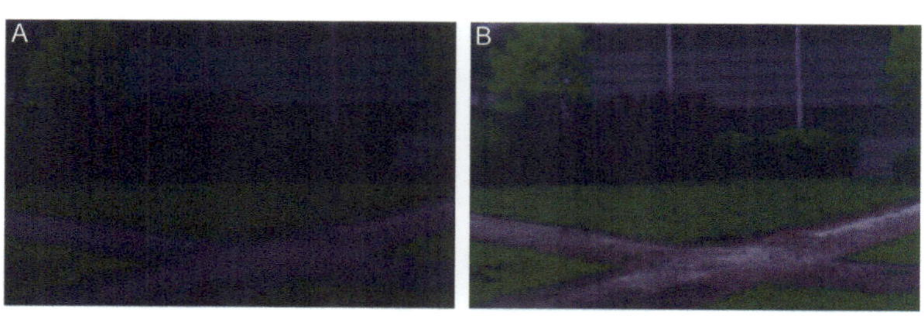

Chapter 4, Fig. 19 Enhancement of a color image (method $f \triangle C_0$). (A) Initial low-light image f "Laboratory" (exposure time 12 ms). (B) Enhanced image of f according to $f \triangle C_0$, with $C_0 = 144$.

Chapter 4, Fig. 20 Enhancement of a color image by maximization of the standard deviation of $GH_f \triangle C$ (method $f \triangle C_1$). (A) Initial color chart f (acquisition time 20 ms). (B) Enhanced image $f \triangle C_1$.

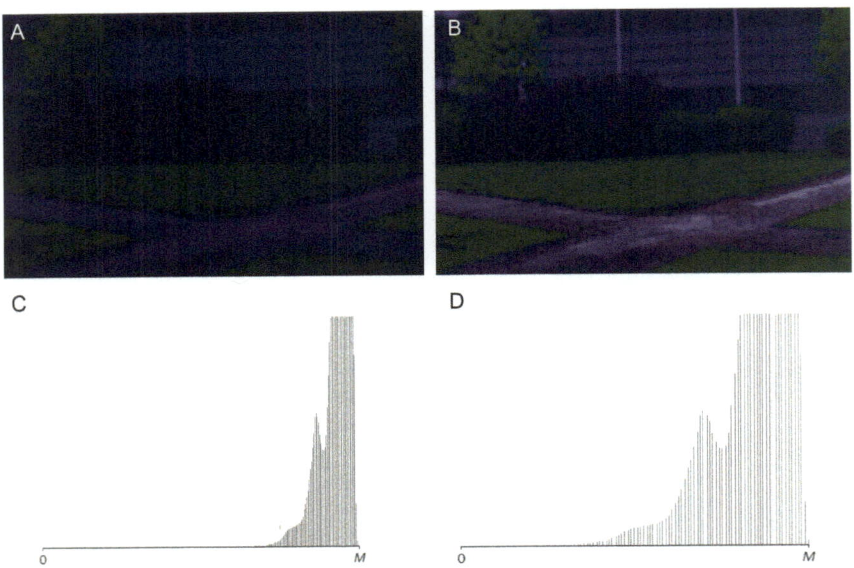

Chapter 4, Fig. 21 Enhancement of a color image (method $f \triangle C_2$ + translation). (A) Initial low-light image f "Laboratory" (exposure time 12 ms). (B) Enhanced image of f according to the method based on C_2 with $C_2 = 145.6$. (C) and (D) Histograms of (A) and (B).

Chapter 4, Fig. 23 First line: images acquired under variable aperture. Second line: corrected images. Third line: corresponding histograms.

Chapter 4, Fig. 24 Brightness enhancement of a low-light image.

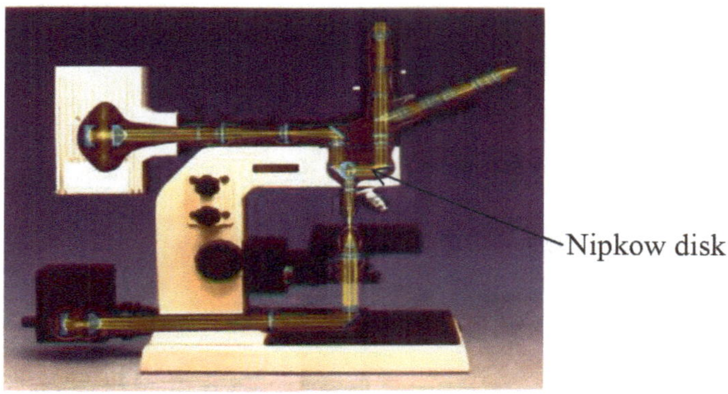

Chapter 4, Fig. 25 A confocal TSM showing light trajectory.

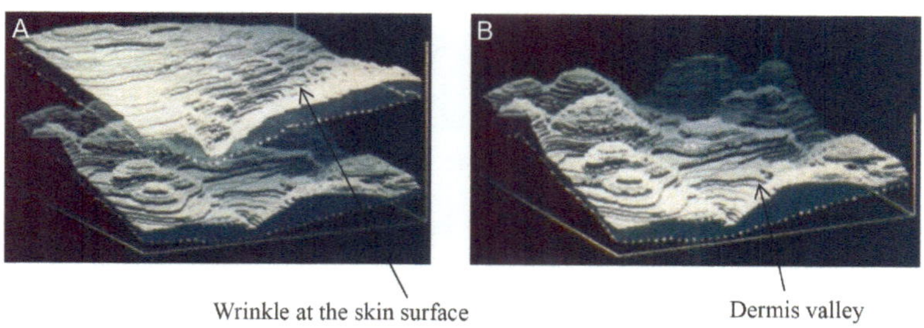

Chapter 4, Fig. 26 (A) 3D imaging of the stratum corneum (upper epidermal layer). The two bright surfaces represent the upper and lower limits of the stratum corneum. (B) The dermis upper surface.

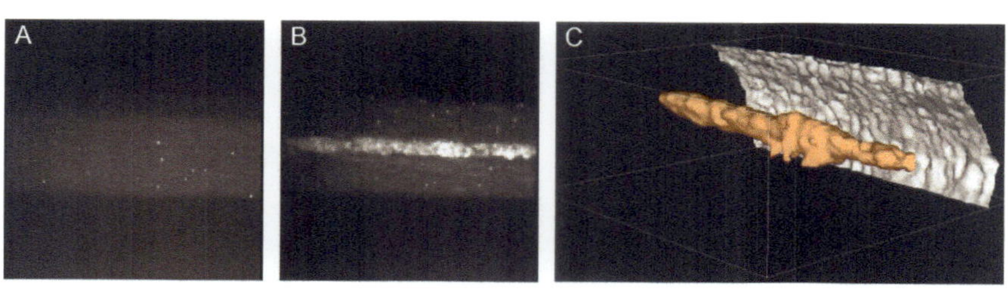

Chapter 4, Fig. 27 (A) An optical cut of the hair. (B) A cut crossing the medulla. (C) 3D reconstruction: here the medulla and the hair surface.

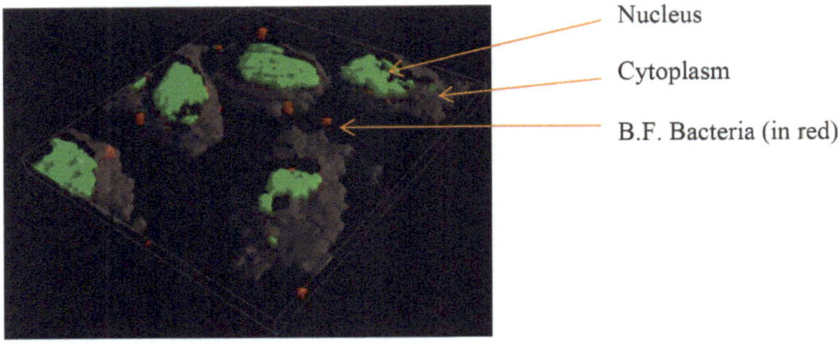

Chapter 4, Fig. 28 3D reconstruction of Hela cells and *Bacteroides fragilis* bacteria.

Chapter 4, Fig. 29 Image of the drone in action.

Chapter 4, Fig. 30 Coolers of a nuclear power station.

Chapter 4, Fig. 31 The initial recomposition (on the *left*) and recomposition with corrected images (on the *right*).

Chapter 4, Fig. 32 Correction on successive images (on the *right*, the transitions between successive images are less visible).

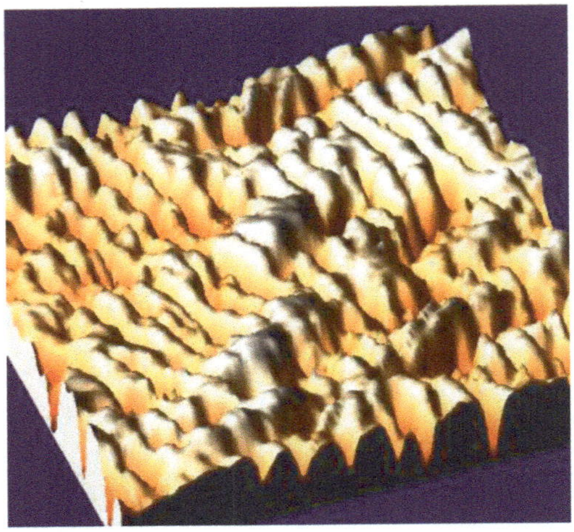

Chapter 4, Fig. 35 Ripples 3D view.

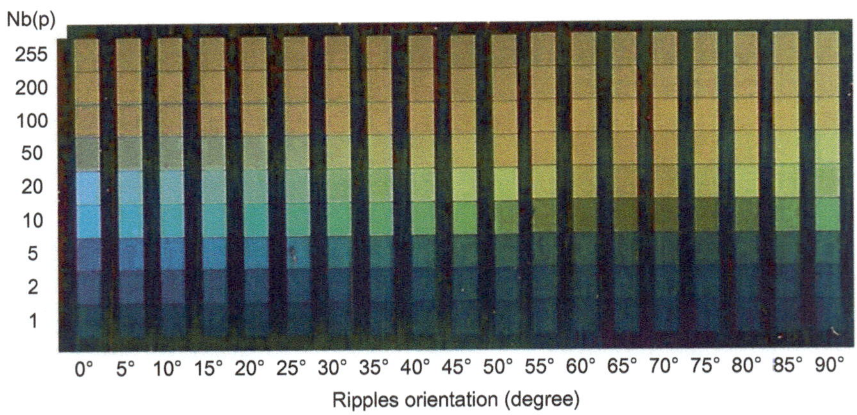

Chapter 4, Fig. 36 Colors resulting of light diffraction due to ripples of various orientations.

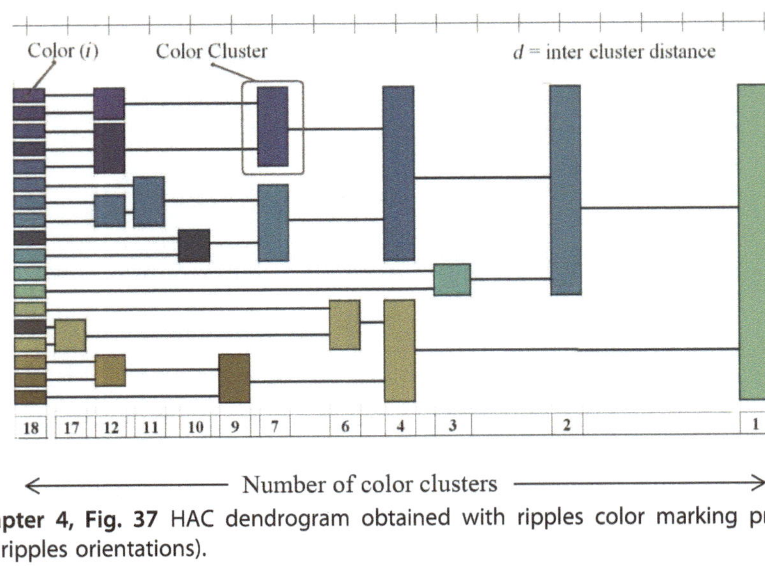

Chapter 4, Fig. 37 HAC dendrogram obtained with ripples color marking process (18 ripples orientations).

Chapter 4, Fig. 38 Examples of images from "Asterix" and "Tintin." On the *right*, the real size (here 17 mm).

Chapter 4, Fig. 42 Examples of local lighting correction on color images.

Chapter 4, Fig. 44 Color image "Laboratory:" Comparison of enhancement methods. (A) Color image "Laboratory": gain = 1, exposure = 100 ms. (B) Color image "Laboratory": gain = 10, exposure = 100 ms. (C) Histogram equalization applied to image (A). (D) LIP subtraction of a constant $C = 245$ applied to (A).

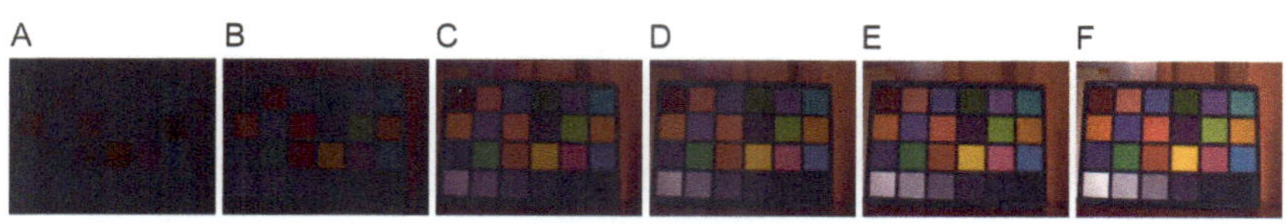

Chapter 5, Fig. 2 Color chart acquired at various exposure times: f_3 (3 ms), f_6 (6 ms), f_{18} (18 ms), f_{24}(33 ms), f_{45}(45 ms). (A) Image f_3: acquired at 3 ms. (B) Image f_6: 6 ms. (C) Image f_{18}: 18 ms. (D) Image f_{24}: 24 ms. (E) Image f_{33}: 33 ms. (F) Image f_{45}: 45 ms.

Chapter 5, Fig. 3 Enhanced images associated to $f_3, f_6, f_{18}, f_{24}, f_{45}$, from left to right.

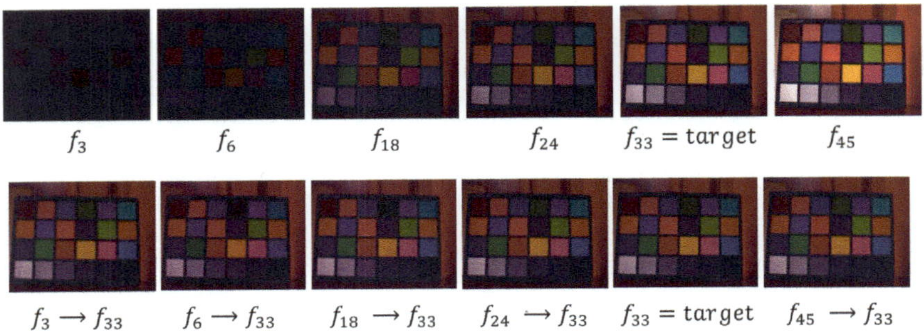

Chapter 5, Fig. 5 Simulations of a target, starting from initial images with various exposure times. *First line*: Initial images and the target. *Second line*: Simulations results.

Chapter 5, Fig. 6 The four colors K_1, K_2, K_3, and K_4 selected in f_{33} after enhancement with C_1-subtraction.

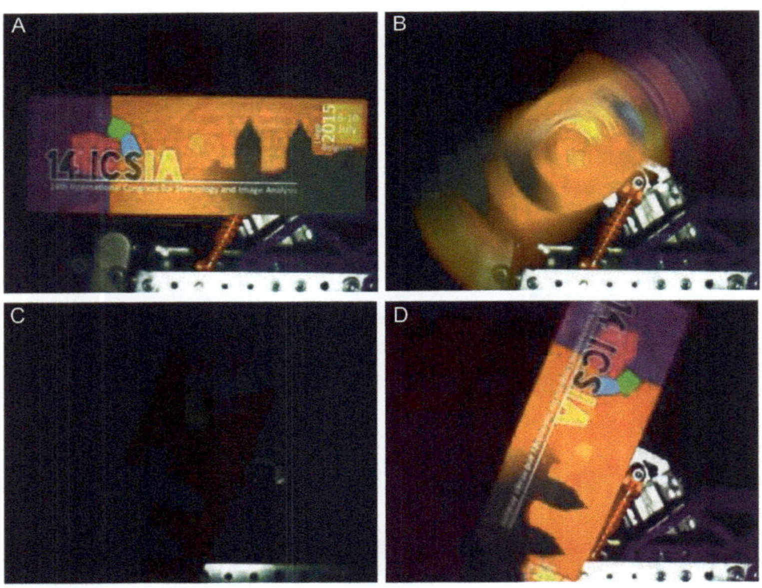

Chapter 5, Fig. 7 Acquisition of a moving object through exposure time simulation. (A) Congress label acquired at 30 ms (motionless wheel). (B) Congress label acquired at 30 ms (wheel rotating at 3.75 revolutions per second). (C) Rotating label acquired at 2 ms. (D) Simulation of an exposure time of 30 ms starting from image (C).

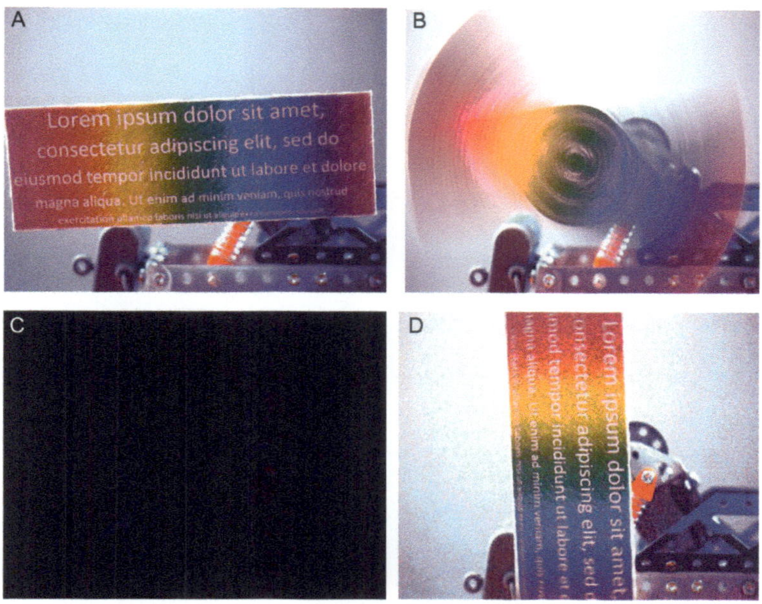

Chapter 5, Fig. 8 Acquisition of a moving object through exposure time simulation. (A) Label acquired at 30 ms (motionless wheel). (B) Label acquired at 30 ms (wheel rotating at 6.4 revolutions per second). (C) Rotating label acquired at 1 ms. (D) Simulation of an exposure time of 30 ms starting from image (C).

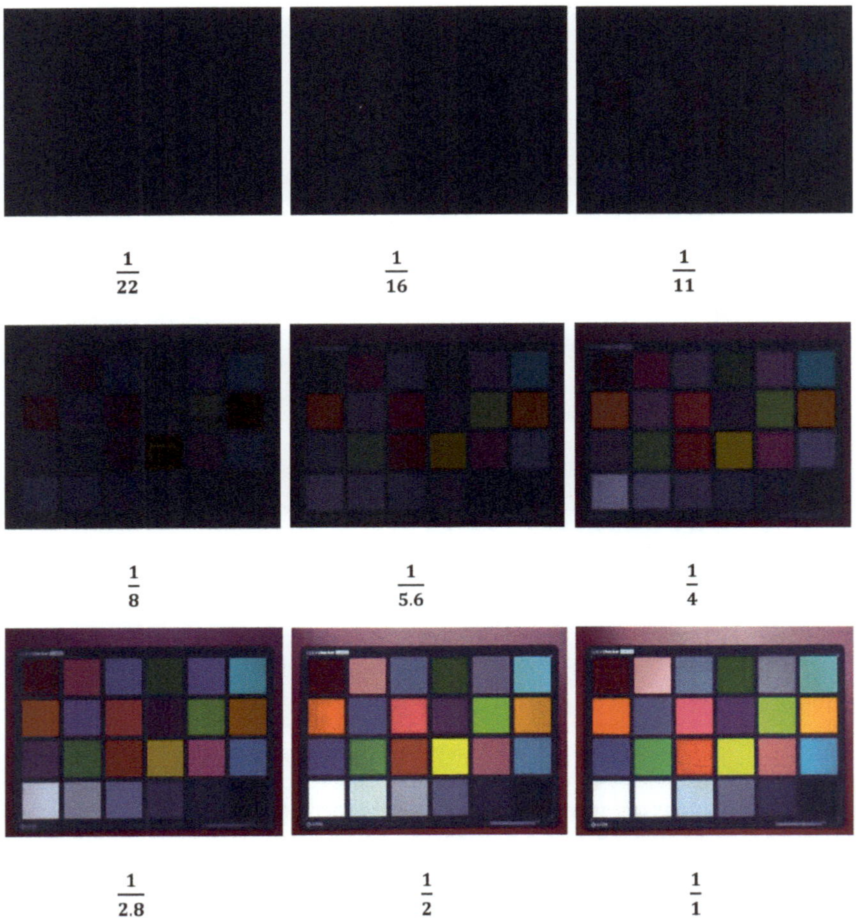

Chapter 5, Fig. 9 Color chart acquired with diaphragm aperture varying from $\frac{1}{22}$ to $\frac{1}{1}$.

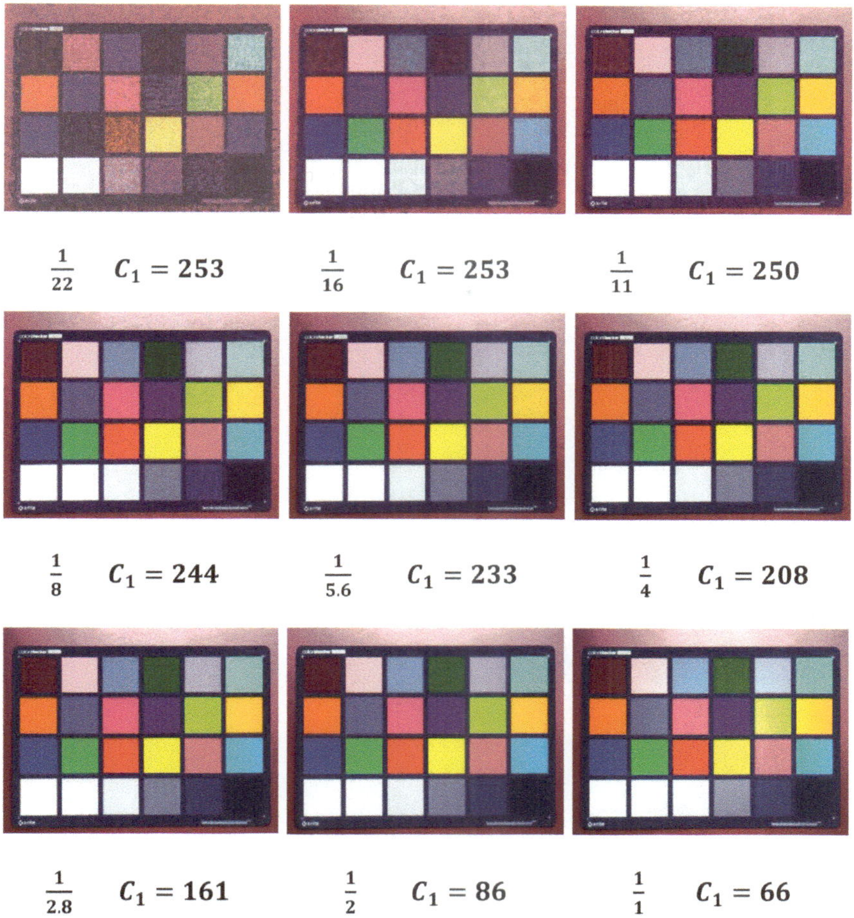

Chapter 5, Fig. 10 Enhancement of initial images presented in Fig. 9.

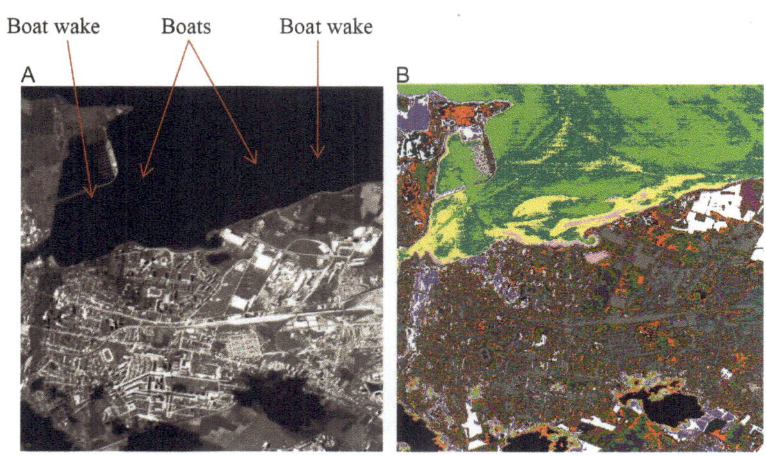

Chapter 6, Fig. 2 Diday's algorithm applied to "La Rochelle port." (A) Initial image: "La Rochelle port." (B) Segmented image (16 classes).

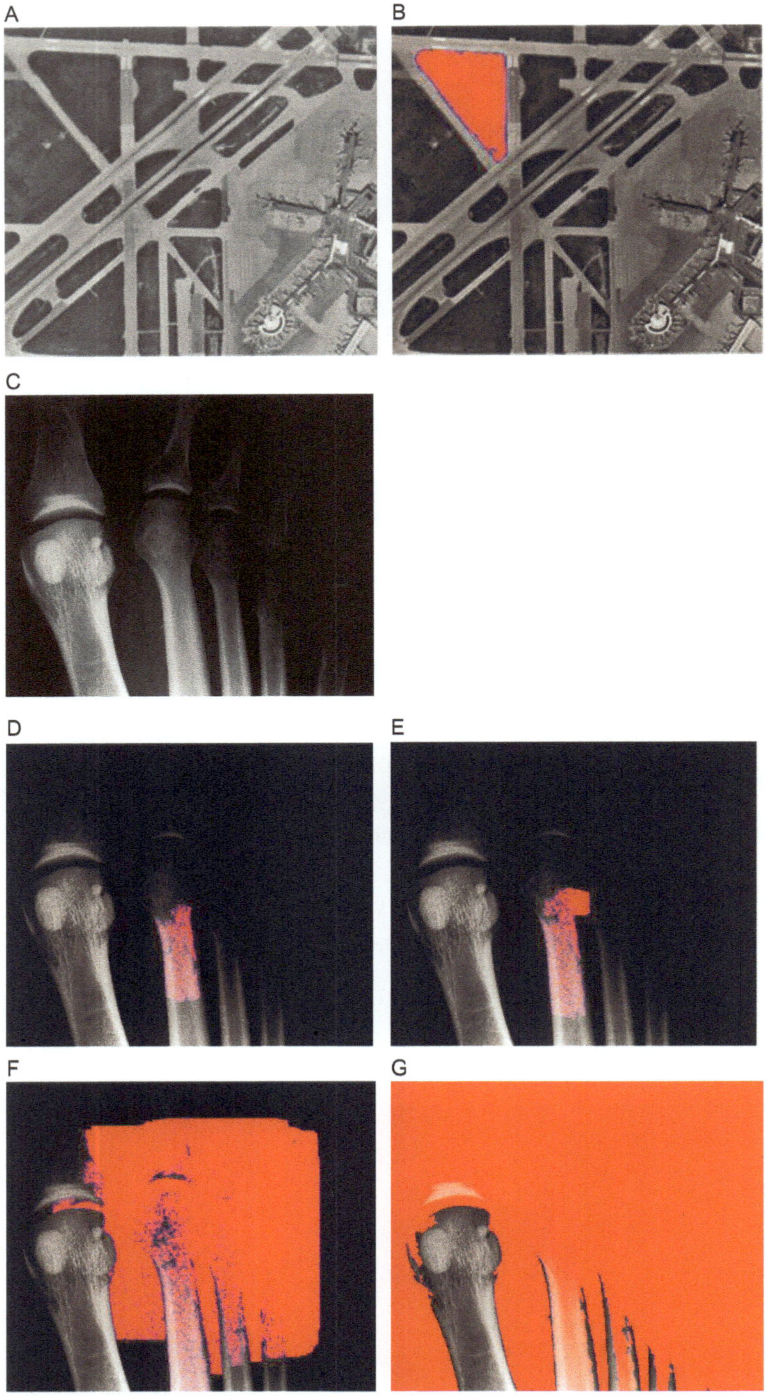

Chapter 6, Fig. 3 Two examples of single linkage effects. (A) Well-contrasted initial image. (B) Single linkage region growing on a part of (A). (C) Initial image. (D) Single linkage region growing on (C) for $\alpha = 5$. (E) Single linkage region growing on (C) for $\alpha > 5$. (F) and (G) Single linkage region growing for increasing values of the "resemblance" parameter α.

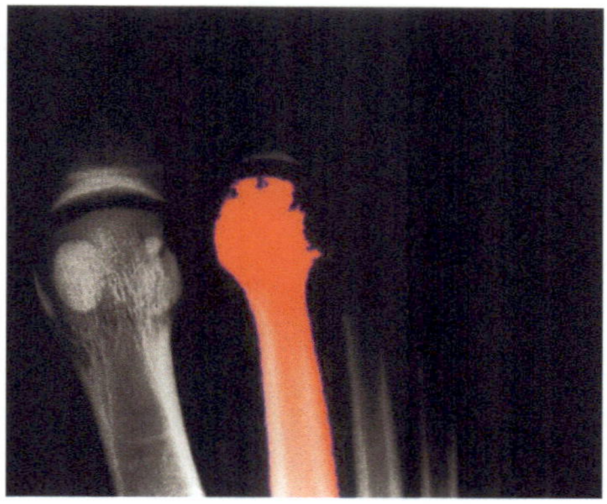

Chapter 6, Fig. 4 Logarithmic single linkage on the initial image (Fig. 3C).

Chapter 6, Fig. 14 Random color LUT on HAC steps. (A) Step 2 of HAC by Euclidean distance d_e. (B) Step 10 of HAC by logarithmic distance $C^{\triangle}_{(x,y)}(f)$.

Chapter 6, Fig. 15 Color HAC by Euclidean metric d_e: steps 0, 1, 5, 8, and 17.

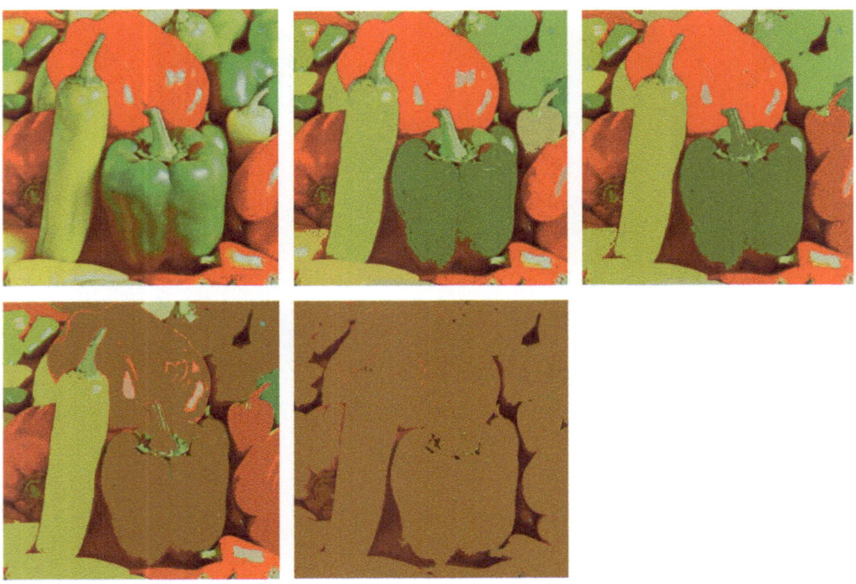

Chapter 6, Fig. 16 Color HAC by logarithmic distance $d^{\triangle_c}$: steps 0, 29, 32, 36, and 57.

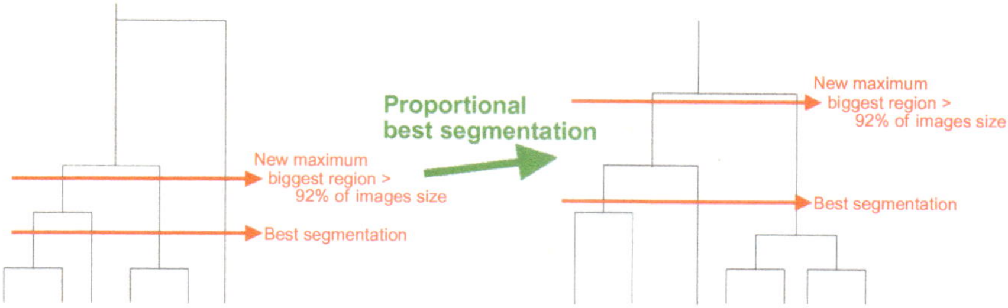

Chapter 6, Fig. 17 Tree's observation for intermetric comparison.

Chapter 6, Fig. 18 HAC comparison between different metrics. (A) Optimal step (5) of HAC for the metric d_e. (B) Corresponding step (32) in the tree for the metric $d^{\triangle_c}$.

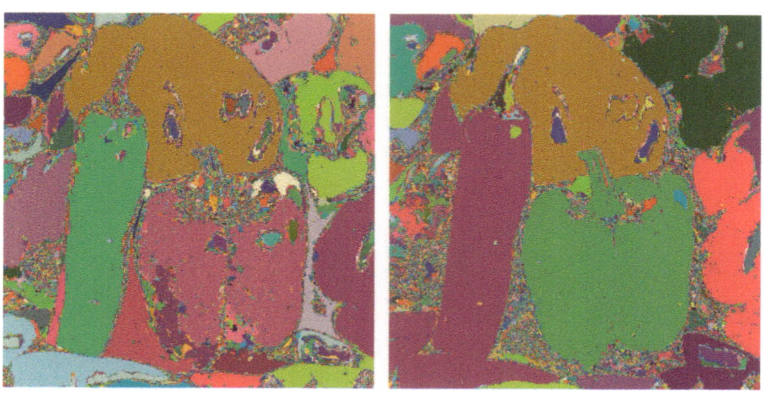

Chapter 6, Fig. 19 Random color LUT on HAC steps. (A) Step (5) of HAC by d_e. (B) Step (32) of HAC by $d^{\triangle_c}$.

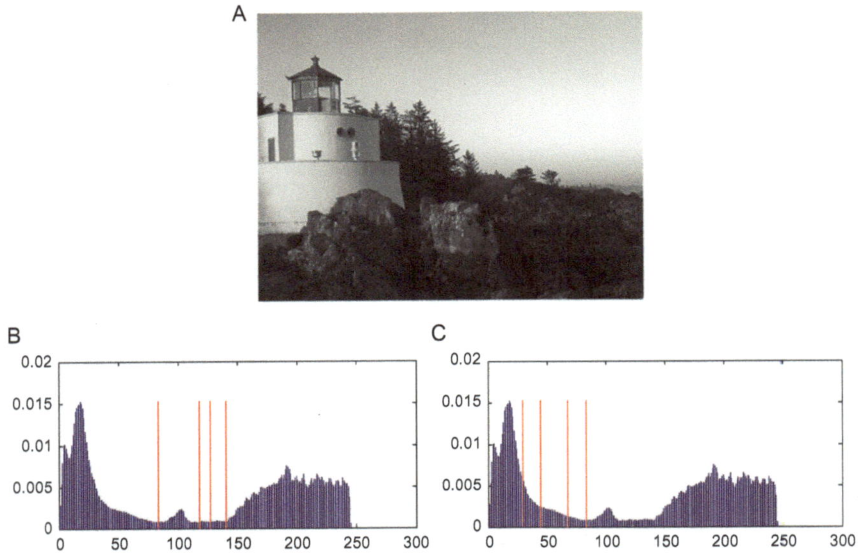

Chapter 6, Fig. 28 For OA algorithm: Initial image and its thresholded histogram (for Euclidean and LIP metrics). (A) Initial image. (B) Histogram with four thresholds in the case of the Euclidean metric. (C) Histogram with four thresholds in the case of the LIP metric.

Chapter 6, Fig. 32 Gravitational clustering (SM algorithm) for a color image driven by Euclidean metric. (A) Initial color image. (B) Classified image (20 colors).

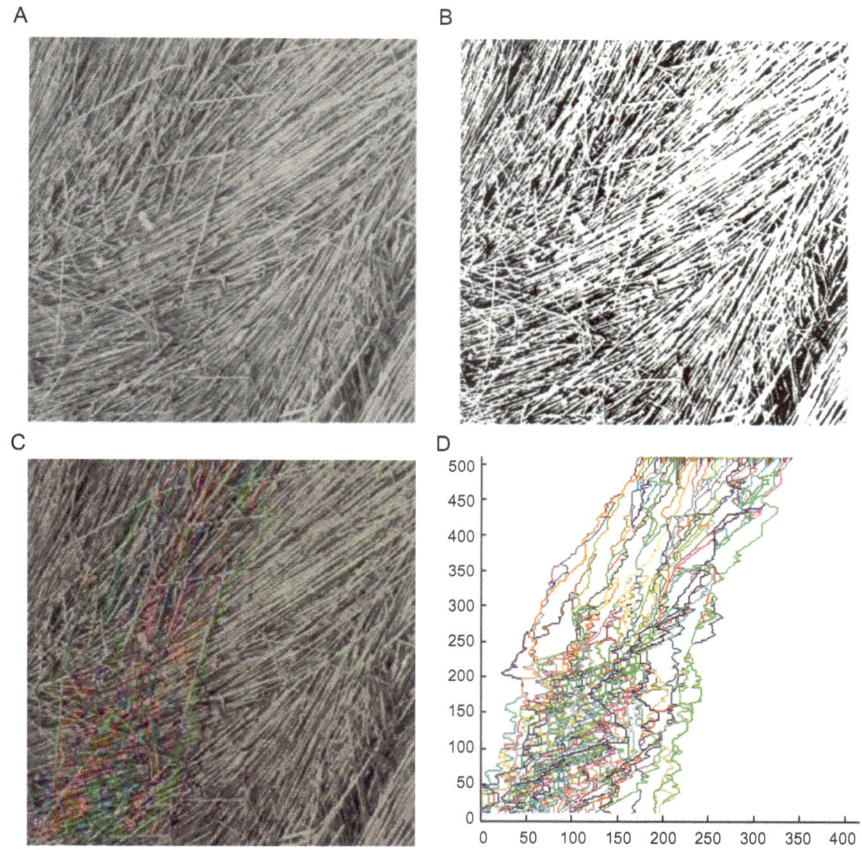

Chapter 6, Fig. 35 Percolation trajectories computed on a textured image. (A) Initial image from Brodatz database. (B) Binary image associated to (A). (C) Original image and its percolation trajectories. (D) Percolation trajectories only.

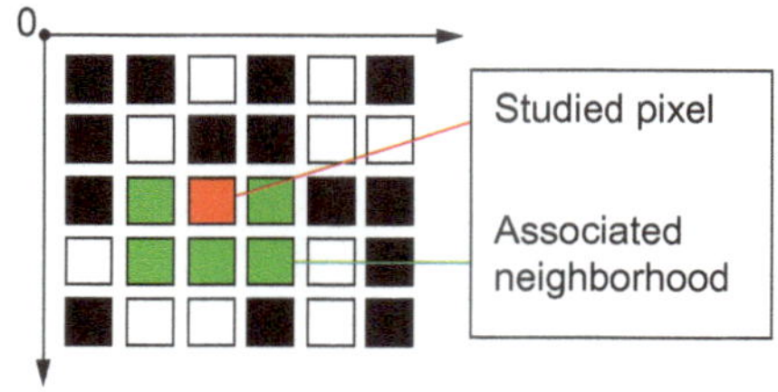

Chapter 6, Fig. 37 Useful part (in *green*) of the neighborhood of the studied pixel (in *red*).

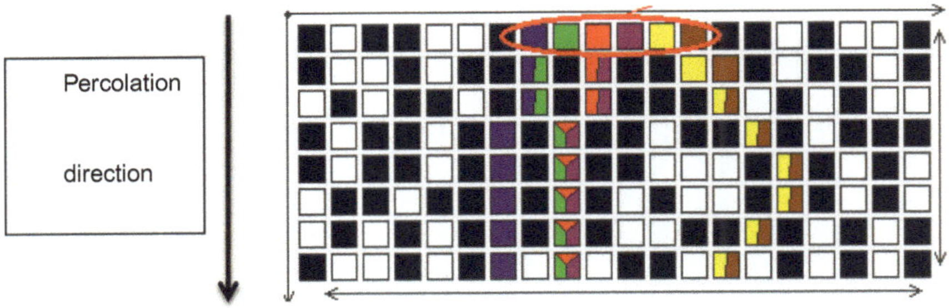

Chapter 6, Fig. 38 Result of the percolation of the example presented in Fig. 36.

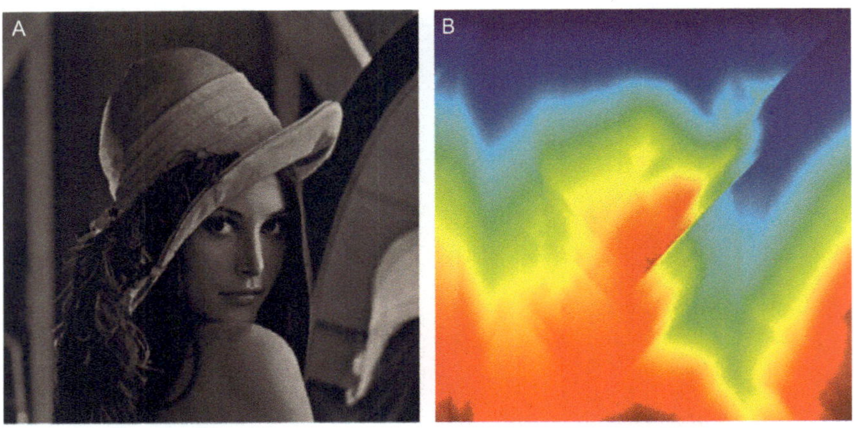

Chapter 6, Fig. 40 Geodesic time propagation on a gray-level image. (A) Initial image. (B) Associated map of geodesic times, in false colors.

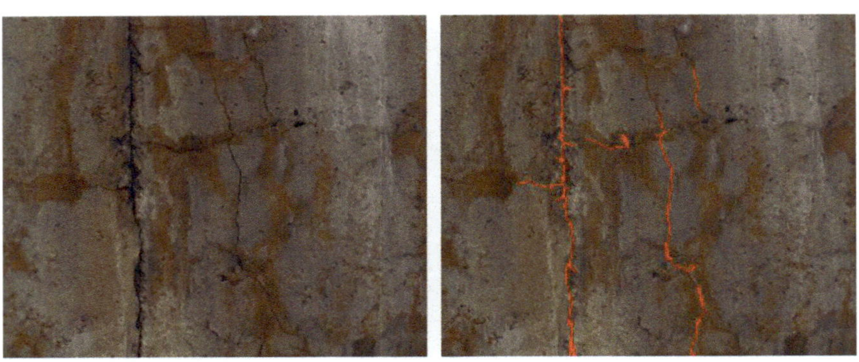

Chapter 6, Fig. 45 Crack detection based on double LIP Top-Hat.

FSC
www.fsc.org
MIX
Papier | Fördert
gute Waldnutzung
FSC® C083411

Kolibri 360 GmbH
Ferdinand-Jühlke-Straße 7
99095 Erfurt, Deutschland
produktsicherheit@kolibri360.de